Mountain
Geomorphology

Edited by Philip N. Owens and Olav Slaymaker

A member of the Hodder Headline Group
LONDON

Distributed in the United S
University Press Inc., New Y

First published in Great Britain in 2004 by
Arnold, a member of the Hodder Headline Group,
338 Euston Road, London NW1 3BH

http://www.arnoldpublishers.com

Distributed in the United States of America by
Oxford University Press Inc.
198 Madison Avenue, New York, NY10016

The advice and information in this book are believed to be true and
accurate at the date of going to press, but neither the editors nor the publisher
can accept any legal responsibility or liability for any errors or omissions.

British Library Cataloguing in Publication Data
A catalogue record for this book is available from the British Library

Library of Congress Cataloging-in-Publication Data
A catalog record for this book is available from the Library of Congress

ISBN 0 340 76417 1

1 2 3 4 5 6 7 8 9 10

Typeset in 10/13pt Gill Sans Light by Servis Filmsetting Ltd, Manchester
Printed and bound in Malta.

What do you think about this book? Or any other Arnold title?
Please send your comments to feedback.arnold@hodder.co.uk

Contents

List of contributors

Helgi Björnsson, Science Institute, University of Iceland, Dunhaga 3, IS-107 Reykjavík, Iceland (hb@raunvis.hi.is)

Nel Caine, Department of Geography and Institute of Artic and Alpine Research, University of Colorado, Boulder University of Colorado CO 80309–0450, USA (cainen@colorado.edu)

Simon Dadson, Department of Earth Sciences, University of Cambridge, Cambridge CB2 3EQ, UK (simon00@esc.cam.ac.uk)

Kenneth Hewitt, Cold Regions Research Centre, Wilfrid Laurier University, Waterloo, Ontario N2L 3C5, Canada (khewitt@wlu.ca)

Niels Hovius, Department of Earth Sciences, University of Cambridge, Cambridge CB2 3EQ, UK (nhovius@esc.cam.ac.uk)

Dimitri Lague, Department of Earth Sciences, University of Cambridge, Cambridge CB2 3EQ, UK (dlay02@esc.cam.ac.uk)

Cliff Ollier, School of Earth and Geographical Sciences, University of Western Australia, Nedlands, WA 6009, Australia (cliffol@geog.uwa.edu.au)

Yuichi Onda, Geomorphology Laboratory, Institute of Geoscience, University of Tsukuba, Tsukuba 305–8571, Japan (onda@atm.geo.tsukuba.ac.jp)

Lewis A. Owen, Department of Earth Sciences, University of California, Riverside, CA 92521, USA (lewis.owen@ucr.edu)

Philip N. Owens, National Soil Resources Institute, Cranfield University, North Wyke Research Station, Okehampton, Devon EX20 2SB, UK (philip.owens@bbsrc.ac.uk)

Olav Slaymaker, Department of Geography, University of British Columbia, 1984 West Mall, Vancouver, British Columbia V6T 1Z2, Canada (olav@geog.ubc.ca)

Jean-Claude Thouret, Laboratoire Magmas et Volcans, Université Blaise-Pascal et CNRS, 5 rue Kessler, 63038 Clermont-Ferrand, France (thouret@opgc.univ-bpclermont.fr)

Tianchi Li, International Centre for Mountain Development (ICIMOD), PO Box 3226, Kathmandu, Nepal (tianchi@icimod.org.np)

Paul W. Williams, School of Geography and Environmental Science, The University of Auckland, Private Bag 92019, Auckland, New Zealand (p.williams@auckland.ac.nz)

Preface

Mountain environments are at the forefront of world attention. The realization in the last few decades of their fragility and the important role that they play as a source of timber, food, water and other natural resources has prompted unparalleled interest in such environments and the people that live within them, and led the United Nations to declare 2002 as International Year of Mountains.

This book has a long history and largely stems from two sources. First, it stems from a book of the same title edited by O. Slaymaker and H.J. McPherson in 1972 (Tantalus Press, Vancouver), and the realization that there has been huge progress since that time. It also arises from a meeting in Edmonton in 1990, organized by Phil Owens, on the theme of 'Mountain geomorphology: a Canadian perspective'. A decade later, Hodder Arnold approached us with the idea of producing a book on a broader theme: one that could be used as a textbook for students while also providing a reference for academics and research scientists. We have tried to achieve this by adopting a simple tripartite structure with three main sections – historical, functional and applied – and by inviting leading experts to write interesting chapters that summarize important research findings and conceptual ideas on specific topics within mountain geomorphology. We have attempted to cover many of the mountain types and ranges in the world, and have tried to include an international authorship. We feel that we have largely achieved this, although we recognize that there will inevitably be some gaps.

As with any project of this size, there are many people to whom thanks are due. We are particularly grateful to Hodder Arnold, in particular, Luciana O'Flaherty, for approaching us in February 2000 with the prospect of producing a book. At Hodder Arnold we would also like to thank Colin Goodlad, Liz Gooster, Lesley Riddle, Tiara Misquitta and Abigail Woodman for their help, patience and understanding. The assistance of Alison Foskett and Sue Thomas with the latter stages of the production process is much appreciated.

We would also like to thank the following people for refereeing chapters and for providing advice at various stages of production:

David Alexander	Jack Ives	Mike Searle
Paul Bishop	Peter Kamp	Graham Smith
Rod Brown	Karna Lidmar-Bergström	Jane Soons
David Chester	Takashi Oguchi	Takasuke Suzuki
David Evans	Tom Pierson	Brian Whalley
Monique Fort	Andy Russell	

Although this book has taken several years to produce, it has been a pleasurable experience and we have enjoyed the opportunity to interact with such a distinguished list of scientists (authors and referees) and enthusiastic publishers. Our greatest hope is that the information contained in this book will encourage and inspire students and scientists (and perhaps even those with a more general interest in mountains) to partake in activities that help us to understand the interactions between society, geomorphological processes and landform evolution (past, present and future) in mountain environments.

Phil Owens, North Wyke
Olav Slaymaker, Vancouver

PART I Introduction

Drepung Monastery, Lhasa, Tibet
Photo: P.N. Owens

1
An introduction to mountain geomorphology

Philip N. Owens and Olav Slaymaker

1 The importance of mountains

Few would argue with the assertion that mountains are some of the most inspiring and attractive natural features on the surface of the Earth. Mountains have attracted mystics, artists, thinkers, scientists and tourists through their aesthetic appeal. Visually, mountains dominate the landscape. Mountains, such as Mt Kailash in western Tibet, have also been a source of religious and spiritual inspiration. They have affected millions of people, both those living in mountains and also those far removed from them (Ives and Messerli, 1999).

The fragility of mountain areas and their likely sensitivity to future changes in land use, management and climate have become increasingly apparent. Three recent events have brought mountain issues to world attention and initiated major scientific research programmes. In 1973, the United Nations Educational, Scientific and Cultural Organization (UNESCO) included mountains as one (Project 6) of the original twelve components in the Man and Biosphere (MAB) programme (Ives and Messerli, 1999). Subsequently, mountains were featured at the UN Conference on Environment and Development (UNCED), or Earth Summit, in Rio de Janeiro, Brazil, in 1992, with Chapter 13 ('Managing fragile ecosystems: sustainable mountain development') becoming part of Agenda 21. Chapter 13 states:

As a major ecosystem representing the complex and interrelated ecology of our planet, mountain environments are essential to the survival of the global ecosystem. Mountain ecosystems are, however, rapidly changing. They are susceptible to accelerated soil erosion, landslides and rapid loss of habitat and genetic diversity . . . As a result, most global mountain areas are experiencing environmental degradation. Hence, proper management of mountain resources and socio-economic development of the people deserves immediate action (United Nations Environment Programme (UNEP), 2003).

The MAB programme and the UN Earth Summit provided a springboard for the development of a variety of research groups and institutes (e.g., the International Centre for Integrated Mountain Development (ICIMOD), the Mountain Forum and the International Mountain Society), reports and publications that reviewed the state of the world's mountains (e.g., Mountain Agenda, 1992, 1997; Stone, 1992; Price, 1995; Messerli and Ives, 1997) and specialist journals (e.g., *Mountain Research and Development*). A useful review is provided by Ives and Messerli (1999).

More recently, 2002 was declared International Year of Mountains by the United Nations General Assembly. This declaration stimulated unparalleled interest in all aspects of mountains and resulted in numerous scientific conferences, workshops and publications (including this book). This momentum is continuing, as 2003 has been declared International Year of Freshwater and mountain environments feature prominently in many of the activities. Thus, with the dawn of the twenty-first century, mountain environments are at the forefront of world attention.

2 What is a mountain?

Although visually the identification of a mountain may appear simple, there is still no single definition of what constitutes a mountain. While the high mountain peaks, such as those in the Himalaya, Andes and European Alps, are clearly mountains, there is uncertainty associated with much smaller mountains (such as the Maya Mountains of Central America) and with high-elevation systems of limited relative relief (i.e., the difference between the maximum elevation of a peak and the elevation of the surrounding land). Thus, many scientists (e.g., Meybeck *et al.*, 2001) would argue that high-elevation plateaux such as the Altiplano of South America and the Tibetan Plateau are not mountains. Similarly, older, and consequently lower, mountain ranges, such as the Appalachians in the USA, the Urals in Europe and the mountains of Scotland, are not included in many classification systems that are based solely on altitude.

Many published works on landforms in general, and mountains specifically, advocate a definition of what constitutes a mountain (e.g., Fairbridge, 1968; Goudie, 1985). Good reviews of early definitions are given by Strahler (1946) and Gerrard (1990). In the German literature there is a clear distinction made between 'hochgebirge' (high mountains), 'gebirge' (mountains) and 'mittelgebirge' (uplands and highlands) (Troll, 1972, 1973). From a geomorphological perspective, Barsch and Caine (1984) propose that there are four characteristics of mountain terrain that are important:

1 elevation;
2 steep, even precipitous, gradients;
3 rocky terrain; and
4 the presence of snow and ice.

Not all of these characteristics apply to high-elevation plateaux. Troll (1972, 1973) and Barsch and Caine (1984) believe that high mountains are also characterized by:

1 diagnostic vegetative-climatic zones;
2 high potential energy for sediment movement;
3 evidence of Quaternary glaciation; and
4 tectonic activity and instability.

High precipitation, low temperatures (and hence typically snow and ice, at least seasonally), internal diversity and variability, infrequent but intense episodic activity, and existence in a metastable state are often associated with mountains in general.

Figure 1.1 shows the altitudinal distribution of environmental variables in relation to prevalent geomorphic processes, based on the Central Karakoram. It illustrates well the relations between elevation, vegetative-climatic zones (and, by association, soils and hydrology), landscapes and geomorphic processes. There are, however, local and regional variations in these relations that reflect, in part, variations in climate (i.e., incoming solar radiation, precipitation) and regional topography and tectonic setting.

There are, no doubt, other attributes that are characteristic of mountains, the significance of which will depend on the scientific questions being asked. It is easy to get bogged down with the issue of establishing an all-encompassing definition, which may ultimately be elusive because of the huge variation of mountain types and forms and the inherent complexity of their features. There are probably greater research questions and environmental concerns in mountain areas to which geomorphologists should turn their attention.

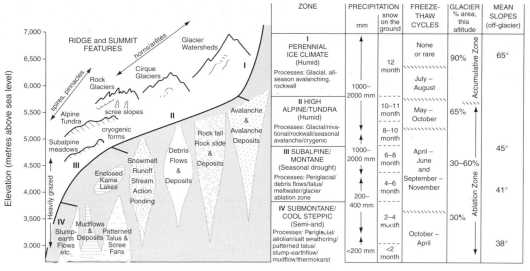

Figure 1.1 *Altitudinal distribution of environmental variables and geomorphic processes in the Central Karakoram*
Source: *from Hewitt (1989), reproduced with the permission of Gebrüder Borntraeger Verlagsbuchhandlung, Stuttgart.*

3 Classification systems

One of the simplest mountain classification systems was developed by Fairbridge (1968) and is based on scale and the degree of continuity:

1 a mountain is a singular, isolated feature or a feature outstanding within a mountain mass;
2 a mountain range is a linear topographic feature of high relief, usually in the form of a single ridge;
3 a mountain chain is a term applied to linear topographic features of high relief, but usually given to major features that persist for thousands of kilometres;
4 a mountain mass, massif, block or group is a term applied to irregular regions of mountain terrain, not characterized by simple linear trends; and
5 a mountain system is reserved for the greatest continent-spanning features.

Table 1.1 further illustrates this classification system according to common North American usage and is based on examples of mountains in British Columbia, Canada. Mountains can also be classified according to, among other things (cf. Slaymaker, 1999; Yamada, 1999):

1 tectonic framework;
2 climate;
3 hydrology;
4 geomorphology;
5 degree of anthropogenic alteration; and
6 morphometry.

Table 1.2 provides an example of a classification system based on tectonic setting. Further information on, and examples of, the relation between mountain environments and tectonic settings is provided by several of the chapters in this volume, including those by Owen, Ollier, Williams and Onda. For reviews of other classical classification systems, which are generally based on tectonic setting and evolution, and denudational history, the reader is directed to Fairbridge (1968) and Gerrard (1990).

A very different, more recent, classification system is that developed by Kapos *et al.* (2000), as part

Table 1.1 – Classification of mountain environments by spatial scale according to common North American usage, with special reference to the mountains of British Columbia

Spatial scales (km²)	Terminology	Examples of mountain region
$>10^6$	Mountain systems	Circum-Pacific orogenic belt
$10^5 - 10^6$	Mountain systems	Canadian Cordillera
$10^4 - 10^5$	Mountain massif (or mass)	Coast Mountains
$10^3 - 10^4$	Mountain ranges (or chains)	Pacific Ranges
$10^2 - 10^3$	Mountain range	Tantalus Range
$<10^2$	Individual mountains	Mt Garibaldi

After Fairbridge (1968); Slaymaker (1999).

Table 1.2 – Classification of mountain environments by tectonic setting

Plate setting	Types	Examples of mountain region
Convergent plates	Oceanic to oceanic	Japanese Alps; Aleutian Arc, Alaska
	Oceanic to continental	South Island, New Zealand; Cascade Ranges, Pacific Northwest
	Continental to continental	Himalaya
	Accreted margins	Coast Mountains, British Columbia
Divergent plates	Oceanic spreading	Iceland; Galapagos Islands
	Intercontinental rifts	Sinai Peninsula; Scottish Highlands
Transform plates	Ridge past ridge	Coast Ranges, California
	Trench past trench	Anatolia, Turkey
	Ridge past trench	Pakistan–Afghanistan
Plate edges	Passive continental margins	Drakensberg; Western Ghats
Plate interiors	Hotspots	Hawaii; Yellowstone National Park
	Continental flood basalts	Deccan Plateau, India; Columbia Plateau, Pacific Northwest
	Shields	Ahaggar Mountains, Sahara
	Intra-cratonic uplift sites	San Rafael Swell, Utah
	Post-tectonic magmatic intrusion sites	Air Mountains, Nigeria
	Evaporite diapirs	Zagros Mountains, Iran

Adapted from Short and Blair (1986); Slaymaker (1999; in press).

of a larger study of mountain forests (cf. Price and Butt, 2001). These workers used the GTOPO30 1-km-grid resolution digital elevation model (DEM) developed by the United States Geological Survey, and a classification system that uses altitude, local elevation range and slope (thereby overcoming potential problems associated with high-elevation areas with little relief and low-elevation areas with significant relief), and the following criteria:

1 a threshold of 2500 m above mean sea level, above which any environment was considered mountainous (thus including high-elevation plateaux);

2 for middle elevations, some degree of slope was a necessary feature, which enabled mid-elevation plateaux to be excluded; and

3 use of local elevation range as a means of including *appropriate* low-elevation and older mountains of regional significance.

Table 1.3 presents a six-fold classification of mountains based on this approach. On the basis of these criteria, the global total area of mountains is 35.8×10^6 km^2 and this represents about 24.3% of the global land surface area. This mountain surface area agrees well with a value of 33.3×10^6 km^2 estimated by Meybeck *et al.* (2001) based on the same dataset but a coarser spatial resolution. Meybeck *et al.* (2001) primarily used relief roughness to determine relief types. As slope is no longer a determining factor above 2500 m in the classification system of Kapos *et al.* (2000), a major difference between the classifications of Kapos *et al.* (2000) and Meybeck *et al.* (2001) lies in the inclusion of very high plateaux,

Table 1.3 – Surface area (km^2) of the world's mountains based on a classification system developed by Kapos *et al.* (2000) and United Nations Environment Programme – World Conservation Monitoring Centre (2001). Values in italics represent the region with the highest value in each of the six classes and the total

Region	≥4500 m	3500–4500 m	2500–3500 m	1500–2500 m and slope≥2°	1000–1500 m and slope≥5° or LER[a] >300	300–1000 m and LER[a] >300	Total
North America	197	11417	200830	*1092881*	*1104529*	1840140	4249994
Central America	38	968	67127	353586	259367	412215	1093301
Caribbean			32	2809	5528	38322	46691
South America	154542	583848	374380	454417	465061	970707	3002955
Europe		225	497886	145838	345255	1222104	2211308
Africa	73	4859	101058	559559	947066	1348382	2960997
Middle East	40363	128790	339954	906461	721135	733836	2870539
Russian Federation	31	1122	31360	360503	947368	*2961976*	4302360
Far East	*1409259*	741876	627342	895837	683221	1329942	5687477
Continental SE Asia	170445	107974	97754	211425	330574	931217	1849389
Insular SE Asia	22	4366	34376	120405	157970	599756	916895
Australia				385	18718	158645	177748
Oceania			41	7745	29842	118010	155638
Antarctica	17	*1119112*	*4530978*	165674	144524	327840	*6288145*
Total	1774987	2704557	6903118	5277525	6160158	12993092	35813437

[a]LER = Local elevation range (m) (7 km radius around a cell, based on a grid spacing of 30 arc-seconds (c.1 km)).

such as the South American Altiplano and the Tibetan Plateau, in the former. If very high plateaux are included in the Meybeck *et al.* (2001) classification, then the total area of mountainous regions increases to 34.7×10^6 km², approximately 24% of the global land surface area. Meybeck *et al.* (2001) also estimated that 26% of the global population live in mountains, although the authors identify that there are several problems associated with this figure and that it is likely to be an overestimate. However, it is likely that a higher proportion of the global population depend on mountains for resources such as water and power.

4 Types of mountains

Barsch and Caine (1984) describe four main relief types within high mountain systems:

1 Alp-type;
2 Rocky Mountain-type;
3 polar mountains; and
4 desert mountains.

The Alp-type relief has been conditioned by intense glacial activity, and has steep rock walls and slopes, distinct glacial features such as cirques, arêtes and troughs, and post-glacial features such as talus slopes and debris fans in valley bottoms. The Rocky Mountain-type relief has been conditioned less by glacial activity and may include interfluve and flat elements in addition to those features that occur in Alp-type relief. Barsch and Caine (1984) point out that the high mountain classification fails in the cases of polar and desert mountains. Polar mountains often have evidence of intense glaciation, but are frequently with a local relief of less than 1000 m. Desert mountains often do not reach timberline and were only lightly glaciated, if at all, during the Pleistocene.

There are, of course, many additions to these relief models presented above. In this respect volcanic mountains deserve a special mention. Volcanic mountains are unique in terms of their morphology (i.e., sometimes cone-shaped), the juxtaposition between construction and erosion processes, the type of volcanic processes that act on them (lahars, pyroclastic flows, etc.) and the rate at which these and other geomorphic processes (fluvial, glacial, etc.) operate, and the hazard that they present to society. The chapter by Thouret (this volume) reviews the various types and characteristics of volcanic mountains and the geomorphic processes that operate on them.

5 A typology of mountain geomorphology

A simple typology of mountain geomorphology makes use of the tripartite division into historical, functional and applied mountain geomorphology, and is based on the original classification of Chorley (1978). This classification system represents a convenient system for examining broader issues within mountain geomorphology, and forms the basis for the structure of this book. The following sections (5.1–5.3) examine this typology in more detail and set the scene for the subsequent chapters. Another useful typology of mountain geomorphic systems was proposed by Slaymaker (1991, see Table 1.4).

5.1 Historical mountain geomorphology

Historical mountain geomorphology focuses on the evolution of mountains and mountain systems over both long and medium timescales. There is generally a distinction between young (of Cenozoic age, cf. Table 1.5) and older mountains, and this is described in more detail in the sections below.

Table 1.4 – A typology of mountain geomorphic systems and appropriate approaches to measurement

System category	Macroscale example	Appropriate approaches to measurement	Mesoscale example	Appropriate approaches to measurement
Morphological	Regional geomorphic and tectonic framework	Remote sensing	Terrain and land system analysis Zero order basins	Mapping and air photos
Morphologic evolutionary	Relief evolution and palaeoenviron-mental reconstruc-tion	Surface chronology Sediment geochronology	Kinematics of landform change	Surface chronology Sediment geochronology
Cascading	Regional water, solute and sediment budgets	Monitoring	Basin water, solute and sediment budgets	Monitoring Pathway identification Storage volumes
Process–response	Energy input and landform response	Physical models Neotectonics	Process studies	Experiments Strength of response
Control	Global change management and prediction	Environmental indicators Global climate models	Geomorphic hazards	Mapping and zoning Magnitude–frequency analysis

Adapted from on Slaymaker (1991).

5.1.1 Plate tectonics and major episodes of mountain building

The evolution of mountain systems is linked to the lithospheric plates and their movement over time. Figure 1.2 shows a cross-section of the outer Earth and illustrates the location of lithospheric plates relative to the asthenosphere. There are six major lithospheric plates (Pacific, Eurasian, Indo-Australian, African, Antarctic and American) and at least 14 minor plates (e.g., Nazca, Cocos, Caribbean, Azores, Scotia and Philippine). Figure 2.1 of this volume illustrates the location of the main plates and many of the minor ones. Plate movements are controlled by the relative motion between the lithosphere and the asthenosphere, and by gradients in lithospheric thickness and density (Windley, 1995). The plates are bordered by three types of seismically and tectonically active boundaries:

1 convergent, including subduction zones and continental–continental collision zones;
2 divergent or spreading zones; and
3 transform faults.

Table 1.2 lists the types of plate setting associated with these three types of boundaries and provides examples of mountains. Convergent boundaries occur where two plates collide and one has overridden the other. One plate is thrust down a subduction zone into the mantle where it melts. Where continental and oceanic plates converge, the greater density of the oceanic plate means that

Table 1.5 – Geological timescale. Nomenclature, division and ages vary slightly between different organizations and authors

Millions of years before present	Epochs	Periods	Periods	Eras	Eons
	Holocene		Quaternary	Cenozoic	Phanerozoic
0.01	Pleistocene				
2	Pliocene	Neogene	Tertiary		
5	Miocene				
24	Oligocene	Palaeogene			
37	Eocene				
58	Palaeocene				
66			Cretaceous	Mesozoic	
144			Jurassic		
208			Triassic		
245			Permian	Palaeozoic	
286			Carboniferous		
360			Devonian		
408			Silurian		
438			Ordovician		
505			Cambrian		
570			Proterozoic	Precambrian/ Proterozoic	Precambrian
2500			Archaean		
ca. 4000			Hadean		
ca. 4600[a]					

[a]Approximate age of the Earth.

Adapted from Press and Siever (1982); Windley (1995); Ollier and Pain (2000).

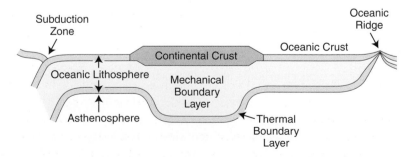

Figure 1.2 *Cross-section of the outer Earth showing the lithospheric plates moving over the asthenospheric mantle. The continental crust (typically 35 km thick), oceanic crust (~6 km) and island arcs form the uppermost part of the lithosphere*
Source: *from* The evolving continents, *Windley, B.F., 1995. © John Wiley & Sons Limited, Chichester. Reproduced with permission.*

this is subducted. Convergent boundaries are the regions of maximum shortening and highest shear motion on the Earth's surface. They are, therefore, zones often associated with maximum seismic activity and the largest earthquakes. Divergent boundaries are associated with spreading, usually at mid-oceanic settings. Examples include the slow-spreading (typically <50 mm a^{-1}) Mid-Atlantic Ridge and the fast-spreading (approximately 0.1–0.2 m a^{-1}) East Pacific Rise, part of which runs along the Californian coast (Windley, 1995). Transform faults are shear boundaries parallel to the direction of movement between two plates. They are not normally associated with the creation of new lithospheric material. Transform faults are the least tectonically active of the three active boundaries, with only shallow earthquakes.

There have been three major mountain-building episodes whose imprint on the present landscape is profound. Useful reviews of these and other mountain-building episodes are provided by Fairbridge (1968), Gerrard (1990) and Windley (1995). Orogeny is a general term usually referring to deformation processes at convergent plate boundaries (however, also see Ollier and Pain, 2000). The Caledonian–Appalachian orogeny occurred in the early Palaeozoic (cf. Table 1.5) when three large continental blocks, Gondwana (Africa, South America, India, Australia and Antarctica), Laurentia (North America, Greenland and northwest Scotland), and Baltica (northwest Europe eastwards to the Ural Mountains), collided to form four segments of the main orogen:

1 the Caledonides of Norway, west Sweden and east Greenland formed between Baltica and Laurentia;
2 the Caledonides of the British Isles between East Avalonia (a small continental block rifted from western Gondwana) and Laurentia;
3 the Appalachians between Laurentia and northwest Africa of northwest Gondwana; and
4 the concealed Caledonian orogen under the North Sea and east Europe formed between north Africa and south Baltica (Windley, 1995).

The Hercynian–Appalachian (Variscan) orogeny occurred in the late Palaeozoic and was even more extensive than the Caledonian–Appalachian orogeny. The Hercynian orogeny involved the collision of north Africa with Europe, and produced a series of NW–SE trending systems which are now isolated blocks in Europe, such as the Central Massif and the Black Forest. The Appalachian mountain system was the result of continent-to-continent collision involving large-scale low-angle thrusting.

The Alpine orogeny was the last major phase of mountain building and this occurred in the Mesozoic and Cenozoic. The mountains associated with this phase of mountain building are discussed further below.

5.1.2 Mountain evolution at active plate boundaries and passive continental margins

At the largest scale, it is common to distinguish between young active mountain belts that have evolved throughout the Cenozoic and are still associated with active lithospheric plate margins, and mountains on passive continental margins. Nearly all the literature on mountain building in the past 40 years has concentrated on active margins where collision and subduction may explain both mountains and the structures within them. Ollier and Pain (2000) describe various mechanisms for the formation of mountains at subduction sites and some of these mechanisms are illustrated in Figure 1.3.

Figure 2.1 of this volume illustrates the distribution of the main mountain chains and systems. In particular, this diagram illustrates that the location of recent global mountain systems of Cenozoic age are primarily associated with lithospheric plate margins (see also Table 1.2). The main global mountain systems at plate boundaries are 'active', in that they are presently deforming due to plate subduction and collision (Owen, this volume). The best examples of this type of mountain system are the

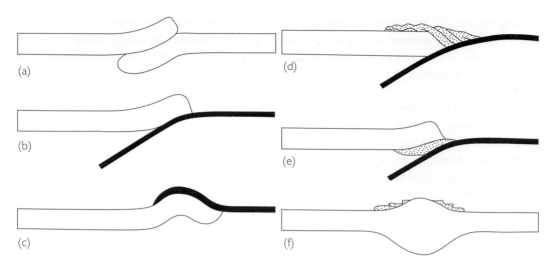

Figure 1.3 *Some mechanisms for the formation of mountains at collision sites: (a) continent–continent collision; (b) continent–ocean collision with buckling-up of the continent and subduction of the ocean floor; (c) obduction of the ocean floor over the continent with later isostatic rise to form mountains; (d) thrusting of sediments on to the continental plate with foreland folding at the front; (e) thrusting of sediments under the continent, where they may melt and form granite; and (f) crustal thickening from plate collision, possibly accompanied by gravity sliding of rocks near the surface*
Source: *from* The origin of mountains, Ollier, C.D. and Pain, C.F., 2000. Reproduced with the permission of Routledge.

Alpine–Himalayan–Tibetan and the Circum-Pacific orogenic systems. As described above, most mid-oceanic ridges are mountain systems also associated with active plate boundaries, albeit in an ocean setting. The distribution and evolution of these recent, active global mountain systems are described in more detail by Owen (this volume). On a smaller scale, Williams (this volume) examines in detail the evolution of the mountains of New Zealand, which are associated with a transform plate boundary due to the motion of the Pacific, Indo-Australian and Antarctic plates.

There are also mountains on passive continental margins. Examples of such mountains include the Eastern Highlands of Australia, the Eastern and Western Ghats of India, and the Appalachians of eastern North America. The chapter by Ollier (this volume) provides a summary of the evolution and distribution of the main mountain ranges that are associated with passive margins. The evolution of these older mountain belts is intrinsically more complex as they do not easily fit into the simple plate tectonic story of mountain building at collisional sites and includes the history of the Earth since the breakup of Gondwanaland during the Mesozoic. With mountains associated with passive margins, a major difference of opinion has emerged between those who place greatest emphasis on the data from fission-track thermochronology (see Gleadow and Brown, 2000, for a useful review on thermochronology) and those who use whatever landform, stratigraphic and geological data that can be found to constrain the interpretation (see Ollier, this volume). Whereas geomorphic models of denudation history can often be difficult to validate, interpretation of fission-track data in terms of denudation history is also contested. Ultimately, however, there is a need for these differences in approaches to meet. Indeed, in historical mountain geomorphology the most exciting developments in recent years have been the trends towards quantifying rates of uplift and denudation with the development of new geochemical, geochronological and geodetic methods. As such, developments in dating techniques such as fission-track thermochronology and cosmogenic dating (cf. Cockburn *et al.*, 2000), and in the use of digital elevation models (cf. Meyer, 2000) offer considerable potential. Similarly, recent developments in the use of numerical models, and in particular coupled tectonic–surface process models (cf. Beaumont et al., 2000), are likely to advance our understanding of the evolution of mountains at a variety of scales. Important here is the potential for examining the relation between tectonics, Earth surface processes and landscape development, and the feedback effects of surface processes on fundamental tectonic mechanisms (Summerfield, 2000).

5.2 Functional mountain geomorphology

Functional mountain geomorphology includes the assessment of processes, rates, and spatial and temporal patterns of mountain belt erosion and deposition, and the chapters by Hovius et al. and Caine (both this volume) provide global perspectives on this. The process framework should ideally involve a consideration of the coupling of uplift and erosion; many geomorphic models have failed to include realistic models of this coupling. In mountain belts, such a consideration is obligatory as both uplift rates and erosion rates achieve maximum values and the coupling of the processes is even more critical than in lowland regions. Improvements in understanding of fluvial bedrock incision processes (e.g., Hartshorn et al., 2002), hillslope mass wasting (e.g., Hovius et al., 2000), glacial valley lowering and sediment routing (e.g., Tucker and Slingerland, 1996) are leading to the development of improved mountain landscape evolution models (see also Hovius et al., this volume). Feedbacks between tectonic, climatic and geomorphic processes have been explored in geodynamic models and solid, solute and organic fluxes from mountain belts have been constrained and considered within a global geochemical context. The topographic evolution of mountain belts can be modelled with increasing realism, but the issue of equilibrium conditions versus transience is still far from resolution (Burbank and Anderson, 2001).

Caine (1974, 1984) proposed a four-fold classification of mountain process systems, based on work in the Rocky Mountains of North America:

1 glacial system;
2 coarse-grained debris system;
3 fine-grained sediment system; and
4 geochemical system.

The various systems have differing controls, responses and activity levels. The systems are not mutually exclusive in terms of processes occurring within them and the temporal and spatial scales over which processes occur, but this classification provides a convenient framework with which to characterize functional mountain geomorphic processes and to investigate material fluxes in mountain systems. The four-fold classification process systems are not exclusive to mountain environments, but this classification lends itself best to systems with a wide range of elevation and distinct vegetative-climatic zones (cf. Figure 1.1). One might expect the relative order of these systems to reflect elevation, with the glacial and coarse-grained systems occurring at higher elevations and the fine-grained sediment and geochemical systems tending to dominate at lower elevations. Table 1.6 gives a qualitative summary of the major hydrologic processes operating in mesoscale alpine systems as identified by Slaymaker (1974), which is also useful in considering material fluxes.

5.2.1 The glacial system

The glacial system is generally found at the highest elevations, although there are situations (e.g., Antarctica, Patagonia) where glaciers enter the sea. The reader is directed to Benn and Evans (1998), who provide an excellent review of the glacial system. The occurrence of water as snow and ice is important in eroding and transporting sediment, and in the delivery and timing of flowing water (i.e., concentrated during the melting season). In glacierized (i.e., those that presently contain active glaciers)

Table 1.6 – Summary of major hydrologic processes operating in mesoscale alpine systems

Characteristic alpine system	>10^3 years	10–10^3 years	10^{-1}–10 years	<10^{-1} years (c.1 month)
1) Glaciers	Water storage term in hydrological cycle	Glacier retreat and advance	Ablation, accumulation	Physics of glacier motion, surging
2) Snowpacks	—	—	Snow metamorphism, firn ice	Melt, storage, accumulation
3) Alpine lakes	Glacial scour and lake formation, sedimentation and meadow formation (see 6)	Sedimentation	Temperature stratification, draining events, ice formation and breakup, sedimentation regime	Seiches, daily level changes, sediment movement into and through lake, evaporation
4) Mountain streams	Glacial erosion	Downcutting, sediment supply (creep)	Discharge regime, sediment supply (talus)	Runoff concentration, 'tumbling' flow, rapid response to precipitation and evaporation
5) Morainic mounds	Glacial deposition	Degradation of morainic slopes	Chemical weathering, frost action, vegetational change	Infiltration, interflow
6) Alpine and sub-alpine meadows; valley bottoms	Sedimentation in glacial lakes (see 3)	Dissection by streams, floodplain development, soil formation	Periodic inundation, jökulhlaups (see 1), vegetational change	Precipitation, infiltration, interflow, baseflow
7) Alpine and sub-alpine meadows; adret (sunny) slope	Post-glacial soil development	Slope degradation, channel dissection	Vegetational change, mass wasting	Precipitation, infiltration, interflow
8) Alpine and sub-alpine tree-covered slopes, adret	Post-glacial soil development, soil creep, gullying	Slope degradation, channel dissection	Frost action, mass wasting	Precipitation, interception, infiltration, interflow
9) Alpine and sub-alpine tree-covered slopes, ubac (shaded) and ridge tops	Post-glacial soil development, slides and flows of earth and mud/sediment	Slope degradation	Frost action, mass wasting	Precipitation, interception, infiltration, overland flow, interflow
10) Alpine barren	Glacial erosion, post-glacial rockfall and talus, slope formation	Slope degradation	Frost action, mass wasting	Precipitation, overland flow,

Adapted from Slaymaker (1974).

mountains, the glacial system is probably the dominant influence on erosional activity (Embleton and King, 1975; Barsch and Caine, 1984; Hallett et al., 1996). Caine (this volume) reviews the literature on denudation rates in glacierized and nonglacierized mountain catchments, and shows that rates are greatest in the former, and indeed represent some of the highest denudation rates in the world. The precise role of glaciers in eroding the land surface and producing large amounts of sediment, thereby increasing river sediment loads, is still uncertain, however (Gurnell, 1987; Harbor and Warburton, 1993), and there are contrasting relations between fluvial specific sediment yield (sediment amount per unit surface area per unit time, i.e., t km^{-2} a^{-1}) and percentage glacial cover (e.g., Hicks et al., 1990; Desloges and Gilbert, 1998). However, part of the problem stems from the fact that percentage glacial cover is a poor measure of glacial activity and sediment production. In addition, the precise form of the relations between glacial cover, glacial erosion and sediment yield is complicated by the degree of slope–glacier–channel coupling and by temporary storage effects in the proglacial zone. The importance of the proglacial zone in controlling sediment yields, by acting either as a sink or as a source of sediment (coarse and fine), has been documented for alpine glacial (cf. Gurnell, 1987; Warburton, 1990, 1999; Harbor and Warburton, 1993) and mountainous arctic glacial (e.g., Hodson et al., 1998; Hasholt et al., 2000) basins (see Figure 1.4).

Although not explicitly part of the glacial system, mountain permafrost (defined as ground with temperatures below 0°C for more than two years) can usefully be considered here. Indeed, the whole cryospheric system (which includes snow and ice, the glacial system and mountain permafrost) is particularly sensitive to changing environmental conditions because of its thermal proximity to melting conditions. Thus, the likely timing, rate and magnitude of response of geomorphic processes and material fluxes within different components of the cryospheric system are strictly dependent on the timing of ice or snow melt.

5.2.2 The coarse-grained sediment system

This system involves the transfer of coarse-grained sediment between cliffs and rock walls and the depositional features they supply, in addition to large-scale mass movement processes. This system may feed into the glacial system, or it may be a closed one. Rockfalls, landslides, avalanches and debris flows are examples of processes that drive the coarse-grained sediment system. Figure 1.5 shows a debris flow channel in the Coast Mountains of British Columbia and illustrates that such processes also provide an important link between steep, high-elevation parts of mountains and lower, flatter parts. Thus, such processes deliver coarse-grained sediment to river channels, thereby coupling hillslopes and channels, and uplands and lowlands. In consequence, bedload transport rates tend to be greatest in mountain streams, particularly steep headwater streams where the abundant supply of coarse-grained sediment coupled with high flow velocities (due to steep gradients, high precipitation amounts, and glacier and snowmelt effects) create optimum conditions. Slaymaker (1974), Gerrard (1990) and Wohl (2000) (see also Table 1.6) provide useful reviews of the characteristics of mountain hydrology and rivers.

The occurrence of landslides, rock avalanches and debris flows is primarily controlled by bedrock and surficial geology in the long and medium term. The chapter by Onda (this volume) illustrates this well, based on examples from the Japanese Alps. In this mountain region, because the soils and surficial materials are relatively thin, the flow of subsurface water also controls the type, location and frequency of mass movement events. There is also often a distinct response due to short-term perturbations: both natural (i.e., climatic) and human-induced (i.e., deforestation). Chang and Slaymaker (2002) examined the frequency and spatial distribution of landslides in a mountainous drainage basin in the Western Foothills of Taiwan. They documented the occurrence of landslides (Figure 1.6), including those due to logging (which started in 1956), road-building activities (early

Figure 1.4 *Proglacial zone, Mittivakkat Glacier, Greenland, showing the sources of sediment (glacial and slope) and the opportunities for storage*
Photo: P.N. Owens.

1980s) and an exceptionally violent typhoon ('Typhoon Herb', which occurred in 1996 with a rainfall total of 1094 mm in 24 hours). This study (cf. Figure 1.6) demonstrates several important issues relating to the occurrence of mass movement events, including: identifying the effects of natural and artificial perturbations; consideration of magnitude and frequency effects; and the timescales associated with response and recovery (the latter estimated to be 20 years for logging activity and 8 years for road construction).

5.2.3 The fine-grained sediment system
This sediment system is more open. Inputs may be both autochthonous (within the basin) and allochthonous (external to the basin). Autochthonous sources include weathering products and

Figure 1.5 *Debris-flow channel in the Coast Mountains, British Columbia, Canada*
Photo: P.N. Owens.

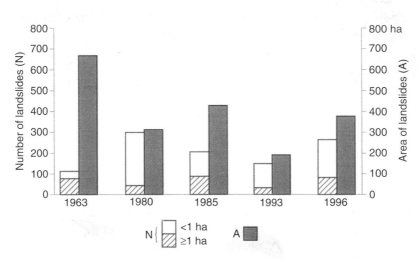

Figure 1.6 *Reconstruction of the number and area of landslides (1963 to 1996) in the Cho-Shui River basin, Taiwan, based on maps and air photographs*
Source: reprinted from Catena, 46, Chang, J.C. and Slaymaker, O., Frequency and spatial distribution of landslides in a mountainous drainage basin: Western Foothills, Taiwan, pp. 285–307. © 2002, with permission from Elsevier.

surface erosion of soils and unconsolidated material. Allochthonous inputs include regional aeolian dustfall. In situations where regional aeolian dust inputs originally derived from outside of the basin are high because of an abundance of local sources of fine material (i.e., nearby moraines – see Figure 12.5 of this volume), such inputs may dominate the fine-grained sediment system (Caine, 1974; Owens and Slaymaker, 1997). Indeed, allochthonous aeolian dustfall to the alpine and subalpine zones of many mountain ranges has been identified as an important pedogenic and geomorphic process (Birkeland, 1973; Kotarba, 1987; Litaor, 1987). Interestingly, while such inputs may provide readily abundant fine-grained sediment for subsequent redistribution by surface erosion and fluvial processes, they may help to protect the bedrock from subaerial weathering (Owens and Slaymaker, 1997). It is also important to identify that aeolian processes become relatively important in eroding, transporting and depositing fine-grained sediment in mountains with limited vegetation cover (i.e., alpine, subalpine, arid and semi-arid) (cf. Thorn and Darmody, 1985).

Fluvial processes assume greater importance here, with rivers also transporting material from the glacier system to lower elevations. As with the coarse-grained (described above) and the geochemical (described below) systems, bedrock geology and the nature of surficial deposits exert an important influence on fluxes of fine-grained sediment. Thus, in a review of sediment (mainly suspended) yields in mountain river basins throughout the world, Dedkov and Moszherin (1992) list the following specific sediment yields (t km^{-2} a^{-1}):

1 loess, 1800;
2 other terrigenic loose rocks, 1300;
3 coarse terrigenic rocks, 500;
4 metamorphic, 420;
5 limestones, 310; and
6 igneous, 100.

Dedkov and Moszherin (1992) describe two contrasting scale-dependent relations between specific sediment yield and drainage basin (i.e., potential contributing) area. In the first, specific sediment yield is positively related to drainage area. This type typically occurs in mountainous areas with forest cover (i.e., montane ecozones, tropical mountain systems), and thus slope erosion is minimal and channel erosion dominates the sediment system. Sediment load (t km^{-2}) is proportional to runoff and thus sediment yield increases with drainage area. In the second type, specific sediment yield is inversely related to increasing basin area. This type commonly occurs in mountainous basins with poor vegetation cover (i.e., subalpine and alpine ecozones, and arid and semi-arid mountain environments), high surface erosion and mechanical denudation, and also in forested areas with human impacts. In these basins, hillslope processes dominate. As there is a greater opportunity for deposition of sediment with increasing transport distance (i.e., on slopes, within channels and on floodplains), specific sediment yield decreases with increasing area, although absolute sediment load may increase. There are, of course, variations to these relations depending on lithology, relief, climate, degree of human activity and geomorphic history. In the mountains of British Columbia, Slaymaker (1987), Church et al. (1989), Church and Slaymaker (1989) and Owens and Slaymaker (1992) have described the increasing specific sediment yield with basin area as a function of the glacial and post-glacial history of the province. Church et al. (1999) have confirmed this effect for almost all regions of Canada.

5.2.4 The geochemical system
Generally, chemical denudation rates are lower than equivalent values for mechanical denudation as represented by river sediment yields (Caine, this volume). However, since the pioneering work of

Rapp (1960) in the Kärkevagge area of northern Scandinavia, the importance of the geochemical system in material fluxes has been increasingly realized (e.g., Darmody et al., 2000). It is intimately linked with solutional weathering, and nivation and fluvial processes. Bedrock and surficial geology and deposits, in addition to the physical characteristics of a basin such as size and relief, exert important influences on the type of geochemical reactions and thus on geochemical fluxes (Sueker et al., 2001). Thus, for example, areas underlain by granitic rocks generally have lower rates of chemical weathering than areas underlain by sedimentary rocks (Bluth and Kump, 1994). The contact time between water (including snow and ice) and material sources, and thus the residence time of water within the basin, are also important in controlling the intensity of weathering and of solutional fluxes within and out of such systems. Increases in temperature, vegetation cover and thickness of deposited and unconsolidated sediments are likely to increase the importance of the geochemical system at lower elevations (Drever and Zobrist, 1992).

The geochemical system is also present within the glacier system, and there has long been debate about the relative importance of chemical weathering in glacierized environments compared with nonglacial mountain environments. Although chemical weathering should be reduced at higher altitudes because of lower temperatures, etc., the effect of glacial activity may counteract the altitude effect by making available abundant fine sediment for subsequent chemical weathering and, in some situations, by increasing specific discharge (water discharge per unit area of basin). Thus, several studies have documented high rates of chemical weathering in alpine glacierized basins (Collins, 1983; Sharp et al., 1995). However, Anderson et al. (1997) were unable to differentiate chemical weathering rates between numerous glacierized and nonglacial catchments throughout the world, with cation and silicate denudation rates often being lower in the former. The role of glacial activity in geochemical systems, they postulated, was primarily in producing very fine and altered sediment (by glacial erosion and grinding) which was subsequently more susceptible to chemical weathering, especially downstream of the glacial zone where environmental conditions are more favourable for rapid weathering activity. The chapter by Caine (this volume) provides further information on this issue.

5.2.5 Glacial lake outburst floods

Although not considered explicitly in Caine's classification of process systems in mountains, glacial lake outburst floods (GLOFs) are worthy of special mention. Such floods represent a particular coupling of glacial and fluvial processes unique to mountain environments. The chapter by Björnsson (this volume) reviews the different types of GLOFs and the mechanisms that cause them. In certain mountain environments (such as parts of Iceland), GLOFs may represent the dominant fluvial process responsible for transporting water, sediment (coarse and fine) and solutes from higher to lower elevations. Owing to the extreme magnitude often associated with GLOFs (cf. Snorrason et al., 2002), such events have major impacts on landscapes both within mountains and also in lowland areas where the floodwaters and associated sediments settle. A major research question here is the role of these floods in long- to medium-term landscape development in mountainous areas. Techniques for estimating the hydrologic and geomorphic magnitude–frequency characteristics of such floods (e.g., Cenderelli and Wohl, 2001), and numerical modelling approaches, may offer ways forward.

GLOFs also represent one of the most extreme geomorphic hazards in mountain environments and adjacent lands. The chapters by Björnsson and Hewitt (this volume) describe examples of GLOFs from different parts of the world.

5.3 Applied mountain geomorphology

An appreciation and understanding of the role of geomorphic processes in mountain environments is perhaps most relevant to the study of the interaction between society and mountains. With an ever-expanding population, society is increasingly interacting with mountains, and with this encroachment comes a need for an understanding of the effect of society on geomorphic processes and, conversely, of geomorphic processes on society. The chapters in this volume by Björnsson, Hewitt, Li and Thouret examine these relations in greater detail, focusing particularly on geomorphic hazards and the response of society to these, while Slaymaker and Owens (this volume) consider the effects of society on geomorphic processes and landscape evolution.

5.3.1 The effect of society on geomorphic processes

Probably the first, most widespread and fundamental change in mountain areas by human activity is that of change in vegetation cover. Deforestation, in particular, has been widespread in mountain areas in order to provide timber for fuel and as a building material, and in order to clear ground for agriculture (see Figure 12.10 of this volume). The removal of the natural vegetation cover exposes the soil to raindrop impact and surface erosion. Vegetation removal also reduces soil binding and alters the hydrological pathways by which water flows through the soil and regolith, and may increase the likelihood of mass failures, such as landslides and debris flows (cf. Onda, this volume), and downstream flooding. Ives and Messerli (1989) developed the 'Himalayan Environmental Degradation Theory' to explain the causes and effects of deforestation in the Himalaya. This theory illustrates well the feedback mechanisms between human activity, geomorphic processes and environmental degradation in mountains (also see Gerrard, 1990).

In addition to the geomorphic effects of widespread changes in land cover to support forestry, agriculture, mining, urban and recreation uses, the infrastructure required to support these activities (such as forest roads, highways, railways, and electricity, power and communication networks) also have major impacts on geomorphic processes by altering drainage pathways and by creating sediment sources. Figure 1.7 shows a large gully on a relatively steep forest road in the Maya Mountains, Belize, caused by excessive erosion of soil.

Figure 1.7 *A large gully on a relatively steep forest road in the Maya Mountains, Belize, caused by excessive erosion of soil on land cleared of the natural forest vegetation*
Photo: P.N. Owens.

There have been many studies which have either qualitatively or quantitatively determined an increase in the rate of operation of geomorphic processes due to changes brought about by human activities (including anthropogenic changes in climate). Most of these changes in the rate and nature of geomorphic processes are detrimental to the successful geomorphological and ecological functioning of mountain environments and the sustainability of such environments to support society. These detrimental effects include the occurrence of excessive soil erosion (Harden, 1991; Wilson and Seney, 1994), soil degradation (including the reduction of soil organic matter, nitrogen, phosphorus, potassium and important trace elements) (Chonghuan and Lixian, 1992; Sandor and Nash, 1995), an increase in the frequency of mass movement events (Montgomery et al., 2000; Barnard et al., 2001), the retreat of mountain glaciers (cf. Beniston, 2000) and degradation of mountain permafrost (Harris et al., 2001; Slaymaker and Owens, this volume). However, human activity does not always result in land degradation (see also Caine, this volume). Several studies (e.g., Harden, 1991; Inbar and Llerena, 2000) have demonstrated that abandonment, neglect and overgrazing of agricultural terraces in the Andes of South America (i.e., generally reduced human activity) may lead to an increase in erosion and sediment delivery to rivers, while other studies (e.g., Quine et al., 1992) have demonstrated how good land management (such as the creation of terraces) can lead to effective protection against erosion and sediment redistribution. Indeed, geomorphologists have much to offer in terms of management advice and technological solutions to reduce the detrimental environmental effects associated with development in mountain areas (cf. Chatwin and Smith, 1992; Clark and Howell, 1992), although the key here is to tailor solutions to local conditions and to recognize local socio-economic and political situations. Li (this volume) describes a variety of appropriate structural and bio-engineering methods for reducing geomorphic hazards in the mountains of China.

5.3.2 The effect of geomorphic processes on society

There is also a need to understand, predict and mitigate the effects of geomorphic processes on society. When such processes impinge on human activities there are often financial and human costs: causing a natural process to become a hazard. There are many different types of geomorphic hazard in mountain areas (Slaymaker, 1995, 1996; Kalvoda and Rosenfeld, 1998) and the chapters in this volume by Björnsson, Hewitt, Li and Thouret provide numerous examples: Hewitt provides an overview with emphasis on the Karakoram Mountains; Li focuses on some of the main geomorphic hazards in Chinese mountains; Thouret describes those hazards which occur on volcanic mountains (many of which are unique to this mountain type); and Björnsson gives examples of hazards associated with GLOFs.

It is important to realize that many geomorphic hazards associated with mountainous areas are not just local. Some processes can be potentially hazardous a considerable distance beyond the source area. Reid et al. (2001) describe a debris flow that originated from Mt Rainier, Washington State, USA, which travelled over 70 km from its source. Similarly, the effects of glacier lake outburst and sediment-dam failure floods, which are initiated in mountainous areas, are often considerable in low-lying areas far removed from mountains. Thus, the flood that occurred when a seismic event caused a massive landslide from Nanga Parbat to block the Indus River, which subsequently failed in June 1841, wreaked havoc for over 400 km downstream of the dam (Owen, 1988; Shroder et al., 1998). Other, similar, examples are described in the chapters by Björnsson, Hewitt, Li and Thouret.

Another consideration with geomorphic hazards in mountains is related to the perception of hazard (see also Hewitt, this volume). In many countries hazard assessment is conditioned not only by likely consequences but also by social, economic and political considerations, especially in less-developed or politically unstable countries. Also important here is the dilemma of allocating resources to dealing with high-magnitude (potentially catastrophic) hazards, and more immediate and pressing hazards of lower magnitude but greater frequency. Alford et al. (2000) describe such a dilemma in the Pamir Mountains of Tajikistan, where a massive earthquake-induced landslide blocked the valley of the Bartang River, causing a lake (Lake Sarez) now 60 km in length. In a worst-case scenario, collapse of the dam could affect an area inhabited by 5 million people. Yet rockfalls, debris flows, avalanches and flooding constantly impinge on local people and their activities. Alford et al. (2000) describe how a programme of lake monitoring and the installation of early warning detection instrumentation offer an appropriate compromise for a disaster that may never happen, in a region that cannot afford an expensive engineering solution.

Of fundamental importance to the reduction of the detrimental impacts of geomorphic processes in mountains are hazard monitoring, hazard mapping and regional risk assessment (also see chapters by Hewitt, Li and Thouret). Hazard mapping, in particular, has become an increasingly important tool in the development of mountain areas, not just in terms of urban development, infrastructure (e.g., road and railway) location and energy facilities (i.e., hydro-electric stations and dams) but also for agricultural land use and tourism (see Aulitzky, 1994; Heuberger and Ives, 1994; Kienholz and Mani, 1994). Hazard mapping usually involves the construction of maps of geomorphic features, underlying geology (solid and drift), soil, hydrological conditions, vegetation cover, land use and management and other anthropogenic factors. Initially 'mountain hazard maps' (maps of both natural and human-induced hazards, Ives and Messerli, 1981) were primarily used in developed countries in Europe and North America (e.g., Kienholz, 1977, 1978; Ives and Bovis, 1978; Dow et al., 1981), but increasingly such maps have become an important planning tool in developing countries (e.g., Ives and Messerli, 1981; Kienholz et al., 1983), particularly in Asia, where slope instability is often a major concern. It is important to recognize, however, that the nature and perception of hazards are likely to vary in contrasting mountain environments, and that the most appropriate solution will similarly vary, often due to local socio-economic and political factors. Thus, Ives and Messerli (1981) identified that soil erosion and loss from agricultural land, which is often not considered in hazard risk assessment, is a serious indirect hazard for people in the Kathmandu–Kakani area of Nepal, which merits consideration in the construction of local hazard maps.

The continuing development of numerical models, remote sensing, DEM and geographical information systems (GIS) techniques (e.g., Keylock and Domaas, 1999; Saito et al., 2001; Clerici et al., 2002; Malet et al., 2002; Marchi et al., 2002; Thouret, this volume) means that the ability of scientists to predict the location and timing of geomorphic hazards in mountain areas is increasing. Thus, Reid et al. (2001) investigated the potential instability of Mt Rainier, USA, and the associated hazard potential, using a three-dimensional slope stability method that combines detailed geological–geomorphological mapping with subsurface geophysical imaging. Despite important technological advances, the key to successful hazard mitigation is always likely to be detailed information on the type and rate of geomorphic processes. Here, the sediment budget approach, sediment fingerprinting techniques, and the use of deposited (i.e., alluvial fan, lake and floodplain) sediments (cf. Dietrich and Dunne, 1978; Warburton, 1990; Jordan and Slaymaker, 1991; Harden, 1993; Slaymaker, 1993; Owens and Walling, 2002) offer considerable potential for geomorphic hazard (including soil erosion) mapping and mitigation in mountainous areas, in that they enable sediment sources and pathways to be identified and quantified, and magnitude–frequency relations to be established.

6 Mountain geomorphology and global environmental change

This is the topic of the last chapter in this book, by Owens and Slaymaker, and the reader is directed to this for more information. One of the most pressing environmental issues of our time concerns the accelerating rate of environmental change, both locally and globally. Slaymaker and Owens (this volume) make the case that mountain environments and their communities are at special risk from global environmental change. Because the livelihoods of mountain peoples are so closely connected to the resources of the land, and because the forces of globalization are marginalizing the priority of mountain peoples, mountain peoples in turn are imposing greater pressures on the land base and land degradation is an almost universal feature of populated mountain environments. The whole set of interdependent processes that has been unleashed to produce global environmental change is still not well understood, and Slaymaker and Owens turn to complexity theory and the interpretation of adaptive cycles (as advocated by Holling, 2001) to suggest a way forward. Mutual dependence of mountain and lowland peoples, mutual dependence of mountain society and environment, and mutual dependence of people of the South and of the North, add up to a situation of increasing vulnerability both for mountain environments and for mountain peoples. It seems clear that little progress can be achieved without a massive increase in investment on monitoring of geomorphic and social change. New ideas and partnerships will be needed. In the context of mountain geomorphology, a radical approach will be necessary to avoid massive human and environmental disaster. Both institutional and technological innovation, based on listening to the needs of the indigenous mountain peoples, will, we believe, be a necessary part of that process.

7 Conclusion and setting

Mountains represent one of the most important morphological environments on the Earth's surface. Most people in some way depend on mountains for resources (e.g., food, water, fuel, minerals and medicine) and quality of life (e.g., recreation, tourism and biodiversity). With increasing pressures on mountains from society, and with concern over the effects of global environmental change, there is a need, now more than ever, to understand how mountains are created and evolve, to determine what processes modify them (under past, present and future environmental conditions), and to establish the interactions between mountains and society (and in particular how to minimize the risks associated with these interactions). Geomorphology lies at the heart of all of these needs. Chapters in this book are intended to contribute to our appreciation and understanding of mountain geomorphology. They are arranged according to the historical (Chapters 2 to 4), functional (Chapters 5 to 8) and applied (Chapters 9 to 11) mountain geomorphology classification system described above. Most of these chapters are concerned with the past and present. Chapter 12 examines the various ways in which past, present and *future* global environmental changes affect geomorphic processes and landscape evolution in mountain areas. Ultimately, the aim of this book is to encourage further research and to stimulate interest in these fascinating environments.

Acknowledgements

We would like to thank Nel Caine, Jack Ives, Cliff Ollier, Lewis Owen and Paul Williams for helpful advice and comments on an early version of this chapter. We would also like to thank Eric Leinberger for drawing some of the figures.

References

Alford, D., Cunha, S.F. and Ives, J.D., 2000. Lake Sarez, Pamir Mountains, Tajikistan: mountain hazards and development assistance. *Mountain Research and Development,* **20**: 20–23.

Anderson, S.P., Drever, J.I. and Humphrey, N.F., 1997. Chemical weathering in glacial environments. *Geology,* **25**: 399–402.

Aulitzky, H., 1994. Hazard mapping and zoning in Austria – methods and legal implications. *Mountain Research and Development*, **14**: 307–13.

Barnard, P.L., Owen, L.A., Sharma. M.C. and Finkel, R.C., 2001. Natural and human-induced landsliding in the Garhwal Himalaya of northern India. *Geomorphology*, **40**: 21–35.

Barsch, D. and Caine, N., 1984. The nature of mountain geomorphology. *Mountain Research and Development*, **4**: 287–98.

Beaumont, C., Kooi, H. and Willett, S., 2000. Coupled tectonic–surface process models with applications to rifted margins and collisional orogens. In Summerfield, M.A. (ed.), *Geomorphology and global tectonics*. Chichester: Wiley, 29–55.

Beniston, M., 2000. *Environmental change in mountains and uplands*. London: Arnold.

Benn, D.I. and Evans, D.J.A., 1998. *Glaciers and glaciation*. London: Arnold.

Birkeland, P.W., 1973. Use of relative age dating methods in a stratigraphic study of rock glacier deposits, Mt Sopris, Colorado. *Arctic and Alpine Research*, **5**: 401–16.

Bluth, G.J.S. and Kump, L.R., 1994. Lithologic and climatologic controls of river chemistry. *Geochemica et Cosmochimica Acta*, **58**: 2341–59.

Burbank, D.W. and Anderson, R.S., 2001. *Tectonic geomorphology*. Oxford: Blackwell Science.

Caine, N., 1974. The geomorphic processes of the alpine environment. In Ives, J.D. and Barry, R.G. (eds), *Arctic and alpine environments*. London: Methuen, 721–48.

Caine, N., 1984. Elevational contrasts in contemporary geomorphic activity in the Colorado Front Range. *Studia Geomorphologica Carpatho-Balcanica*, **18**: 5–31.

Cenderelli, D.A. and Wohl, E.E., 2001. Peak discharge estimates of glacial-lake outburst floods and 'normal' climatic floods in the Mount Everest region, Nepal. *Geomorphology*, **40**: 57–90.

Chang, J.-C. and Slaymaker, O., 2002. Frequency and spatial distribution of landslides in a mountainous drainage basin: Western Foothills, Taiwan. *Catena*, **46**: 285–307.

Chatwin, S.C. and Smith, R.B. 1992. Reducing soil erosion associated with forestry operations through integrated research: an example from coastal British Columbia, Canada. In Walling, D.E., Davies, T.R. and Hasholt, B. (eds), *Erosion, debris flows and environment in mountain regions*. IAHS Publication 209, Wallingford: IAHS Press, 377–85.

Chonghuan, N. and Lixian, W., 1992. A preliminary study of soil erosion and land degradation. In Walling, D.E., Davies, T.R. and Hasholt, B. (eds), *Erosion, debris flows and environment in mountain regions*. IAHS Publication 209, Wallingford: IAHS Press, 439–45.

Chorley, R.J., 1978. Bases for theory in geomorphology. In Embleton, C., Brunsden, D. and Jones, D.K.C. (eds), *Geomorphology: present problems and future prospects*. Oxford: Oxford University Press, 1–13.

Church, M. and Slaymaker, O., 1989. Disequilibrium of Holocene sediment yield in glaciated British Columbia. *Nature*, **337**: 452–54.

Church, M., Ham, D., Hassan, M. and Slaymaker, O., 1999. Fluvial sediment yield in Canada: a scaled analysis. *Canadian Journal of Earth Sciences*, **36**: 1267–80.

Church, M., Kellerhals, R. and Day, T.J., 1989. Regional clastic sediment yield in British Columbia. *Canadian Journal of Earth Sciences*, **26**: 31–45.

Clark, J.E. and Howell, J.H., 1992. Development of bioengineering strategies in rural mountain areas. In Walling, D.E., Davies, T.R. and Hasholt, B. (eds), *Erosion, debris flows and environment in mountain regions*. IAHS Publication 209, Wallingford: IAHS Press, 387–97.

Clerici, A., Perego, S., Tellini, C. and Vescovi, P., 2002. A procedure for landslide susceptibility zonation by the conditional analysis method. *Geomorphology*, **48**: 349–64.

Cockburn, H.A.P., Brown, R.W., Summerfield, M.A. and Seidl, M.A., 2000. Quantifying passive margin denudation and landscape development using a combined fission-track thermochronology and cosmogenic isotope analysis approach. *Earth and Planetary Science Letters*, **179**: 429–35.

Collins, D.N., 1983. Solute yield from a glacierized high mountain basin. In Webb, B.W. (ed.), *Dissolved loads of rivers and surface water quantity/quality relationships*. IAHS Publication 141, Wallingford: IAHS Press, 41–49.

Darmody, R.G., Thorn, C.E., Harder, R.L., Schlyter, J.P.L. and Dixon, J.C., 2000. Weathering implications of water chemistry in an arctic-alpine environment, northern Sweden. *Geomorphology*, **34**: 89–100.

Dedkov, A.P. and Moszherin, V.I., 1992. Erosion and sediment yield in mountain regions of the world. In Walling, D.E., Davies, T.R. and Hasholt, B. (eds), *Erosion, debris flows and environment in mountain regions*. IAHS Publication 209, Wallingford: IAHS Press, 29–36.

Desloges, J.R. and Gilbert, R., 1998. Sedimentation in Chilko lake: a record of the geomorphic environment of the eastern Coast Mountains of British Columbia, Canada. *Geomorphology*, **25**: 75–91.

Dietrich, W.E. and Dunne, T., 1978. Sediment budget for a small catchment in mountainous terrain. *Zeitschrift für Geomorphologie, Supplementband*, **29**: 191–206.

Dow, V., Kienholz, H., Plam, M. and Ives, J.D., 1981. Mountain hazard mapping: the development of a prototype combined hazards map, Monarch Lake Quadrangle, Colorado, U.S.A. *Mountain Research and Development*, **1**: 55–64.

Drever, J.I. and Zobrist, J., 1992. Chemical weathering of silicate rocks as a function of elevation in the southern Swiss Alps. *Geochemica et Cosmochimica Acta*, **56**: 3209–16.

Embleton, C. and King, C.A.M., 1975. *Glacial geomorphology*. London: Edward Arnold.

Fairbridge, R.W., 1968. Mountain systems. In Fairbridge, R.W. (ed.), *The encyclopaedia of geomorphology*. New York: Rheinhold.

Gerrard, A.J., 1990. *Mountain environments: an examination of the physical geography of mountains*. Cambridge, MA: MIT Press.

Gleadow, A.J.W. and Brown, R.W., 2000. Fission-track thermochronology and the long-term denudational responses to tectonics. In Summerfield, M.A. (ed.), *Geomorphology and global tectonics*. Chichester: Wiley, 57–75.

Goudie, A. (ed.), 1985. *The encyclopaedic dictionary of physical geography*. Oxford: Blackwell.

Gurnell, A., 1987. Suspended sediment. In Gurnell, A.M. and Clark, M.J. (eds), *Glacio-fluvial sediment transfer: an alpine perspective*. Chichester: Wiley, 305–54.

Hallet, B., Hunter, L. and Bogen, J., 1996. Rates of erosion and sediment evacuation by glaciers: a review of field data and their implications. *Global and Planetary Change*, **12**: 213–35.

Harbor, J. and Warburton, J., 1993. Relative rates of glacial and nonglacial erosion in alpine environments. *Arctic and Alpine Research*, **25**: 1–7.

Harden, C.P., 1991. Andean soil erosion. *National Geographic Research and Exploration*, **7**: 216–31.

Harden, C.P., 1993. Upland erosion and sediment yield in a large Andean drainage basin. *Physical Geography*, **14**: 254–71.

Harris, C., Haeberli, W., Mühll, D.V. and King, L., 2001. Permafrost monitoring in the high mountains of Europe: the PACE project in its global context. *Permafrost and Periglacial Processes*, **12**: 3–11.

Hartshorn, K., Hovius, N., Wade, W.B. and Slingerland, R.L., 2002. Climate driven bedrock incision in an active mountain belt. *Science*, **297**: 2036–38.

Hasholt, B., Walling, D.E. and Owens, P.N., 2000. Sedimentation in arctic proglacial lakes: Mittivakkat Glacier, south-east Greenland. *Hydrological Processes*, **14**: 679–99.

Heuberger, H. and Ives, J.D., 1994. Mountain hazard geomorphology – preface. *Mountain Research and Development*, **14**: 271–72.

Hewitt, K., 1989. The altitudinal organisation of Karakoram geomorphic processes and depositional environments. *Zeitschrift für Geomorphologie, Supplementband*, **76**: 9–32.

Hicks, D.M., McSaveney, M.J. and Chinn, T.J.H., 1990. Sedimentation in proglacial Ivory Lake, Southern Alps, New Zealand. *Arctic and Alpine Research*, **22**: 26–42.

Hodson, A., Gurnell, A., Tranter, M., Bogen, J., Hagen, J.O. and Clark, M., 1998. Suspended sediment yield and transfer processes in a small high arctic glacier basin, Svalbard. *Hydrological Processes*, **12**: 73–86.

Holling, C.S., 2001. Understanding the complexity of economic, ecological and social systems. *Ecosystems*, **4**: 390–405.

Hovius, N., Stark, C.P., Chu, H.-T. and Lin, J.C., 2000. Supply and removal of sediment in a landslide dominated mountain belt: Central Range, Taiwan. *Journal of Geology*, **108**: 73–89.

Inbar, M. and Llerena, C.A., 2000. Erosion processes in high mountain agricultural terraces in Peru. *Mountain Research and Development*, **20**: 72–79.

Ives, J.D. and Bovis, M.J., 1978. Natural hazard maps for land-use planning, San Juan mountains, Colorado, USA. *Arctic and Alpine Research*, **10**: 185–212.

Ives, J.D. and Messerli, B., 1981. Mountain hazard mapping in Nepal: introduction to an applied mountain research project. *Mountain Research and Development*, **1**: 223–30.

Ives, J.D. and Messerli, B., 1989. *The Himalayan dilemma: reconciling development and conservation*. London: Routledge.

Ives, J.D. and Messerli, B., 1999. AD 2002 declared by United Nations as 'International Year of the Mountains'. *Arctic, Antarctic and Alpine Research*, **31**: 211–13.

Jordan, P. and Slaymaker, O., 1991. Holocene sediment production in Lillooet River basin, British Columbia: a sediment budget approach. *Géographie physique et Quaternaire*, **45**: 45–57.

Kalvoda, J. and Rosenfeld, C.L. (eds), 1998. *Geomorphological hazards in high mountain areas*. Dordrecht: Kluwer.

Kapos, V., Rhind, J., Edwards, M., Price, M.F. and Ravilious, C., 2000. Developing a map of the world's mountain forests. In Price, M.F. and Butt, N. (eds), *Forests in sustainable mountain development: a state of knowledge report for 2000*. Task Force on Forests in Sustainable Mountain Development. Wallingford: CAB International, 4–9.

Keylock, C. and Domaas, U., 1999. Evaluation of topographic models of rockfall travel distance for use in hazard applications. *Arctic, Antarctic and Alpine Research*, **31**: 312–20.

Kienholz, H., 1977. Kombinierte Geomorphologische Gefahrenkarte 1:10,000 von Grindelwald. *Catena*, **3**: 265–94.

Kienholz, H., 1978. Maps of geomorphology and natural hazards of Grindelwald, scale 1:10,000. *Arctic and Alpine Research*, **10**: 169–84.

Kienholz, H. and Mani, P.J., 1994. Assessment of geomorphic hazards and priorities for forest management on the Rigi North Face, Switzerland. *Mountain Research and Development*, **14**: 321–28.

Kienholz, H., Hafner, H., Schneider, G. and Tamraker, R., 1983. Mountain hazards mapping in Nepal's Middle Mountains with maps of land use and geomorphic damage (Kathmandu–Kakani area). *Mountain Research and Development*, **3**: 195–220.

Kotarba, A., 1987. *High-mountain denudational system of the Polish Tatra Mountains*. Geographical Studies Special Issue No. 3, Ossolineum, Warsaw: Polish Academy of Sciences.

Litaor, M.I., 1987. Influence of eolian dust on the genesis of alpine soils in the Front Range, Colorado. *Soil Science Society of America Journal*, **51**: 142–47.

Malet, J.-P., Maquaire, O. and Calais, E., 2002. The use of Global Positioning System techniques for the continuous monitoring of landslides: application to the Super-Sauze earthflow (Alpes-de-Haute-Provence, France). *Geomorphology*, **43**: 33–54.

Marchi, L., Arattano, M. and Deganutti, A.M., 2002. Ten years of debris-flow monitoring in the Moscardo Torrent (Italian Alps). *Geomorphology*, **46**: 1–17.

Messerli, B. and Ives, J.D. (eds), 1997. *Mountains of the world: a global priority*. London: Parthenon.

Meybeck, M., Green, P. and Vörösmarty, C., 2001. A new typology for mountains and other relief classes: an application to global continental water resources and population distribution. *Mountain Research and Development*, **21**: 34–45.

Meyer, L., 2000. Application of digital elevation models to macroscale tectonic geomorphology. In Summerfield, M.A. (ed.), *Geomorphology and global tectonics*. Chichester: Wiley, 16–27.

Montgomery, D.R., Schmidt, K.M., Greenberg, H.M. and Dietrich, W.E., 2000. Forest clearing and regional landslides. *Geology*, **28**: 311–14.

Mountain Agenda, 1992. *An appeal for mountains*. Berne: Geographical Institute, University of Berne, Switzerland.

Mountain Agenda, 1997. *Mountains of the world: challenges for the 21st century*. Berne: Geographical Institute, University of Berne, Switzerland.

Ollier, C.D. and Pain, C.F., 2000. *The origin of mountains*. London: Routledge.

Owen, L.A., 1988. Neotectonics and glacial deformation in the Karakorum Mountains and Nanga Parbat Himalaya. *Tectonophysics*, **163**: 227–65.

Owens, P.N. and Slaymaker, O., 1992. Late Holocene sediment yields in small alpine and subalpine drainage basins, British Columbia. In Walling, D.E., Davies, T.R. and Hasholt, B. (eds), *Erosion, debris flows and environment in mountain regions*. IAHS Publication 209, Wallingford: IAHS Press, 147–54.

Owens, P.N. and Slaymaker, O., 1997. Contemporary and post-glacial rates of aeolian deposition in the Coast Mountains of British Columbia, Canada. *Geografiska Annaler*, **79A**: 267–76.

Owens, P.N. and Walling, D.E., 2002. Changes in sediment sources and floodplain deposition rates in the catchment of the River Tweed, Scotland, over the last 100 years: the impact of climate and land use change. *Earth Surface Processes and Landforms*, **27**: 403–23.

Press, F. and Siever, R., 1982. *Earth*. San Francisco, CA: W.H. Freeman and Company.

Price, M.F., 1995. *Mountain research in Europe: an overview of MAB research from the Pyrenees to Siberia*. Paris and Casterton: UNESCO and Parthenon.

Price, M.F. and Butt, N., 2001. *Forests in sustainable mountain development: a state of knowledge report for 2000*. Task Force on Forests in Sustainable Mountain Development. Wallingford: CAB International.

Quine, T.A., Walling, D.E., Zhang, X. and Wang, Y., 1992. Investigation of soil erosion on terraced fields near Yanting, Sichuan province, China, using caesium-137. In Walling, D.E., Davies, T.R. and Hasholt, B. (eds), *Erosion, debris flows and environment in mountain regions*. IAHS Publication 209, Wallingford: IAHS Press, 155–68.

Rapp, A., 1960. Recent developments of mountain slopes in Kärkevagge and surroundings, northern Scandinavia. *Geografiska Annaler*, **42A**: 73–200.

Reid, M.E., Sisson, T.W. and Brien, D.L., 2001. Volcano collapse promoted by hydrothermal alteration and edifice shape, Mount Rainier, Washington. *Geology*, **29**: 779–82.

Saito, T., Eguchi, T., Takayama, K. and Taniguchi, H., 2001. Hazard predictions for volcanic explosions. *Journal of Volcanology and Geothermal Research*, **106**: 39–51.

Sandor, J.A. and Nash, N.S., 1995. Ancient agricultural soils in the Andes of Peru. *Soil Science Society of America Journal*, **59**: 170–79.

Sharp, M., Tranter, M., Brown, G.H. and Skidmore, M., 1995. Rates of chemical denudation and CO_2 breakdown in a glacier-covered alpine catchment. *Geology*, **23**: 61–64.

Short, N.M. and Blair, R.W. (eds), 1986. *Geomorphology from space*. Washington, DC: NASA.

Shroder, J.F., Jr, Bishop, M.P. and Scheppy, R., 1998. Catastrophic flood flushing of sediment, western Himalaya, Pakistan. In Kalvoda, J. and Rosenfeld, C.L. (eds), *Geomorphological hazards in high mountain areas*. Dordrecht: Kluwer, 27–48.

Slaymaker, O., 1974. Alpine hydrology. In Ives, J.D. and Barry, R.G. (eds), *Arctic and alpine environments*. London: Methuen, 134–55.

Slaymaker, O., 1987. Sediment and solute yields in British Columbia and Yukon: their geomorphic significance re-examined. In Gardiner, V. (ed.), *International geomorphology*. Chichester: Wiley, 925–45.

Slaymaker, O., 1991. Mountain geomorphology: a theoretical framework for measurement programmes. *Catena*, **18**: 427–37.

Slaymaker, O., 1993. The sediment budget of the Lillooet River basin, British Columbia. *Physical Geography*, **14**: 304–20.

Slaymaker, O. (ed.), 1995. *Steepland geomorphology*. Chichester: Wiley.

Slaymaker, O. (ed.), 1996. *Geomorphological hazards*. Chichester: Wiley.

Slaymaker, O., 1999. Mountain environments. In Alexander, S.E. and Fairbridge, R.W. (eds), *Encyclopedia of environmental science*. Dordrecht: Kluwer, 413–16.

Slaymaker, O., in press. Mountain geomorphology. In Goudie, A.S. (ed.), *Encyclopedia of geomorphology*. London: Routledge.

Snorrason, A., Finnsdóttir, H.P. and Moss, M.E. (eds), 2002. *The extremes of the extremes: extraordinary floods*. IAHS Publication 271, Wallingford: IAHS Press.

Stone, P.B. (ed.), 1992. *The state of the world's mountains: a global report*. London: Zed Books.

Strahler, A.N., 1946. Geomorphic terminology and classification of masses. *Journal of Geology*, **54**: 32–42.

Sueker, J.K., Clow, D.W., Ryan, J.N. and Jarrett, R.D., 2001. Effect of basin physical characteristics on solute fluxes in nine alpine/subalpine basins, Colorado, USA. *Hydrological Processes*, **15**: 2749–69.

Summerfield, M.A., 2000. Geomorphology and global tectonics: introduction. In Summerfield, M.A. (ed.), *Geomorphology and global tectonics*. Chichester: Wiley, 3–12.

Thorn, C.E. and Darmody, R.G., 1985. Grain size sampling and characterisation of eolian lag surfaces within alpine tundra, Niwot Ridge, Front Range, Colorado, USA. *Arctic and Alpine Research*, **17**: 443–50.

Troll, C., 1972. Geoecology and the world-wide differentiation of high-mountain ecosystems. In Troll, C. (ed.), *Geoecology of the high mountain regions of Eurasia*. Wiesbaden: Franz Steiner, 1–13.

Troll, C., 1973. High mountain belts between the polar caps and the equator: their definition and lower limit. *Arctic and Alpine Research*, **5**: 19–27.

Tucker, G.E. and Slingerland, R., 1996. Predicting sediment flux from fold and thrust belts. *Basin Research*, **8**: 329–50.

United Nations Environment Programme, 2003. Managing fragile ecosystems: sustainable mountain development. Chapter 13. http://www.unep.org/Documents/Documents.htm (last accessed 20 January 2003).

United Nations Environment Programme – World Conservation Monitoring Centre, 2001. Mountains and mountain forests. http://www.unep-wcmc.org/habitats/mountains/background.htm (last accessed 16 September 2001).

Warburton, J., 1990. An alpine proglacial fluvial sediment budget. *Geografiska Annaler*, **72A**: 261–72.

Warburton, J., 1999. Environmental change and sediment yield from glacierised basins: the role of fluvial processes and sediment storage. In Brown, A.G. and Quine, T.A. (eds), *Fluvial processes and environmental change*. Chichester: Wiley, 363–84.

Wilson, J.P. and Seney, J.P., 1994. Erosional impact of hikers, horses, motorcycles, and off-road bicycles on mountain trails in Montana. *Mountain Research and Development*, **14**: 77–88.

Windley, B.F., 1995. *The evolving continents*. Chichester: Wiley.

Wohl, E.E., 2000. *Mountain rivers*. Water Resources Monograph 14, Washington, DC: American Geophysical Union Press.

Yamada, S., 1999. Mountain ordering: a method for classifying mountains based on their morphometry. *Earth Surface Processes and Landforms*, **24**: 653–60.

PART 2 Historical Mountain Geomorphology

Cascade Mountain, Banff, Alberta, Canada
Photo: P.N. Owens

2
Cenozoic evolution of global mountain systems

Lewis A. Owen

1 Introduction

Mountain systems are among the most prominent geomorphic features on the Earth. Tectonically, they are major belts of pervasive deformation that include thick sequences of shallow-water sandstones, limestones and shales deposited on continental crust, and oceanic deposits characterized by deep-water turbidites and pelagic sediments, commonly with volcaniclastic sediments and volcanic rocks. Typically, mountain systems have been deformed and metamorphosed to varying degrees and intruded by plutonic rocks, chiefly of granitic affinity (Moores and Twiss, 1995).

The geologic evolution of these orogenic belts is complex and may span hundreds of millions of years. During the latter part of the twentieth century the application of plate tectonic theories to the study of orogenic belts revolutionized the understanding of the dynamics and evolution of these systems. Furthermore, the rapid development of geophysical and geochemical techniques has aided the measurement, monitoring and modelling of the evolution of mountain systems on local, regional and global scales. Contemporary research on the evolution of mountain systems involves most branches of geology, particularly geodesy, geophysics, geochemistry, structural geology, sedimentology, stratigraphy, geomorphology and palaeoclimatology (Zeitler *et al.*, 2001; Bishop *et al.*, 2002).

A casual comparison of topographic and tectonic maps of the world clearly shows that the major high mountain systems occur along or are parallel to lithospheric plate boundaries (Figure 2.1). The majority of these mountain systems began to form and largely evolved during the Cenozoic (~65 Ma to present). These are commonly referred to as 'young' mountain systems and 'active' if they are presently deforming. Of particular note are the Alpine–Himalayan–Tibetan and the Circum-Pacific orogenic systems, and the ocean ridges. Closer inspection reveals regionally extensive and significant mountain belts of lesser relief. These 'ancient' mountain systems generally have little or no relationship to the present lithospheric plate boundaries and may have begun to have formed many hundreds of millions of years ago. Despite their age and distance from plate margins these mountain systems may still experience deformation, albeit not so dramatic as young active mountain belts. Often their major pervasive geologic structures are zones of discontinuity along which earthquakes may occur. The Appalachian–Caledonide system is one of the best examples. This mountain system stretched for some 6000 km and now includes the Caledonides of east Greenland, Svalbard, Ireland, Britain and Scandinavia, the Appalachians of the USA and Canada, the Innuitian Mountains of Arctic Canada and Greenland, the Ouachita Mountains of south-central USA, the Cordillera Oriental in Mexico, the Venezuelan Andes and the West African fold belt. The evolution of this mountain system began in the late Precambrian (>~570 Ma), and the deformation and mountain building occurred during three major orogenies, during the early, middle and late Palaeozoic (between ~570 Ma and ~250 Ma). The mountain system

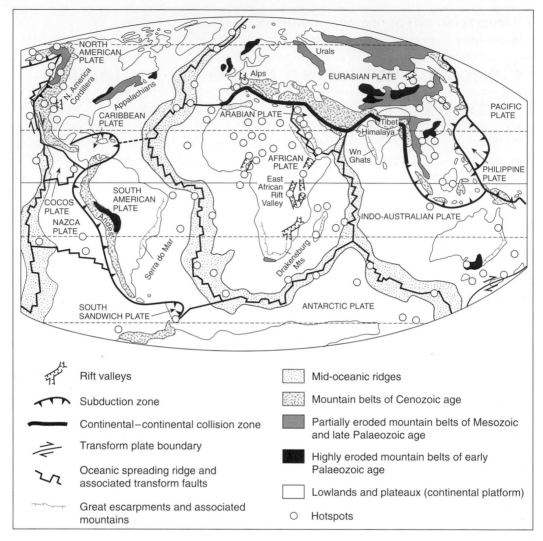

Figure 2.1 *The distribution of mountain systems showing their relationship to plate boundaries and tectonic settings*
Adapted from Uyeda (1978), Vogt (1981) and Summerfield (1991a).

was subsequently broken up with the opening of the Atlantic and is extensively covered by Mesozoic (~250 Ma to ~65 Ma) continental margins and the Atlantic Ocean. Nevertheless, its remains still constitute impressive mountain ranges.

The study of young active mountain systems provides knowledge and understanding of the dynamics of mountain building that may be used to understand contemporary and ancient systems, and can aid in effective management and hazard mitigation in mountainous regions. The aim of this chapter is to provide a framework for understanding the evolution of Cenozoic mountain systems that can be applied to help explain contemporary landscapes and the evolution of ancient and young orogens. Particular emphasis is placed on the Alpine–Himalayan–Tibetan orogen that constitutes part of the highest and greatest mountain mass on Earth and is hence one of the best natural laboratories to study the nature and dynamics of orogenic processes.

2 Geographic extent of global mountain belts

The association of young active mountains with plate boundaries reveals that major mountain systems occur in three main tectonic settings: continental–continental collision zones; subduction related settings (oceanic–oceanic and continental–oceanic collision zones); and oceanic spreading ridges. Other young mountains, however, are associated with transform plate boundaries, hotspots, rift systems and passive margins.

The longest mountain system is associated with the oceanic spreading ridges and extends for >40000 km (Figure 2.1). Although these mountains may rise in elevation by >5 km from the ocean floors, they only occur above sea level where an oceanic spreading ridge astrides a hotspot. The Icelandic hotspot that is broadly coincident with the mid-Atlantic ridge and helps to form Iceland provides a contemporary example (Gudmundsson, 2000).

The largest mountain mass on Earth, however, is the Alpine–Himalayan–Tibetan system. This stretches from the Betic Mountains in Southern Spain through the European Alps, the Turkish–Iranian Plateau, the Zagros Mountains, the Himalaya, the Tibetan Plateau, to the Sumatra arc of Indonesia and is some 7000 km long and exceeds 2000 km at its widest part (Figure 2.1). Mountain ranges such as the Tien Shan and Gobi Altai Mountains are also part of this orogen. These mountains are associated with the collision of the African and Indian continental lithospheric plates with the Eurasian continental lithospheric plate.

The Circum-Pacific oceanic–oceanic and continental–oceanic collision zones constitute the next major mountain systems of note. These include the Antarctic Peninsula, Andes, Western Cordillera of North America, and the volcanic island arcs of the Aleutians through to Japan and the Philippines and on to New Guinea (Figure 2.1).

Mountain systems that are associated with other tectonic settings and include transform plate boundaries, passive margins and hotspots are not really of continental/global scale but are impressive topographic features (Figure 2.1). These include the Alps of New Zealand, which provide one of the best examples of a mountain system associated with a transform plate boundary. This is the result of the relative motion between the Antarctic, Indian Australian and Pacific plates (Tippett and Hovius, 2000; Williams, this volume). The Transverse Ranges of Southern California within the San Andreas–Gulf of California transform system provide another example of a mountain system within a transform plate boundary (Cox et al., 2003). These essentially form within the double bend of the San Andreas fault system and they rise from a few hundred metres to 3500 m above sea level (asl) within little more than 10 km.

The Western Ghats of India and Drakensberg Mountains of South Africa are impressive examples of mountain ranges that have formed along passive margins (cf. Ollier, this volume). These are thought to be the result of uplift due to denudational unloading and isostatic flexuring as the adjacent plateau regions are eroded along their margins (Gilchrist and Summerfield, 1990, 1994; Summerfield, 1991a, b; Brown et al., 2000; Gunnell and Fleitout, 2000).

Mountains produced by hotspots are a consequence of regional warping and associated volcanism and rifting. The Grand Tetons in Wyoming provide a spectacular example of uplift along a rifted margin associated with a hotspot, in this case related to the Yellowstone hotspot (Love and Reed, 1971; Pierce and Morgan, 1992). The Hawaiian Islands–Emperor Seamount chain provide an example of volcanic mountains that have grown over the Hawaiian hotspot as the Pacific plate has moved progressively northwestwards and then westwards over time. However, such mountains subside as they are tectonically transported away from the hotspot and as their mass increases and causes isostatic subsidence (Watts and ten Brink, 1989).

3 Characteristics of Cenozoic mountain belts

The greatest mountain systems traverse many climatic belts. As a consequence they include along their length nearly every environmental and geomorphic setting. For example, they may include tropical rainforest, deciduous forest, alpine meadows, tundra, desert and glacial environments (Troll, 1973a, b). Since most Cenozoic mountains exceed 5000 m asl, they are extensively glacierized. They commonly have a precipitation gradient across their ranges and rainshadows on their leeward slopes. The steep slopes and glacierized catchments result in high river discharges and extensive landsliding.

The geomorphic processes within these environments play a major role in shaping the landscapes. Furthermore, it is becoming increasingly apparent that denudation influences the tectonism in these regions by such processes as denudational unloading and basin subsidence resulting from the thick piles of sediments that are deposited in the forelands (Montgomery, 1994; Gilchrist *et al.*, 1994; Shroder and Bishop, 2000; Bishop *et al.*, 2002).

Dramatic climatic changes have taken place throughout the Cenozoic, and particularly throughout the Quaternary. This has caused major fluctuations in the magnitude and frequency of Earth surface processes in mountain regions. Moreover, the mountain uplift may have also contributed to climate change throughout the Cenozoic by affecting global atmospheric circulation, deflecting jetstreams, initiating and enhancing monsoons and altering biogeochemical cycles (Ruddiman and Kutzbach, 1989; Raymo and Ruddiman, 1992; Ruddiman, 1997, 1998; Ramstein *et al.*, 1997). Such are the links and feedbacks between tectonism, climate, Earth surface processes and biology that research in the evolution of Cenozoic mountain systems is becoming increasingly multidisciplinary.

Despite the variety of tectonic and geomorphic settings for mountain systems, the two largest subaerial mountain systems, the Alpine–Himalayan–Tibetan and the Circum-Pacific systems, have a number of similarities in their evolution and geologic characteristics. In the mature stages of the orogen, the mountain system may be broadly divided into geologic and topographic belts. These are illustrated in Figure 2.2(A) and include:

1 an outer foredeep or foreland basin;
2 a foreland fold-and-thrust belt;
3 a crystalline core complex that includes: sedimentary rocks and their basement; volcanic and igneous rocks and associated sediments; metamorphosed ocean crust (ophiolites); gneissic terranes with abundant ultramafic bodies; and granitic batholiths;
4 rectilinear (high-angle) fault zones.

The Himalayan–Tibetan region illustrates this well. It exhibits all these belts, although they are developed to varying degrees along different transects of the orogen (Figure 2.2(B)–(D)).

Geologic observations of orogenic belts suggest that a sequence of events occurs as part of an orogenic cycle (Moores and Twiss, 1995). These events are summarized in Table 2.1. Dilek and Moores (1999) illustrate some of these similarities in their comparative study of the early Tertiary Western United States Cordillera and the modern Tibetan and Turkish–Iranian Plateau. They stressed that, as a consequence of an orogenic belt becoming overthickened, the mountains become the loci of lithospheric extension and experience tectonic collapse during their late-stage post-collisional evolution. It follows that the hinterland of major orogenic belts share a common taphrogenic (rifting) evolutionary path. This is related to rapid increase in the geothermal gradient and thus rapid isobaric heating, prograde high-temperature metamorphism, intrusion of post-tectonic granites and the extrusion of ignimbrites and associated minor extension. This phase is commonly followed by a further increase in the geothermal gradient, accelerated lithospheric extension and thinning with erosional denudation, superposition of high-temperature/low-pressure metamorphic assemblages, mantle partial melting

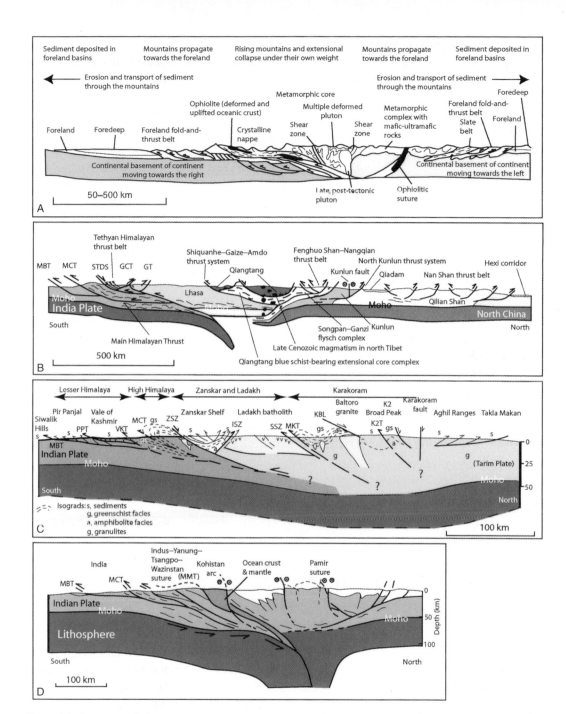

Figure 2.2 *Comparison of selected cross-sections across the Himalayan–Tibetan orogenic belt with a schematic cross-section across a model composite orogenic belt. (A) Model composite orogenic belt showing the major structures and tectonic components (adapted from Hatcher and Williams, 1986, and Moores and Twiss, 1995). Schematic sections across (B) the Himalaya, Tibet and Qilian Shan from Nepal to the Hexi Corridor (after Yin and Harrison, 2000); (C) the western Himalaya and central Karakoram (after Searle, 1991); and (D) the Himalaya, Kohistan and Pamir (adapted from Mattauer, 1986). Figure 2.5 shows the locations of sections (B), (C) and (D). GCT, Greater Counter Thrust; GT, Gangdese Thrust; ISZ, Indus Suture Zone; K2T, K2 Thrust; KBL, Karakoram Batholith Lineament; MBT, Main Boundary Thrust; MCT, Main Central Thrust; MKT, Main Karakoram Thrust; MMT, Main Mantle Thrust; PPT, Pir Panjal Thrust; STDS, South Tibet Detachment System; SSZ, Shyok Suture Zone; VKT, Vale of Kashmir Thrust; XF, Xianshuihe Fault; ZSZ, Zanskar Shear Zone. The Moho marks the boundary between the crust and the mantle*

Table 2.1 – Sequence of events in an orogenic cycle

1	Accumulation in separate areas of thick deposits of both shallow-water and deep-water marine sediments, the latter in association with intrusions or extrusions of mafic or intermediate magmatic rocks
2	Commencement of deformation in the foreland fold-and-thrust belt together with the emplacement of ophiolitic rocks and the subsequent isostatic rise of the ophiolite and the deformed sediments beneath it
3	Continued deformation in the fold-and-thrust belt – and metamorphism, deformation, and intrusion of granitic batholiths in the core zone – together with deposition of synorogenic sediments
4	Further isostatic rise of the orogenic region and the deposition and partial deformation of post-orogenic continental sediments in the outer foredeep
5	Block faulting, the development of fault-bounded basins, and the intrusion of scattered alkalic dykes and intrusive bodies

Source: Moores and Twiss (1995).

and mafic magmatism, and rapid subsidence and deposition of nonmarine sediments. This sequence of events and the similarity of tectonic structures for the Tibetan Plateau and Himalaya, Turkish–Iranian Plateau, and Western US Cordillera and Great Basin are summarized in Table 2.2 and Figure 2.3. In Figure 2.3(C) it should be noted that the North American craton is underplating the Sevier thrust belt and the overall morphology and tectonics of the high plateau and the Great Basin are analogous to the Tibetan Plateau and Turkish–Iranian Plateau. Furthermore, the Great Basin, Himalayan–Tibetan and Turkish–Iranian plateaux all adjoin a suture zone (union of lithospheric scale units), where continental apposition occurred and where major shortening and imbrication took place resulting in crustal overthickening and surface uplift.

Clearly, these observations and sequence of events are somewhat simplistic and the evolution of each individual orogen varies spatially and temporally. This model, however, does provide a working framework to help understand the evolution of Cenozoic and ancient mountain systems. Some of these differences and the detailed evolution of several of the major mountain ranges will now be discussed in more detail.

4 Alpine–Himalayan–Tibetan orogenic belt

The Alpine–Himalayan–Tibetan orogenic belt incorporates the Betic Mountains, European Alps, Zagros, Himalaya, Trans-Himalaya, Tibetan Plateau and its ranges, Tien Shan and the Gobi Altai. Several major zones of continental–continental collision are evident along its length and these include: the Alps (African–European collision); the Turkish–Iranian Plateau (Arabian–Asian collision); and the Himalayan–Tibetan orogen (Indian–Asian collision) (Figure 2.1). These major zones of convergence, for most of the orogen, are shown in Figure 2.3(C). Until the beginning of the 1980s little attention had been given to the orogen outside of the Alps. This was mainly due to political and logistical problems. However, during the last two decades considerable efforts have been made to study the evolution of Tibet and its bordering mountains. Unfortunately, studies of the Turkish–Iranian Plateau are still few because of the difficulties of fieldwork in this politically sensitive part of the world.

Some of the first orogenic studies were undertaken in the European Alps and their influence still persists in modern geology (Hsu, 1995). For example, the concept of fold nappes (sheet-like units of deformed rock that have moved on a predominant horizontal surface as a result of thrust faulting, recumbent folding or both mechanisms) was first introduced in 1841 by an Alpine geologist, Escher

Table 2.2 – Nature and chronology of tectonic and magmatic events during the taphrogenic evolution of orogenic belts with comparisons from the Tibetan Plateau and Himalaya, Turkish–Iranian Plateau, and Western US Cordillera and Great Basin

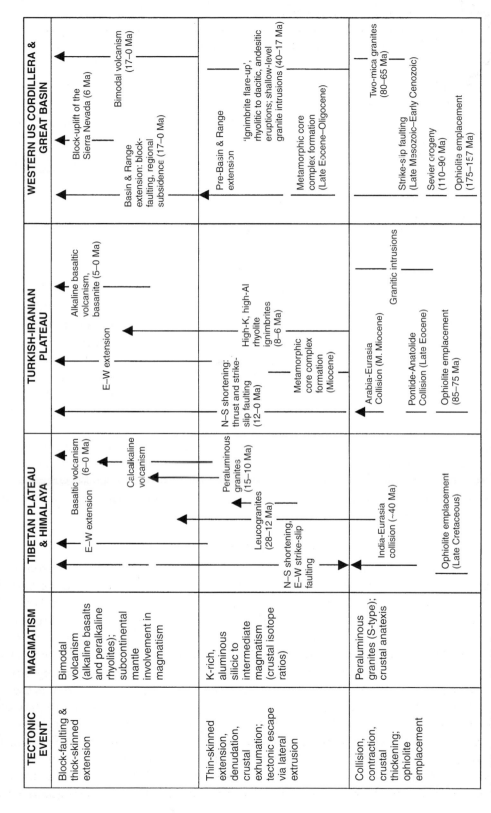

TECTONIC EVENT	MAGMATISM	TIBETAN PLATEAU & HIMALAYA	TURKISH-IRANIAN PLATEAU	WESTERN US CORDILLERA & GREAT BASIN
Block-faulting & thick-skinned extension	Bimodal volcanism (alkaline basalts and peralkaline rhyolites); subcontinental mantle involvement in magmatism	Basaltic volcanism (6–0 Ma); E–W extension; Calcalkaline volcanism	Alkaline basaltic volcanism, basanite (5–0 Ma); E–W extension	Block-uplift of the Sierra Nevada (6 Ma); Bimodal volcanism (17–0 Ma); Basin & Range extension: block-faulting, regional subsidence (17–0 Ma)
Thin-skinned extension, denudation, crustal exhumation; tectonic escape via lateral extrusion	K-rich, aluminous silicic to intermediate magmatism (crustal isotope ratios)	Peraluminous granites (15–10 Ma); Leucogranites (28–12 Ma); N–S shortening, E–W strike-slip faulting	High-K, high-Al rhyolite ignimbrites (8–6 Ma); Metamorphic core complex formation (Miocene); N–S shortening: thrust and strike-slip faulting (12–0 Ma)	Pre-Basin & Range extension; 'Ignimbrite flare-up', rhyolitic to dacitic, andesitic eruptions; shallow-level granite intrusions (40–17 Ma); Metamorphic core complex formation (Late Eocene–Oligocene)
Collision, contraction, crustal thickening; ophiolite emplacement	Peraluminous granites (S-type); crustal anatexis	India-Eurasia collision (~40 Ma); Ophiolite emplacement (Late Cretaceous)	Arabia-Eurasia Collision (M. Miocene); Granitic intrusions; Pontide-Anatolide Collision (Late Eocene); Ophiolite emplacement (85–75 Ma)	Two-mica granites (80–65 Ma); Strike-s ip faulting (Late Mesozoic–Early Cenozoic); Sevier orogeny (110–90 Ma); Ophiolite emplacement (175–157 Ma)

After Dilek and Moores (1999). Reproduced with the permission of the Geographical Society of London.

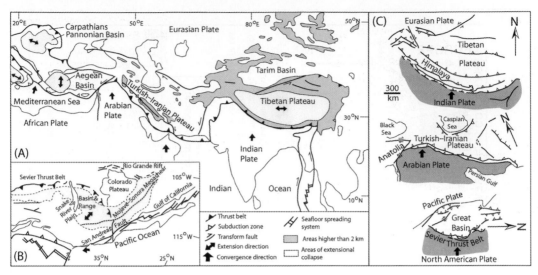

Figure 2.3 *Simplified tectonic maps of the (A) Alpine–Himalayan–Tibetan orogenic belt and (B) the Western US Cordillera showing areas of high elevation (>2000 m asl), extensional orogenic collapse, and plate convergence. (C) Simplified tectonic maps of the Himalaya–Tibet, Turkish–Iranian Plateau, and the pre-Basin and Range Western US Cordillera at the same scale. The plate boundaries, geometry of the collision zones, and the direction of relative plate motion are shown. Large arrows show the direction of relative plate motion*
Adapted from Dilek and Moores (1999).

van der Linth (Ryan, 2000). Despite the Alps being the most studied of all orogenic belts, its history has still to be fully understood because of its complex evolution involving a combination of subduction of Mesozoic oceanic crust, ophiolite emplacement, back-arc spreading, volcanism, metamorphism, thrusting and nappe emplacement, denudation and foreland basin sedimentation.

The Alpine sector of the Alpine–Himalayan orogenic belt is extensive, stretching from Gibraltar to Turkey, and includes the Betic Mountains, European Alps, Dinarides, Hellenides and Carpathians, while the Turkish–Iranian sector includes the Turkish–Iranian and Zagros Mountains (Figures 2.3 and 2.4(A)). The Alpine–Iranian belt developed on late Palaeozoic Hercynian (~345–225 Ma) and late Proterozoic–early Palaeozoic Pan-African (~800–500 Ma) orogenic belts as the continental plates of Africa and Arabia advanced into the Eurasian continental plate. The movement of Africa and Arabia into Eurasia is the consequence of the opening of the Atlantic and Indian Oceans. The convergence history is therefore complex and this has resulted in an orogen that varies considerably along its length. It has also resulted in abrupt curves and large changes in the strike of fold-and-thrust belts along its length. Bends of 90° to 180°, for example, characterize the Gibraltar region, the Alps, the Carpathians and the Balkanides. These bends, together with the complex fold-and-thrust vergences, suggest that considerable rotation and/or strike-slip deformation must have taken place. Furthermore, as illustrated in Figure 2.4(B), the orogen is more complex than the simple bilateral model shown in Figure 2.2(A). This section through the Swiss Alps shows two outward-directed thrust sequences on either side of a metamorphic core, associated with an apparent offset of the Moho (the Mohorovičić Discontinuity, which marks the boundary between the crust and mantle). In comparison, the orogen is more symmetrical in the Dinarides and the Carpathians where thrusts verge in opposite directions away from the volcanic rich Pannonian Basin (Figure 2.4(A)). Similarly in Turkey, the Tauride and Pontide thrust complexes bound either side of a central core of deformed and metamorphosed rock and younger volcanic rock that underlie the central Anatolian Plateau (Moores and Twiss, 1995).

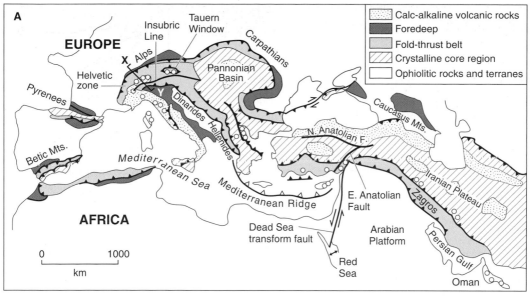

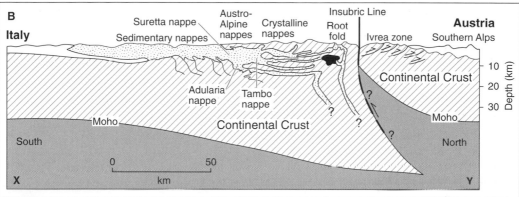

Figure 2.4 *Characteristics of the Alpine–Mediterranean sector of the Alpine–Himalayan orogenic belt. (A) Simplified geologic map showing the main structural features (adapted from Dewey et al., 1973, and Moores and Twiss, 1995). (B) Cross-section through the Swiss Alps showing recumbent nappes and root fold in the crystalline core zone of the Alps (after Laubscher, 1982)*

No one model explains the tectonic evolution of the whole of the Alpine–Iranian sector. Nevertheless, much of the alpine sector can be explained by developing a model that involves the formation and deformation of island arcs and associated basins, and the collision of a microcontinent (Penninic) with the Apulia (eastern Italy, the Ionian Sea, Slovenia, Croatia, Bosnia, Albania, Montenegro, Greece and western Turkey), and ultimately Europe (Roeder, 1977). Simplified, the Alpine sector really began to form in the middle Cretaceous (~120 Ma) with the subduction of the southern ocean basin beneath an island arc that was separated from the passive margin of the Apulian terrane by a marginal basin. This was followed (~110 Ma) by the collision of the Penninic microcontinent and deformation of the overriding island arc and marginal basin. During the late Cretaceous (~90 Ma), the Apulian continental margin overrode the marginal basin intensifying the collision zone. Following this, the northern basin closed as it was subducted under an arc separated from the rifted passive margin (Helvetic miogeocline) by a back-arc basin. The southern continental mass collided and overrode the northern continental mass deforming the arc, back-arc basin and Helvetic miogeocline during the early Miocene (~20 Ma). Since the late Neogene (5–0 Ma) there has been continued convergence and shortening resulting in nappe emplacement and backfolding (Roeder, 1977; Moores and Twiss, 1995).

Although the Himalayan–Tibetan orogen is commonly considered to be one of the youngest mountain belts, its history spans far beyond the beginning of the Cenozoic. Initially, throughout the early Palaeozoic, this involved the sequential accretion of microcontinents and island arcs onto the southern margin of Eurasia (Hsu et al., 1995; Sengor and Natal'in, 1996). This was followed by the collision of the Indian continental lithospheric plate with the Eurasia continental lithospheric plate between 50 and 70 Ma (Yin and Harrison, 2000). During the past 40–50 Ma the Indian plate has been moving at a nearly constant rate of ~50 mm a^{-1} northward with respect to stable Eurasia, resulting in between 1400 and 2000 km of crustal shortening (Molnar and Tapponnier, 1975; Patriat and Achache, 1984; DeMets et al., 1994). Ultimately, this led to the formation of the present Tibetan Plateau and the adjacent mountains. The collision of India into Asia helped to rejuvenate the Tien Shan orogen and has affected regions as far north as the Gobi Altai Mountains and Baikal rift, and may have played a role in the opening of the South China Sea (Molnar and Tapponnier, 1975, 1978; Tapponier et al., 1986; Hendix et al., 1994; Abdrakhmatov et al., 1996; Cunningham et al., 1996).

The region is, seismically, one of the most active in the world (Holt et al., 1995; Chen and Kao, 1996; Chen et al., 1999). The partitioning of the post-collisional crustal shortening is complex and is essentially divided between crustal thickening and lateral extrusion along strike-slip fault systems (Avouac and Tapponier, 1993; Houseman and England, 1996). Estimating the post-collisional shortening is difficult because of the uncertainty associated with estimating the initial crustal thickness before the Indian–Asian collision and because the shortening is distributed beyond the Himalayan–Tibetan region, with a substantial amount occurring in the Tien Shan (Murphy et al., 1997). Yin and Harrison (2000) suggest that the shortening since the Indian–Asian collision is distributed as follows: >360 km across the Himalaya, >60 km across the Gangdese thrust system, ~250 km along the Shiquanhe–Gaize–Amdo thrust system, >60–80 km across the Fenghuo Shan–Nangqian fold-and-thrust belt, ~270 km across the Qimen Tagh–North Kunlun thrust system, and ~360 km across the Nan Shan thrust belt. Furthermore, they suggest that the shortening is expressed in two modes at the surface: (a) discrete thrust belts with relatively narrow zones of contraction or regional décollement (a detachment structure resulting from deformation), and (b) distributed shortening over a wide region involving basement rocks.

A compilation of the rates of shortening and strike-slip faulting on Holocene and late Pleistocene timescales is summarized in Figure 2.5. These data are based on measuring and dating offset landforms and displaced outcrops. The current rate of deformation is beginning to be quantified by Global Positioning System (GPS) measurements (King et al., 1997; Larson et al., 1999; Wang et al., 1999; Chen et al., 2000). These studies show relatively good agreement with the geologic data, yet they are somewhat limited by the short duration over which the measurements have been undertaken.

Numerous models have been constructed to help understand the geodynamics of the Indian–Asian collision and they involve numerical simulation of indentation of a viscous thin-sheet (England and McKenzie, 1982; Vilotte et al., 1982; England and Houseman, 1989; Ellis, 1996; Yang and Lui, 2000), analogue models of indentation of a plasticine plane (Tapponnier and Molnar, 1976; Tapponnier et al., 1986; Peltzer and Tapponnier, 1988) and three-dimensional (3-D) finite element modelling (Lui et al., 2000). Such modelling studies add to the knowledge and understanding of the deep structure of the Himalayan–Tibetan orogen and they complement the deep crustal research (Nelson et al., 1996; Owens and Zandt, 1997).

The timing of the Tibetan plateau uplift has been difficult to quantify because of the uncertainty in

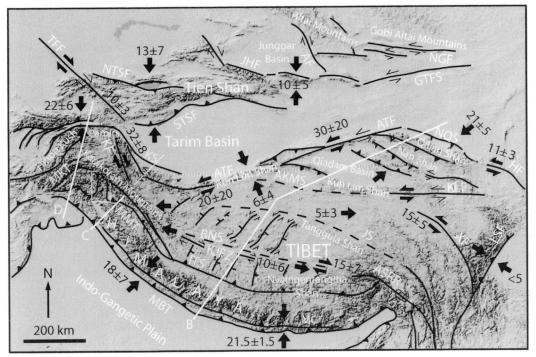

Figure 2.5 *Digital elevation model of Tibet and the bordering mountains showing the major faults and sutures. Estimates of late Quaternary strike-slip, convergence and extension rates are shown in millimetres per annum (after Larson* et al.'s *(1999) compilation of recent data). The sections B, C and D are shown in Figure 2.2. AF, Altai fault; AKMS, Ayimaqin–Kunlun–Mutztagh suture; ASRR, Ailao Shan-Red River shear zone; ATF, Altyn Tagh fault; BNS, Bangong Nujiang suture; GTFS, Gobi–Tien Shan fault system; HF, Haiyuan fault; ITS, Indus Tsangpo suture; JHF, Junggar Hegen fault; JS, Jinsha suture; KF, Karakoram fault; KJFZ, Karakoram Jiali fault zone; KLF, Kunlun fault; KS, Kudi suture; LSF, Longmen Shan fault; MBT, Main Boundary Thrust; MCT, Main Central Thrust; MKT, Main Karakoram Thrust (Shyok suture zone); MMT, Main Mantle Thrust; NGF, North Gobi fault; NQS, North Qilian suture; NTSF, North Tien Shan fault; STSF, South Tien Shan fault; TFF, Talus–Fergana fault; XF, Xianshuihe Fault Adapted from Searle (1991); Cunningham* et al. *(1996); Chung* et al. *(1998); Yin* et al. *(1999); Yin and Harrison (2000); Blisniuk* et al. *(2001); Hurtado* et al. *(2001).*

determining palaeoaltitudes (cf. Gregory and Chase, 1992). Several uplift patterns have been proposed (Harrison *et al.*, 1992, 1998), but recent geologic data suggest that the initiation and rates of uplift varied considerably across the orogen (Chung *et al.*, 1998). Furthermore, Murphy *et al.* (1997) suggested that a significant portion of southern Tibet was elevated before the Indian–Asian collision and Chung *et al.* (1998) suggested that northeastern Tibet had uplifted by 40 Ma, while in western Tibet the uplift occurred at about 20 Ma. These observations are consistent with sedimentation records from the Ganges–Brahmaputra delta and the Bengal fan (Chung *et al.*, 1998). The uplift history also helps to explain the nature of the strontium isotope evolution of the oceans and global cooling over the past 20 Ma (Chung *et al.*, 1998).

By about 14 Ma, the Tibetan Plateau had become sufficiently thick that it began to extend gravitationally (Coleman and Hodges, 1995). Two types of extensional structures are apparent: the south Tibetan fault system, a family of east-striking shallow to moderate north-dipping normal faults exposed near the crest of the Himalaya from Bhutan to northwest India; and numerous north-trending rift systems that largely dictate the topographic pattern of the southern Tibetan Plateau (Armijo *et al.*, 1986; Wu *et al.*, 1998; Yin *et al.*, 1999; Yin, 2000; Blisniuk *et al.*, 2001; Hurtado *et al.*,

2001). These structures are summarized on Figure 2.5 and the relationship to the evolution of the Himalayan–Tibetan orogen is reviewed in Table 2.2 and Figure 2.2.

The uplift and subsequent denudation of the Himalayan–Tibetan orogen resulted in a varied topography and geology. This is summarized in Figures 2.2 and 2.5. Several pervasive structures are present along the length of the Himalaya. These include: the Main Boundary Thrust that delimits the southern margin of the Himalaya; the Main Central Thrust that forms a major crustal suture zone within the Indian plate; and the Main Mantle Thrust (Indus Tsangpo Suture) that marks the main boundary between the Indian and Asian continental plates. Other major thrusts and sutures are present, but they are not so regionally pervasive; they include the K2 Thrust, Karakoram Batholith Lineament, Pir Panjal Thrust and the Vale of Kashmir Thrust. Several major sutures traverse Tibet and include the Bangong Nujiang, Jinsha and Ayimaqin–Kunlun–Mutztagh sutures, which started to form during the Palaeozoic. In addition, continental-scale strike-slip fault systems transverse Tibet and include the Karakoram, Altyn Tagh and Kunlun faults, and the Ailao Shan–Red River Shear Zone. These are considered to be important in allowing the regional shortening to be accommodated as eastward lateral extrusion (Tapponnier and Molnar, 1976). For example, the total slip along the Altyn Tagh fault during the Cenozoic probably exceeds 600 km (Yin and Harrison, 2000) and along the Karakoram fault it is >100 km (Searle and Owen, 1999). The Altyn Tagh and Karakoram faults act as major transfer faults linking major thrust belts and extensional systems, respectively (Figure 2.5).

The Trans-Himalayan Batholith is an important component of the Himalayan orogen. It is discontinuous along the entire length of the Trans-Himalaya, some 2500 km. Along the eastern stretch it occurs north of the Indus–Tsangpo suture and it was emplaced into an Andean-type margin during the mid-Cretaceous and in the Palaeocene–lower Eocene (England and Searle, 1986; Debon et al., 1986). In the west, in northern India and Pakistan, it forms the Kohistan–Ladakh arc. This was an island arc that grew on the northern side of the Neo-Tethys Ocean that separated India from Eurasia during the mid-Cretaceous. The arc collided with the Karakoram plate at between 102 and 85 Ma to become the leading edge of an active continental margin under which the Neo-Tethys was subducted (Petterson and Windley, 1985; Coward et al., 1987; Reuber, 1989) (Figures 2.2(C) and 2.2(D)). This arc was intruded by an Andean-type granodiorite batholith between 78 and 75 Ma, and 48 and 45 Ma (Sullivan et al., 1993). The Indian plate eventually collided with the arc during the earliest Eocene and the continuous underthrusting of the Indian plate below the arc led to crustal thickening and melting and the intrusion of leucogranites at ~30 Ma and subsequent deformation (Petterson and Windley, 1985).

The occurrence of syn-collisional igneous activity is an important characteristic of the Himalayan–Tibetan orogen (Figure 2.2(B)–(D); Table 2.2). Yin and Harrison (2000) listed five different mechanisms that may have been responsible for the generation of syn-collisional igneous activity. These are: (i) an early crustal thickening followed by slip along a shallow dipping décollement (Himalayan leucogranites); (ii) slab break-off during the early stage of the Indian–Asian collision (Linzizong volcanic sequence in southern Tibet); (iii) continental subduction in southern and central Tibet, which generated calcalkaline magmatism; (iv) formation of releasing bends and pull-apart structures that serve both as a possible mechanism to generate decompressional melting and as conduits to trap melts (Pulu basalts and other late Neogene–Quaternary volcanic flows along the Altyn Tagh and the Kunlun faults); (v) viscous dissipation in the upper mantle and subduction of Tethyan flysch complexes to mantle depths may be the fundamental cause of widespread and protracted partial melting in the Himalayan–Tibetan orogen in the Cenozoic.

The presence of calcalkaline type volcanism in southern and central Tibet suggests that some portion of the continental crusts from both the north and south must have been subducted into the mantle beneath Tibet (Yin and Harrison, 2000).

The role of denudation in shaping the Himalayan–Tibetan region is a subject of intense debate. Molnar and England (1990), for example, hypothesized that Cenozoic climatic change would have increased glaciation throughout the Himalaya and this, and its associated processes, would have increased erosion creating deeply incised valleys. They argued that high isolated mountain peaks would have been isostatically uplifted because of the denudation unloading caused by the deep valley incision. This helps increase the maximum elevation of the mountains. Others argue that the geometry of the valleys and the erosion rates are not significant to allow such uplift to occur (Harbor and Warburton, 1992; Whittington, 1996; Whipple and Tucker, 1999).

Zeitler et al. (2001) proposed an interesting model relating erosion, geomorphology and metamorphism in the Nanga Parbat Himalaya in northern Pakistan. Nanga Parbat is the ninth highest mountain in the world and is essentially defined by the Main Mantle Thrust that forms a syntaxis around Nanga Parbat and Haramosh massifs (Figure 2.6(A)). The core of the Nanga Parbat massif is characterized by very young (<3 Ma) granites, low-P cordierite-bearing granulites, low seismic velocities, resistive lower crust and shallow microearthquakes implying shallow brittle-ductile transition bowed upwards by ~3 km. Incision rates for the Indus River in this region are in the order of 2–12 mm a^{-1} (Burbank et al., 1996) and tributary valley incision rates around Nanga Parbat are 22 ± 11 mm a^{-1} (Shroder and Bishop, 2000). Zeitler et al. (2001) proposed that the incision that produced the deep river gorge of the Indus helps weaken the crust in this region. This, in turn, encourages failure and helps draw in advective flow toward the topographic gap (Figure 2.6(B)). This builds elevation and, together with the incising river, builds relief and leads to high erosion rates. The result is a steepened thermal gradient, which raises the brittle-ductile transition, and further weakens the crust. Deep and mid-crustal material can then experience decompression melting and low-P–high-T metamorphism as it is moved rapidly to the surface. They called this process a 'tectonic aneurysm' and they believe that this is an important orogenic process in continental–continental collision zones.

The study of these continental–continental collision zones provides an insight into the evolution of the continents and helps in understanding and explaining the nature and distribution of ancient mountain systems. It is necessary, however, to examine active oceanic–oceanic and oceanic–continental

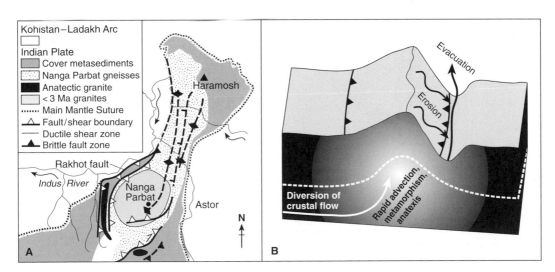

Figure 2.6 *The geology of Nanga Parbat massif, northern Pakistan, illustrating the relationship between erosion, crustal processes and uplift. (A) Geologic sketch map of the Nanga Parbat massif (after Schneider et al., 1999, and Zeitler et al., 2001). (B) Schematic representation illustrating the dynamics of a tectonic aneurysm, shown at a mature stage (see text for explanation) (after Zeitler et al., 2001)*

collision zones to fully understand the early evolution of continental–continental collision zones. The Circum-Pacific orogenic belt provides such an opportunity and, ultimately, it may itself become a continental–continental collision zone in the distant future.

5 Circum-Pacific orogenic belt

The Circum-Pacific orogenic belt can be broadly divided into eastern and western sectors (Figure 2.1). The western sector of the Circum-Pacific orogenic belt is the result of convergence of oceanic plates including the Pacific, Philippine and Indian–Australian plates, and the eastern margin of the Eurasian plate (Figure 2.1). This sector, however, is discontinuous and includes volcanic island arcs and arc-collision zones. The associated mountains are not very geographically extensive, but nevertheless are impressive in terms of their relative relief and rates of erosion.

Taiwan, the Philippines, New Guinea and the Vanuatu arc in the southeast Pacific provide the best examples of arc–continental and arc–arc collisions. The convergence in this region is complex, with the interaction of the Pacific, Indian–Australian, Eurasian and Philippine plates, and two major trench–trench–trench triple junctions. Landforms include volcanic chains, fold-and-thrust belts and accretionary wedges. Taiwan provides one of the best examples of an area of rapid mountain uplift that is a consequence of arc–continental collision. The island rises to 3997 m asl and formed during the past 4.5 Ma as the Philippine Sea plate moved northwest into the Eurasian continental plate at a rate of ~70 km Ma^{-1} (Seno, 1977; Angelier et al., 1986; Lee and Wang, 1987; Figure 2.7). Intense internal deformation and metamorphism has resulted in tectonic uplift rates of between 1 and 10 mm a^{-1} (Lin, 1991; Wang and Burnett, 1991). This uplift, together with rates of denudation of between 1 and 5 mm a^{-1} (Li, 1976) that are a consequence of the extreme monsoonal climate with its frequent tropical cyclones, has resulted in one of the youngest and most dynamic landscapes on Earth.

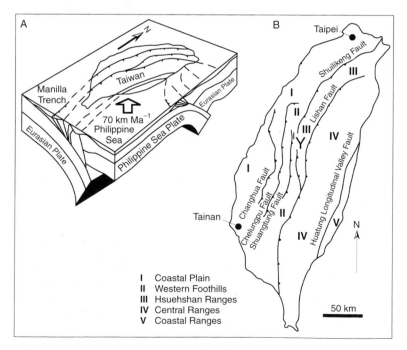

Figure 2.7 *The geologic setting of Taiwan and its associated mountain ranges. (A) Schematic plate tectonic setting (after Lin, 2000). (B) Major faults, and geologic and geomorphic units (after Chang, 2000)*

The Andes chain and North American Cordillera are the two greatest mountain ranges in the Circum-Pacific orogenic belt and stretch almost continuously for >20000 km. These constitute the eastern sector of the Circum-Pacific orogenic belt. Their evolution is essentially the consequence of the convergence of the oceanic and continental plates. Today this includes the collision of the Pacific, Juan Fuca, Cocos and Nazca oceanic plates with the North and South American continental plates (Figure 2.1). Presently, the margin is consumed beneath Alaska, the US Pacific Northwest and southwestern Canada, Central and South America, the Scotia Arc and the Antarctica Peninsula. Transform margins are present, connecting the trenches of Alaska and the Pacific Northwest and connecting the Mendocino triple junction to the Gulf of California, which comprises the San Andreas fault system (Moores and Twiss, 1995). The mountains along this sector of the Circum-Pacific orogenic belt have a long and complex history beginning in the late Precambrian. Most of the mountain building that produced the present landscapes, however, has occurred during the last 200 Ma (Figures 2.1 and 2.8). Structures verge towards the forelands on the eastern and western sides of the mountain belts, but there is a strong asymmetry within the orogens (Figure 2.8).

The Andean chain has been a site of continental accretion, crustal growth, and both compressional and extensional deformation throughout the Phanerozoic. Palaeozoic subduction and accretion resulted in the amalgamation of various terrains, associated with regional compression events (Ramos, 1988). Since the Triassic (~225 Ma) the southern Andes have formed a classic continental-type subduction margin and with no further terrain accretion. The northern Andes are more complex, influenced by Caribbean tectonics and the relative motion of the North and South America plates. This resulted in the accretion of island-arcs during the latest Mesozoic and early Tertiary. During the Jurassic and Cretaceous there was extensive rifting in fore-arc and back-arc basins, and magmatic activity along the length of the Andes that included the emplacement of massive granite batholiths (McCourt et al., 1984; Jaillard et al., 1990; Kay et al., 1991). Increased plate convergence occurred during the early Cenozoic and middle to late Cenozoic resulting in major regional deformation (Allmendinger et al., 1983; Jordan et al., 1983). During this time the eastern Andes flexed downwards in response to deformation and crustal loading. This resulted in a series of Cenozoic foreland basins that contain thick (≤5 km) sequences of terrestrial sediments. This general pattern of events is similar throughout the Andes, but, as illustrated in Table 2.3, the timing of orogenic events is diachronous along the mountain belt.

Like the Andes, the North American Cordillera has a complex history of continental accretion, crustal growth, and both compressional and extensional deformation. In addition, however, the southern stretch has also experienced the development of a continental transform plate boundary. This formed during the latter part of the Cenozoic, probably as a consequence of the subduction of the Pacific–Farallon ridge-transform system under North America (Atwater and Molnar, 1973; Atwater, 1989) (Figure 2.8). Several belts of deformation of different ages are present throughout the orogen. Palaeozoic deep-water rocks of the so-called 'eugeocline' were deformed by the Antler and Sonoma orogenies during the Devonian–Mississippian and Permo-Triassic, respectively (Speed et al., 1988). A phase of major deformation during the Sevier orogeny in the late Jurassic to late Cretaceous produced an extensive fold-and-thrust belt that extends from southeast California to Canada (Allmendinger and Jordan, 1981) (Figures 2.3 and 2.8). A complex hinterland of thick Palaeozoic shallow-water rocks of the 'miogeosyncline' is present east of this belt. These are involved in major Mesozoic nappes, Tertiary low-angled denudational faulting and metamorphic core complexes of Mesozoic and Tertiary age (Dilek and Moores, 1999). The easternmost segment of the North American Cordillera was deformed during the Cretaceous–Tertiary Laramide orogeny. This involves Precambrian crystalline crust and Palaeozoic–Mesozoic platform sedimentary rocks (Hamilton, 1988).

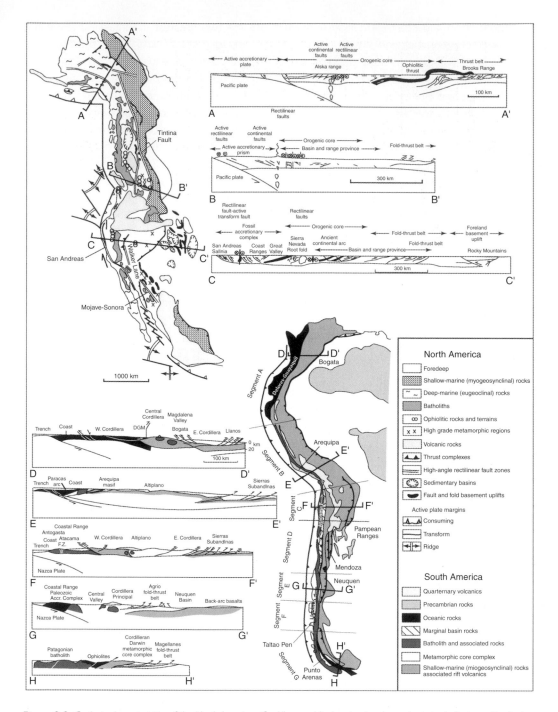

Figure 2.8 *Geologic characteristics of the North American Cordillera and Andes showing the major tectonic features. The Andes are divided into seven segments (A to G) and their geologic history is summarized in Table 2.3*
After King (1977), Megard (1989), Mpodozis and Ramos (1989) and Moores and Twiss (1995), and the geologic cross-sections are adapted from Moores and Twiss (1995), after Maxwell (1974), Roeder and Mull (1978), Csejtey et al. (1982), Potter et al. (1986), Allmendinger et al. (1987), Roeder (1988), Mpodozis and Ramos (1989) and Vicente (1989).

Table 2.3 – Time-space diagram showing the principal tectonic events along the length of the Andes

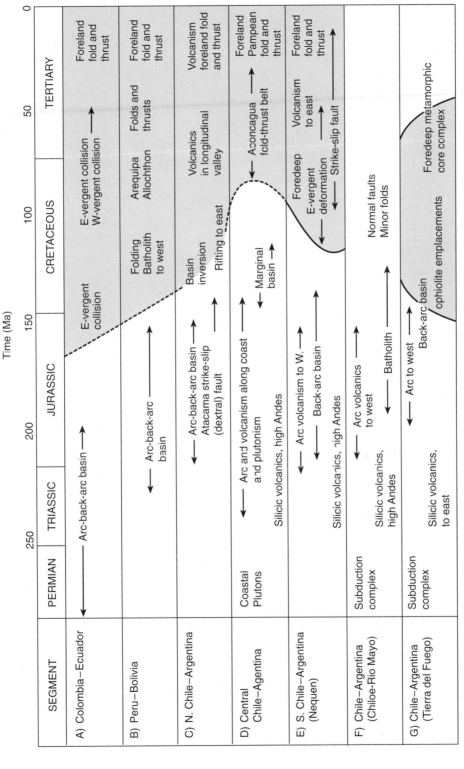

Notes: The heavy dashed line indicates the onset of the main Andean deformation. The location of each segment is shown in Figure 2.6.

After Moores and Twiss (1995).

The prolonged Mesozoic orogeny produced a north-trending crustal high that had a maximum thickness of about 60 km and an elevation of >3 km (Wolfe *et al.*, 1997). By the mid-Tertiary, this highland region had begun to undergo orogenic extension resulting in the exhumation of metamorphic cores and widespread calcalkaline volcanism. The early stage of orogenic collapse was followed by the Basin and Range extension at between 18 and 16 Ma and associated volcanism (Coney, 1987). This ultimately produced the Great Basin with a mean elevation of ~1.5 km asl and a crustal thickness of ~30 km (Thompson and Burke, 1974; Wolfe *et al.*, 1997). The succession of events that is important in producing the present orogen is summarized in Table 2.2 and discussed above in comparison with the Tibetan Plateau and the Turkish–Iranian Plateau.

The mountains of the Circum-Pacific orogenic belt help illustrate the variety of tectonic settings that can produce substantial relief along different types of convergent and transform plate boundaries. Furthermore, they provide valuable models for understanding the orogenic evolution of orogens that ultimately become continental–continental collision zones such as the Alpine–Himalayan–Tibetan orogen.

6 Ocean ridges

As a type of global mountain system, oceanic ridges are commonly neglected. This is probably because they are the least well studied owing to their inaccessibility. Ocean ridges occur in mid-ocean settings associated with divergent oceanic plates and back-arc spreading centres behind volcanic arcs of subduction zones. Mid-ocean ridges are between 1000 and 4000 km wide, they rise 2–3 km above the surrounding ocean floors and their crests have an average depth of 2500 m below sea level (Nicholas, 1995) (Figure 2.1). Back-arc spreading centres are considerably smaller than the mid-oceanic ridges and therefore little attention is given to them in this section.

Oceanic ridges are elevated because they consist of rock that is hotter and less dense than the adjacent oceanic crust. Furthermore, hot mantle material rises beneath the ridges to fill the gap created by the spreading plates and this helps to increase their elevation. As the mantle rises it decompresses and undergoes partial melting at depths that can exceed 100 km and over a broad region of several hundred kilometres. Gabbros form within magma chambers, and magma may be intruded into dykes and may erupt at the ocean floor to form basaltic shield volcanoes and lava flows. All this contributes to form new oceanic crust. As time progresses the new oceanic crust moves away from the spreading centre, cools, contracts and subsides. The spreading rates vary from a few millimetres per annum in the Gulf of Aden to 10 mm a^{-1} in the North Atlantic near Iceland and 60 mm a^{-1} for the East Pacific Rise, although the rates may vary over the duration of the ocean ridge's history (Reading and Mitchell, 2000).

Ridges with slow spreading rates have a well-defined (1.5–3 km deep) symmetrical axial rift valley. In contrast, the fastest spreading ridges have subdued topography more reminiscent of Hawaiian volcanoes, with a small summit ridge or graben (Macdonald, 1982). Well-developed axial valleys may drop to depths below that of the surrounding ocean floor. Hydrothermal activity is associated with ridges producing extremely hot springs that may form columnar structures known as chimneys.

Ocean ridges are broken into segments by transverse fractures (transform faults) which displace the ridges by tens, or even hundreds, of kilometres (Figure 2.1). These transform faults are sub-vertical and may produce fault scarps that exceed 500 m in height (Collette, 1986). Complex stress patterns are associated with the transform faults and transpressional and transtensional zones are common. Such stresses help to produce landforms analogous to those seen along continental strike-slip faults and include pressure ridges, pull-apart basins and shutter ridges.

The coincidence of the Icelandic hotspot with the mid-Atlantic ridge, that helped produce Iceland, provides an opportunity to examine some of the geologic aspects of oceanic ridges above sea level. However, the evolution of the ocean ridge at this location clearly differs from true oceanic ridges. This is because, for much of its history, it evolved by the successive subaerial and subglacial eruptions and a considerable portion of its uplift history and rifting is related to the hotspot (Gudmundsson, 2000).

As a consequence of the spreading, oceanic ridges are geologically young. Even with the slowest spreading rates, the rocks that comprise them are rarely more than a few tens of millions of years old. Nevertheless, they are among some of the world's most impressive geomorphic and tectonic features.

7 Conclusions

The above descriptions of the global mountain systems help illustrate their complex history, structure and morphology. Strong contrasts exist between global mountain systems that develop along mid-oceanic ridges, continental–continental collision zones and oceanic–oceanic/continental convergence zones. Furthermore, there is considerable variability within a single mountain system along any one plate boundary setting. This is really well illustrated along the Circum-Pacific and Alpine–Himalayan–Tibetan orogenic systems. Despite this, global mountain systems share a number of similar characteristics, both in their evolutionary path and the resultant forms. These are summarized in Figure 2.2(A) and Tables 2.1 and 2.2.

The evolution of individual mountain ranges may be in excess of hundreds of millions of years. Moreover, most Cenozoic mountain belts began their evolution long before the onset of the Cenozoic. Most orogenic belts grow outwards from a central core and may be diachronous along their lengths. Furthermore, uplift is not simple. It may propagate through a mountain system as the orogen evolves. In addition, as mountains grow in height, the denudation increases as a consequence of steeper slopes, increased river power and more prevalent mass movement, and possibly as a result of glaciation. High denudation rates may, in turn, contribute to uplift as a result of denudational unloading. The transfer of sediment to foreland regions may also contribute to uplift because of crustal flexuring associated with basin subsidence.

The growth of mountain ranges may also affect local, regional and even global climatic conditions. This, in turn, affects the rates and magnitudes of Earth surface processes that help shape the evolving orogen. This complex interaction between tectonic processes, climate and geomorphology needs quantifying to fully understand the links and interactions, and hence the evolution of global mountain systems. Fortunately, new analytical and computational methods and techniques are beginning to be applied to help explore and examine orogenic systems. Furthermore, much can be learned by applying space–time substitutions and by comparing ancient and modern mountain belts using tectonic, geomorphic and palaeoclimatological techniques. This is useful in helping to provide a fuller picture of the evolution of mountain belts and an understanding of their dynamics. It is encouraging that the study of orogenesis is becoming increasingly multidisciplinary, allowing for a better understanding of ancient and modern mountain systems. Such knowledge is also essential for sustainable development and hazard mitigation in mountain regions, especially as these regions become more populated, exploited and utilized by the world's growing population (cf. Hewitt, this volume).

Acknowledgements

I would like to thank Phil Owens, Mike Searle and Brian Whalley for very useful and constructive comments on an earlier version of this chapter.

References

Abdrakhmatov, K.Ye., Aldazhanov, S.A., Hager, B.H., Hamburger, M.W., Herring, T.A., Kalabaev, K.B., Makarov, V.I., Molnar, P., Panasyuk, S.V., Prilepin, M.T., Reilinger, R.E., Sadybakasov, I.S., Souter, B.J., Trapeznikov, A., Tsurkov, V.Y. and Zubovich, A.V., 1996. Relatively recent construction of the Tian Shan inferred from GPS measurements of present-day crustal deformation rates. *Nature*, **384**: 450–53.

Allmendinger, R.W. and Jordan, T.E., 1981. Mesozoic evolution, hinterland of the Sevier orogenic belt. *Geology*, **9**: 308–13.

Allmendinger, R.W., Hauge, T.A., Hauser, E.C., Potter, C.J., Klemperer, S.L., Nelson, K.D., Knuepfer, P. and Oliver, J., 1987. Overview of the COCORP 40°N transect, western USA. *Geological Society of America Bulletin*, **98**: 364–72.

Allmendinger, R.W., Ramos, V.A., Jordan, T.E., Palma, M. and Isacks, B.L., 1983. Paleogeography and Andean structural geometry, northwest Argentina. *Tectonics*, **2**: 1–16.

Angelier, J., Barrier, E. and Chen, H.T., 1986. Plate collision and paleostress trajectories in a fold-thrust belt: the foothills of Taiwan. *Tectonophysics*, **125**: 161–78.

Armijo, R., Tapponnier, P., Mercier, J. and Han, T., 1986. Quaternary extension in southern Tibet: field observations and tectonic implications. *Journal of Geophysical Research*, **91**: 13803–72.

Atwater, T., 1989. Plate tectonic history of the northeast Pacific and western North America. In Winterer, E.L., Hussong, D.M. and Decker, R.W. (eds), *The geology of North America*. Boulder, CO: Geological Society of America, volume N, 21–72.

Atwater, T. and Molnar, P., 1973. Relative motion of the Pacific and North American plates deduced from seafloor spreading in the Atlantic, Indian, and south Pacific oceans. In Kovach, R.L. and Nur, A. (eds), *Proceedings of the conference on tectonic problems of the San Andreas Fault System*. Stanford, CA: Stanford University Press, 314–24.

Avouac, J.P. and Tapponnier, P., 1993. Kinematic model of active deformation in central Asia. *Geophysical Research Letters*, **20**: 895–98.

Bishop, M.P., Shroder, J.F., Bonk, R. and Olsenholler, J., 2002. Geomorphic change in high mountains: a western Himalayan perspective. *Global and Planetary Change*, **32**: 311–29.

Blisniuk, P.M., Hacker, B.R., Glodny, J., Ratschbacher, L., Siwan B., Zhenhan W., McWilliams, M.O. and Calvert, A., 2001. Normal faulting in central Tibet since at least 13.5 Myr ago. *Nature*, **412**: 628–32.

Brown, R.W., Gallagher, K., Gleadow, A.J.W. and Summerfield, M.A., 2000. Morphotectonic evolution of the South Atlantic margins of Africa and South America. In Summerfield, M.A. (ed.), *Geomorphology and global tectonics*. Chichester: Wiley, 255–82.

Burbank, D., Leland, J., Fielding, E., Anderson, R.S., Brozovik, N., Reid, M.R. and Duncan, C., 1996. Bedrock incision, rock uplift and threshold hillslopes in the northwestern Himalaya. *Science*, **276**: 571–74.

Chang Hui-Cheng (ed.), 2000. *Geological studies of the Chi-Chi (921) earthquake*. Taipei: Central Geological Survey.

Chen, W.P. and Kao, H., 1996. Seismotectonics of Asia: some recent progress. In Yin, A. and Harrison M. (eds), *The tectonic evolution of Asia*. New York: Cambridge University Press, 37–62.

Chen, W.P., Chen, C.Y. and Nabelek, J.L., 1999. Present-day deformation of the Qaidam Basin with implications for intra-continental tectonics. *Tectonophysics*, **305**: 165–81.

Chen, Z., Burchfield, B.C., Liu, Y., King, R.W., Royden, L.H., Tang, W., Wang, E., Zhao, J. and Zhang, X., 2000. Global positioning system measurements from eastern Tibet and their implications for India–Eurasia intercontinental deformation. *Journal of Geophysical Research*, **105**: 16215–27.

Chung, S.L., Lo, C.H., Lee, T.Y., Zhang, Y., Xie, Y., Li, X., Wang, K.L. and Wang, P.L., 1998. Diachonous uplift of the Tibet Plateau starting 40 Myr ago. *Nature*, **394**: 769–73.

Coleman, M. and Hodges, K., 1995. Evidence for Tibetan Plateau uplift before 14 Myr ago from a new minimum age for east-west extension. *Nature*, **374**: 49–52.

Collette, B.J., 1986. Fracture zones in the North Atlantic: morphology and a model. *Journal of the Geological Society of America*, **143**: 763–74.

Coney, P.J., 1987. The regional tectonic setting and possible causes of Cenozoic extension in the North America Cordillera. In Coward, M.P., Dewey, J.F. and Hancock, P.L. (eds), *Continental extensional tectonics*. Geological Society Special Publication 28. London: Geological Society of London, 177–86.

Coward, M.P., Butler, R.W.H., Khan, M.A. and Knipe, R.J., 1987. The tectonic history of Kohistan and its implications for Himalayan structure. *Journal of the Geological Society of London*, **144**: 377–91.

Cox, B.F., Hillhouse, J.W. and Owen, L.A., 2003. Pliocene and Pleistocene evolution of the Mojave River, and associated tectonic development of the Transverse Ranges and Mojave Desert, based on borehole stratigraphy studies and mapping of landforms and sediments near Victorville, California. In Enzel, Y., Wells, S. and Lancaster, N. (eds), *Paleoenvironment and paleohydrology of the Mojave and southern Great Basin deserts*. Geological Society of America Special Paper, 368, pp. 1–42.

Csejtey, B., Cox, D.P., Evarts, R.C., Stricker, G.D. and Foster, H.L., 1982. The Cenozoic Denali fault system and the Cretaceous accretionary development of southern Alaska. *Journal of Geophysical Research*, **87**: 3742–54.

Cunningham, W.D., Windley, B.F., Dorjnamjaa, D., Badamgarov, G. and Saander, M., 1996. Late Cenozoic transpression in southwestern Mongolia and the Gobi Altai-Tien Shan connection. *Earth and Planetary Science Letters*, **140**: 67–82.

Debon, F., Le Fort, P., Sheppard, S.M.F. and Sonet, J., 1986. The four plutonic belts of the Transhimalaya–Himalaya: a chemical, mineralogical and chronological synthesis along a Tibet–Nepal section. *Journal of Petrology*, **27**: 219–50.

DeMets, C., Gordon, R.G., Argus, D.F. and Stein, S., 1994. Effect of recent revisions to the geomagnetic reversal time scale on estimates of current plate motion. *Geophysical Research Letters*, **21**: 2191–94.

Dewey, J.F., Pitman, III, W.C., Ryan, W.B.F. and Bonnin, J., 1973. Plate tectonics and the evolution of the Alpine system. *Geological Society of America Bulletin*, **84**: 3137–80.

Dilek, Y. and Moores, E.M., 1999. A Tibetan model for the early Tertiary western United States. *Journal of the Geological Society of London*, **156**: 929–41.

Ellis, S., 1996. Forces driving continental collision: reconciling indentation and mantle subduction tectonics. *Geology*, **24**: 699–702.

England, P. and Houseman, G., 1989. Extension during continental convergence, with application to the Tibetan Plateau. *Journal of Geophysical Research*, **94**: 17561–69.

England, P. and McKenzie, D.P., 1982. A thin viscous sheet model for continental deformation. *Geophysical Journal of the Royal Astronomical Society*, **70**: 295–321.

England, P. and Searle, M.P., 1986: The Cretaceous–Tertiary deformation of the Lhasa block and its implications for crustal thickening in Tibet. *Tectonics*, **5**: 1–14.

Gilchrist, A.R. and Summerfield, M.A., 1990. Differential denudation and flexural isostasy in formation of rifted margin upwarps. *Nature*, **346**: 739–42.

Gilchrist, A.R. and Summerfield, M.A., 1994. Tectonic models of passive margin evolution and their implications for theories of long-term landscape development. In Kirkby, M.J. (ed.), *Process models and theoretical geomorphology*. Chichester: Wiley, 55–84.

Gilchrist, A.R., Summerfield, M.A. and Cockburn, H.A.P., 1994. Landscape dissection, isostatic uplift, and the morphologic development of orogens. *Geology*, **22**: 963–66.

Gregory, K.M. and Chase, C.G., 1992. Tectonic significance of paleobotanically estimated climate and altitude of the late Eocene erosion surface. *Geology*, **20**: 581–85.

Gudmundsson, A., 2000. Dynamics of volcanic systems in Iceland: example of tectonism and volcanism at juxtaposed hot spot and mid-ocean ridge systems. *Annual Review of Earth and Planetary Sciences*, **28**: 107–40.

Gunnell, Y. and Fleitout, L., 2000. Morphotectonics evolution of the Western Ghats, India. In Summerfield, M.A. (ed.), *Geomorphology and global tectonics*. Chichester: Wiley, 321–38.

Hamilton, W.B., 1988. Laramide crustal shortening. In Schmidt, C.J. and Perry, W.J. (eds), Interaction of the Rocky Mountain Foreland and the Cordilleran Thrust Belt. *Geological Society of America Memoirs*, **171**: 27–39.

Harbor, J. and Warburton, J., 1992. Glaciation and denudation rates. *Nature*, **356**: 751.

Harrison, T.M., Copeland, P., Kidd, W.S.F. and Yin, A., 1992. Raising Tibet. *Science*, **255**: 1663–70.

Harrison, T.M., Yin, A. and Ryerson, F.J., 1998. Orographic evolution of the Himalaya and Tibetan plateau. In Crowley, T.J. and Burke, K.C. (eds), *Tectonic boundary conditions for the climate reconstruction*. New York: Oxford University Press, 21–71.

Hatcher, R. and Williams, R.T., 1986. Mechanical model for single thrust sheets. *Geological Society of America Bulletin*, **97**: 975–85.

Hendix, M.S., Dumitru, T.A. and Graham, S.A., 1994. Late Oligocene–early Miocene unroofing in the Chinese Tien Shan: an early effect of the India–Asia collision. *Geology*, **22**: 487–90.

Holt, W.E., Li, M. and Haines, A.J., 1995. Earthquake strain rates and instantaneous relative motions within central and eastern Asia. *Geophysical Journal International*, **122**: 569–93.

Houseman, G. and England, P., 1996. A lithospheric thickening model for the Indo-Asian collision. In Yin, A. and Harrison, T.M. (eds), *The tectonic evolution of Asia*. New York: Cambridge University Press, 3–17.

Hsu, K.J., 1995. *The geology of Switzerland: an introduction to tectonic facies*. Princeton, NJ: Princeton University Press.

Hsu, K.J., Sengor, A.M.C., Briegel, U., Chen, H., Chen, C., Harris, N., Hsu, P., Li, J., Luo, J., Typhon, L., Li, Z.X., Chiayu, L., Powell, C., Wang, Q. and Winterer, E.L., 1995. Tectonic evolution of the Tibetan Plateau: a working hypothesis based on the archipelago model of orogenesis. *International Geological Reviews*, **37**: 473–508.

Hurtado, J.M., Hodges, K.V. and Whipple, K.X., 2001. Neotectonics of the Thakkhola graben and implications for recent activity on the South Tibetan fault system in the central Himalaya. *Geological Society of America Bulletin*, **113**: 222–40.

Jaillard, E., Soler, P., Carlier, G. and Mourier, T., 1990. Geodynamic evolution of the northern and central Andes during early to middle Mesozoic times: a Tethyan model. *Journal of the Geological Society of London*, **147**: 1009–22.

Jordan, T.E., Isacks, B.L., Allmendinger, R.W., Brewer, J.A., Ramos, V.A. and Ando, C.J., 1983. Andean tectonics related to geometry of subducted Nazca plate. *Geological Society of America Bulletin*, **94**: 341–61.

Kay, S.M., Mpodozis, C., Ramos, V.A. and Munizaga, F., 1991. Magma source variations for mid–late Tertiary magmatic rocks associated with a shallowing subduction zone and a thickening crust in the central Andes (28–33°S). *Geological Society of America Special Paper*, **265**: 113–37.

King, P.B., 1977. *Evolution of North America*. Princeton, NJ: Princeton University Press.

King, R.W., Shen, F., Burchfield, B.C., Royden, L.H., Wang, E., Chen, Z., Liu, Y., Zhang, X., Zhao, J. and Li, Y., 1997. Geodetic measurement of crustal motion in southwest China. *Geology*, **25**: 1279–82.

Larson, K., Burgmann, R., Billham, R. and Freymueller, J.T., 1999. Kinematics of the India–Eurasian collision zone from GPS measurements. *Journal of Geophysical Research*, **104**: 1077–94.

Laubscher, H.P., 1982. Detactment, shear, and compression in the central Alps. *Geological Society of America Memoir*, **158**: 191–211.

Lee, C.T. and Wang, Y., 1987. Palaeostress change due to the Pliocene–Quaternary arc-continent collision in Taiwan. *Memoir of the Geological Society of China*, **9**: 63–86.

Li, Y.H., 1976. Denudation of Taiwan Island since the Pliocene Epoch. *Geology*, **4**: 105–107.

Lin, J.C., 1991. The structural landforms of the Coastal Range of eastern Taiwan. In Cosgrove, J. and Jones, M. (eds), *Neotectonics and resources*. London: Belhaven Press, 65–74.

Lin, J.C., 2000. Morphotectonic evolution of Taiwan. In Summerfield, M.A. (ed.), *Geomorphology and global tectonics*. Chichester: Wiley, 135–46.

Love, J.D. and Reed, J.C., 1971. *Creation of the Teton landscape, the geologic story of the Grand Teton National Park*. Moose, WY: Grand Teton Natural History Association.

Lui, M., Shen, Y. and Yang, Y., 2000. Gravitational collapse of orogenic crust: a preliminary three-dimensional finite element study. *Journal of Geophysical Research*, **105**: 3159–73.

Macdonald, K.C., 1982. Mid-ocean ridges: fine scale tectonic, volcanic and hydrothermal processes within the plate boundary zone. *Annual Review of Earth and Planetary Science*, **10**: 155–90.

Mattauer, M., 1986. Intracontinental subduction, crust–mantle decollement and crustal-stacking wedge in the Himalayas and other collision belts. In Coward, M.P. and Ries, A.C. (eds), *Collision tectonics*. Geological Society Special Publication, 19. London: Geological Society of London 37–50.

Maxwell, J.C., 1974. Anatomy of an orogen. *Geological Society of America Bulletin*, **85**: 1195–204.

McCourt, W.J., Aspden, J.A. and Brook, M., 1984. New geological and geochronological data from the Colombian Andes: continental growth by multiple accretion. *Journal of the Geological Society of London*, **141**: 831–45.

Megard, F., 1989. The evolution of the Pacific Ocean margin in South America north of the Africa elbow (18°S). In Ben-Avrahem, Z. (ed.), *The evolution of the Pacific Ocean margins*. New York: Oxford University Press, 208–30.

Molnar, P. and England, P., 1990. Late Cenozoic uplift of mountain ranges and global climate change: chicken or egg? *Nature*, **346**: 29–34.

Molnar, P. and Tapponnier, P., 1975. Cenozoic tectonics of Asia: effects of a continental collision. *Science*, **189**: 419–26.

Molnar, P. and Tapponnier, P., 1978. Active tectonics in Tibet. *Journal of Geophysical Research*, **85**: 5361–75.

Montgomery, D.R. 1994. Valley incision and the uplift of mountain peaks. *Journal of Geophysical Research*, **99**: 913–21.

Moores, E.M. and Twiss, R.J., 1995. *Tectonics*. New York: W.H. Freeman and Company.

Mpodozis, C. and Ramos, V., 1989. The Andes of Chile and Argentina. In Erickson, G.E., Canas, P.M.T. and Reinemund, J.A. (eds), *Geology of the Andes and its relation to hydrocarbon and mineral resources*. Houston, TX: Circum-Pacific Council for Energy and Mineral Resources Earth Science Series.

Murphy, M.A., Yin, A., Harrision, T.M., Durr, S.B., Chen, Z., Ryerson, F.J., Kidd, W.S.F., Wang, X. and Zhou, X., 1997. Did the Indo-Asian collision alone create the Tibetan plateau? *Geology*, **25**: 719–22.

Nelson, K.D., Zhao, W., Brown, L.D., Kuo, J., Che, J., Liu, X., Klemperer, S.L., Makovosky, Y., Meissner, R., Mechie, J., Kind, R., Wenzel, F., Ni, J., Nabelek, J., Leshou, C., Tan, H., Wei, W., Jones, A.G., Brooker, J., Unsworth, M., Kidd, W.S.F., Huack, M., Alsdorf, D., Ross, A., Rogan, M., Wu, C., Sandvol, E. and Edwards, M., 1996. Partially molten middle crust beneath southern Tibet: synthesis of project INDEPTH results. *Science*, **274**: 1684–88.

Nicholas, A., 1995. *The mid-oceanic ridges: mountains below sea level*. Berlin: Springer-Verlag.

Owens, T.J. and Zandt, G., 1997. Implications of crustal property variations for models of Tibetan plateau evolution. *Nature*, **387**: 37–43.

Patriat, P. and Achache, J., 1984. India–Eurasia collision chronology has implications for crustal shortening and driving mechanism of plates. *Nature*, **311**: 615–21.

Peltzer, G. and Tapponnier, P., 1988. Formation and evolution of strike-slip faults, rifts, and basins during the India–Asia collision: an experimental approach. *Journal of Geophysical Research*, **93**: 15085–117.

Petterson, M.G. and Windley, B.F., 1985. Rb-Sr dating of the Kohistan batholith in the Trans-Himalaya of North Pakistan, and tectonic implications. *Earth and Planetary Science Letters*, **74**: 45.

Pierce, K.L. and Morgan, L.A., 1992. The track of the Yellowstone hot spot: volcanism, faulting and uplift. *Geological Society of America Memoir*, **179**: 1–53.

Potter, C.J., Sanford, W.E., Yoos, T.R., Prussen, E.I., Keach, II, R.W., Oliver, J.E., Kaufman, S. and Brown, L.D., 1986. COCORP transect of the Washington–Idaho Cordillera. *Tectonics*, **5**: 1007–26.

Ramos, V.A., 1988. Late Proterozoic–early Palaeozoic of South America. *Episodes*, **11**: 168–74.

Ramstein, G., Fluteau, F.F., Besse, J. and Joussaume, S., 1997. Effect of orogeny, plate motion, and land–sea distribution on Eurasian climate over the past 30 million years. *Nature*, **286**: 788–95.

Raymo, M.E. and Ruddiman, W.F., 1992. Tectonic forcing of late Cenozoic climate. *Nature*, **359**: 117–22.

Reading, H.G. and Mitchell, N.C., 2000. Mid-ocean ridges. In Hancock, P.L. and Skinner, B.J. (eds), *The Oxford companion to the Earth*. Oxford: Oxford University Press, 683–87.

Reuber, I., 1989. The Dras arc: two successive volcanic events in eroded oceanic crust. *Tectonophysics*, **161**: 93–106.

Roeder, D.H., 1977. Continental convergence in the Alps. *Tectonophysics*, **40**: 339–50.

Roeder, D.H., 1988. Andean-age structure of eastern Cordillera (Province of La Paz, Bolivia). *Tectonics*, **7**: 23–40.

Roeder, D.H. and Mull, C.G., 1978. Ophiolites of the Brooks Range. *American Association of Petroleum Geologists Bulletin*, **62**: 1692–702.

Ruddiman, W.F., 1997. *Tectonic uplift and climate change*. New York: Plenum Press.

Ruddiman, W.F., 1998. Early uplift in Tibet? *Nature*, **394**: 723–35.

Ruddiman, W.F. and Kutzbach, J.E., 1989. Forcing of Late Cenozoic northern hemisphere climate by plateau uplift in southern Asia and the American west. *Journal of Geophysical Research*, **94**: 18409–27.

Ryan, P.D., 2000. Alpine orogeny. In Hancock, P.L. and Skinner, B.J. (eds), *The Oxford companion to the Earth*. Oxford: Oxford University Press, 14–16.

Schneider, D.A., Edwards, M.A., Kidd, W.S.F., Khan, M.A., Seeber, L. and Zeitler, P.K., 1999. Tectonics of Nanga Parbat, western Himalaya: synkinematic plutonism within the double vergent shear zones of crustal-scale pop-up structure. *Geology*, **27**: 999–1002.

Searle, M.P., 1991. *Geology and tectonics of the Karakoram Mountains*. Chichester: Wiley.

Searle, M.P. and Owen, L.A., 1999. The evolution of the Indus River in relation to topographic uplift, erosion, climate and geology of western Tibet, the Transhimalayan and High Himalayan Ranges. In Meadows, A. and Meadows, P. (eds), *The Indus River*. Oxford: Oxford University Press, 210–30.

Sengor, A.M.C. and Natal'in, B.A., 1996. Paleotectonics of Asia: fragments of a synthesis. In Yin, A. and Harrison, M. (eds), *The tectonics of Asia*. New York: Cambridge University Press, 486–640.

Seno, T., 1977. The instantaneous rotation vector of the Philippine Sea Plate relative to the Eurasian Plate. *Tectonophysics*, **42**: 209–26.

Shroder, Jr, J.F. and Bishop, M.P., 2000. Unroofing of the Nanga Parbat Himalaya. In Khan, M.A., Treloar, P.J., Searle, M.P. and Jam, M.Q. (eds), *Tectonics of the Nanga Parbat Syntaxis and the Western Himalaya*. Geographical Society Special Publication, 170. London: Geological Society of London, 163–79.

Speed, R., Ellison, M.W. and Heck, F.R., 1988. Phanerozoic tectonic evolution of the Great Basin. In Ernst, W.G. (ed.), *Metamorphism and crustal evolution of the Western United States*. Ruby Volume VII. Englewood Cliffs NJ: Prentice Hall, 572–605.

Sullivan, M.A., Windley, B.F., Saunders, A.D., Haynes, J.R. and Rex, C.C., 1993. A palaeogeographic reconstruction of the Dir Group: evidence for magmatic arc migration within Kohistan, N. Pakistan. In Treloar, P.J. and Searle, M.P. (eds), *Himalayan tectonics*. Geographical Society Special Publication, 74. London: Geological Society of London 139–60.

Summerfield, M.A., 1991a. *Global geomorphology*. London: Longman.

Summerfield, M.A., 1991b. Sub-aerial denudation of passive margins: regional elevation versus local relief models. *Earth and Planetary Science Letters*, **102**: 460–69.

Tapponnier, P. and Molnar, P.J., 1976. Slip-line field theory and large-scale continental tectonics. *Nature*, **264**: 319–24.

Tapponnier, P., Peltzer, G. and Armijo, R., 1986. On the mechanics of the collision between India and Asia. In Coward, M.P. and Ries, A.C. (eds), *Collision tectonics*. Geographical Society Special Publication, 19. London: Geological Society of London, 115–57.

Thompson, G.A. and Burke, D.B., 1974. Regional geophysics of the Basin and Range province. *Annual Reviews of Earth and Planetary Sciences*, **2**: 213–38.

Tippett, J.M. and Hovius, N., 2000. Geodynamic processes in the Southern Alps, New Zealand. In Summerfield, M.A. (ed.), *Geomorphology and global tectonics*. Chichester: Wiley, 109–34.

Troll, C., 1973a. The upper timberlines in different climatic zones. *Arctic and Alpine Research*, **5**: 3–18.

Troll, C., 1973b. High mountain belts between the polar caps and the equator: their definition and lower limit. *Arctic and Alpine Research*, **5**: 19–21.

Uyeda, S., 1978. *The new view of the Earth*. San Francisco, CA: W.H. Freeman and Company.

Vicente, J.-C., 1989. Early late Cretaceous overthrusting in the western Cordillera of southern Peru. In Erickson, G.E., Canas, P.M.T. and Reinemund, J.A. (eds), *Geology of the Andes and its relation to hydrocarbon and mineral resources*. Houston, TX: Circum-Pacific Council for Energy and Mineral Resources Earth Science Series, Volume 11.

Vilotte, J.P., Daignieres, M. and Madariaga, R., 1982. Numerical modeling of intraplate deformation: simple mechanical models of continental collision. *Journal of Geophysical Research*, **87**: 19709–28.

Vogt, P.R., 1981. On the applicability of thermal conduction models to mid-plate volcanism; comments on a paper by Gass et al. *Journal of Geophysical Research*, **86**: 950–60.

Wang, C.-H. and Burnett, W.C., 1991. Holocene mean uplift rates across an active plate-collision boundary in Taiwan. *Science*, **248**: 204–206.

Wang, M., Shen, Z.K., Jackson, D., Yin, A., Li, Y., Zhao, C., Dong, D. and Fang, P., 1999. GPS-derived deformation along the northern rim of the Tibetan Plateau and southern Tarim basin. *EOS, Transactions of the American Geophysical Union*, **80**: F1009.

Watts, A.B. and ten Brink, U.S., 1989. Crustal structure, flexure and subsidence history of the Hawaiian Islands. *Journal of Geophysical Research*, **94**: 10473–500.

Whipple, K.X. and Tucker, G.E., 1999. Dynamics of the stream-power river incision model: implications for height limits of mountain ranges, landscape response timescales, and research needs. *Journal of Geophysical Research*, **104**: 17661–74.

Whittington, A.G., 1996. Exhumation overrated. *Tectonophysics*, **206**: 215–26.

Wolfe, J.A., Schorn, H.E., Forest, C.E. and Molnar, P., 1997. Paleobotanical evidence for high altitudes in Nevada during the Miocene. *Nature*, **276**: 1672–75.

Wu, C., Nelson, K.D., Wortman, G., Samson, S.D., Yue, Y., Li, J., Kidd, W.S.F. and Edwards, M.A., 1998. Yadong cross structure and South Tibetan detachment in the east central Himalaya (89°–90°E). *Tectonics*, **17**: 28–45.

Yang, Y. and Liu, M., 2000. The rise and fall of the Tibetan Plateau: results of 3D numeric modelling. *EOS, Transactions of the American Geophysical Union*, **80**: F1230.

Yin, A., 2000. Modes of Cenozoic east–west extension in Tibet suggesting a common origin of rifts in Asia during the Indo-Asian collision. *Journal of Geophysical Research*, **105**(B9): 21745–59.

Yin, A. and Harrison, T.M., 2000. Geologic evolution of the Himalayan–Tibetan orogen. *Annual Reviews of Earth and Planetary Sciences*, **28**: 211–80.

Yin, A., Kapp, P.A., Murphy, M.A., Manning, C.E., Harrison, T.M., Grove, M., Ding, L., Deng, X. and Wu, C., 1999. Significant late Neogene east–west extension in northern Tibet. *Geology*, **27**: 787–90.

Zeitler, P.K., Meltzer, A.S., Koons, P.O., Craw, D., Hallet, B., Chamberlain, C.P., Kidd, W.S.F., Park, S.K., Seeber, L., Bishop, M. and Shroder, J., 2001. Erosion, Himalayan geodynamics and the geomorphology of metamorphism. *GSA Today*, January: 4–9.

3
The evolution of mountains on passive continental margins

Cliff Ollier

1 Introduction

In plate tectonic terms, the Earth's surface is covered by a number of plates. These can grow at spreading sites, and be consumed by subduction where plates collide. Continental margins without collision are called passive margins, in contrast to the active margins where plates collide. Nearly all the literature on mountain building in the past 40 years has concentrated on active margins where collision and subduction may explain both mountains and the structures within them, but in reality there are also mountains on passive margins. The popular concept of a mountain range is a tent-shaped ridge with a sharp top, but the name 'range' has often been given to escarpments. Many of the world's great mountain 'ranges' are plateaux bounded by escarpments.

In this chapter I first describe the main features of passive margin landscapes. After a brief review of different approaches to the study of these landscapes, the major features are reviewed by geographical regions. Finally a number of general topics are reviewed, possible mountain-building mechanisms are discussed, and general conclusions are drawn.

2 Features of the geomorphology of passive continental margins

The basic geomorphic features of passive margins with mountains are shown in Figure 3.1.

2.1 Plateaux

Plateaux are upland areas with relatively flat topography. A few may be original depositional surfaces, but most are erosion surfaces. In areas of folded sedimentary rock they are recognized by bevelled cuestas and accordant, level strike ridges. They may be extensive or dissected until only fragments are left on the hardest rocks.

2.2 Marginal swells

The marginal swell is a widespread swell or bulge along the edges of a continent (*Randschwellen* in German; *bourrelets marginaux* in French: Jessen, 1943; Godard, 1982; Bremer, 1985; Ollier, 1985). The marginal swell is formed after the planation surface of the plateau, and after formation of major valleys. The whole land surface has been warped into an asymmetrical bulge, with the steeper slope to the coast, though the 'steep' slope is usually only about 2°.

2.3 Great Escarpments

Great Escarpments are landforms on the grand scale, thousands of kilometres long and up to 1000 m high. They occur on many types of rocks including horizontal or folded sedimentary rocks, granites,

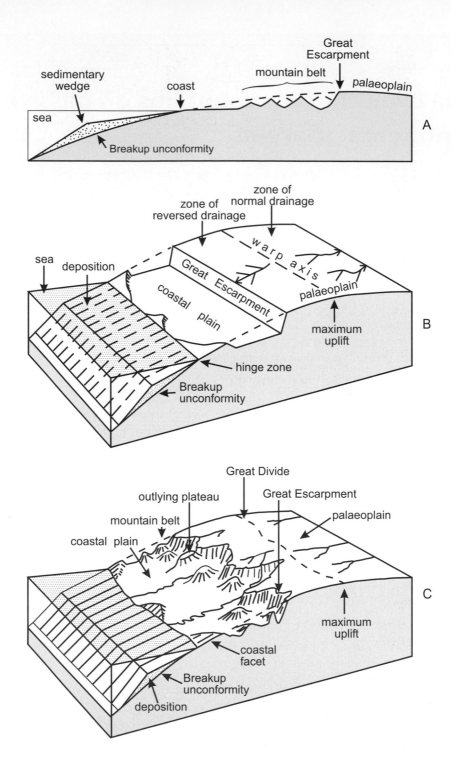

Figure 3.1 *The basic morphotectonic features of a passive continental margin with mountains. (A) Diagrammatic cross-section showing the range of features of a passive continental margin; (B) simplified block diagram showing the major morphotectonic features of a passive continental margin; (C) a more realistic block diagram of the topographic features. The Great Escarpment is embayed by major valleys, and a belt of mountains lies between it and the coastal plain*
Source: *reprinted from* Geomorphology, *19, Ollier, C.D. and Pain, C.F.,* Equating the basal unconformity with the palaeoplain: a model for passive margins. *pp. 1–15. © 1997, with permission from Elsevier.*

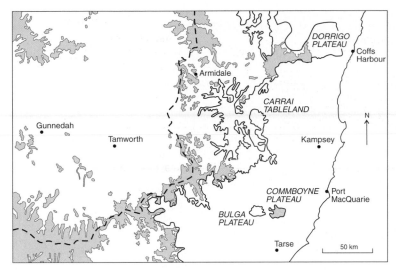

Figure 3.2 *Part of northern New South Wales, Australia, showing the highly indented Great Escarpment, with peninsulas such as the Dorrigo Plateau and Carrai Tableland, outlying plateaux such as the Comboyne and Bulga plateaux, the separation of the Great Divide (dashed line) and the Great Escarpment (solid line). Relationships with basalt (shaded areas) are also shown. The large patch west of the Dorrigo Plateau is the Ebor Volcano After Ollier (1982). Published with the permission of the Geological Society of Australia.*

basalts and metamorphic rocks. Great Escarpments run roughly parallel to the coast, and they separate a high plateau from a coastal plain. The top of a Great Escarpment can be very abrupt. Great Escarpments are undoubtedly erosional, as is clear where scarps are deeply embayed along rivers (Figure 3.2, and see Lidmar-Bergström et al., 2000), but in some places the escarpments are fairly straight for hundreds of kilometres (Ollier and Stevens, 1989).

Rugged mountainous areas form below the Great Escarpment, where the old plateau surface has been deeply dissected. Occasionally a patch of plateau is isolated from the main tableland to form a peninsula or tableland as an outlier from the main plateau. Many of the world's large waterfalls are found where rivers cross a Great Escarpment, such as Wollomombi Falls, Australia (Figure 3.3), described later.

2.4 Coastal plains

Seaward of the Great Escarpments lie lowland plains. These are basically erosional, but may have a cover of sediment or volcanic rocks that help to date them.

2.5 Breakup unconformity

A wedge of sediments is deposited offshore from the continental margin. At the base of the sediments is an unconformity, sloping seaward, which might be termed the breakup unconformity, meaning the one that is related to the breakup of a supercontinent and formation of new continental margins. Ollier and Pain (1997) believe the breakup unconformity is the downwarped equivalent to the palaeoplain on land.

2.6 Drainage patterns and drainage divides

Some passive margins have simple drainage patterns with streams flowing in opposite directions away from a ridge at the top of the Great Escarpment as in Namibia, the Western Ghats of India and

Figure 3.3 *Wollomombi Falls, New South Wales, Australia. Here the Great Escarpment is retreating as a wall, and old valleys are perched on the top of the cliffs, similar to hanging valleys at the top of marine cliffs Photo: C.D. Ollier.*

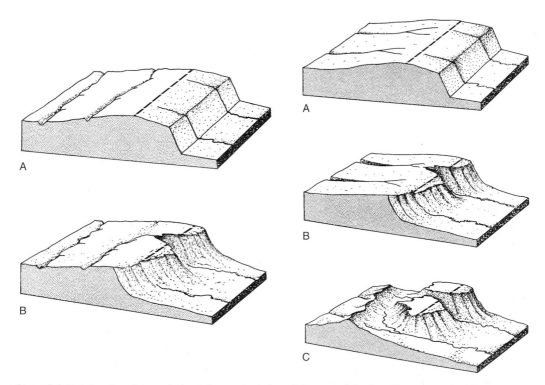

Figure 3.4 *Evolution of passive margins in relation to prior drainage. Left: retreat of the Great Escarpment where the palaeodrainage is roughly age-parallel to the line of uplift and continental margin. Streams developed on the steep, ocean-facing side cut back by headward erosion and capture rivers flowing parallel to the coast. Right: Retreat of the Great Escarpment where the divide is roughly perpendicular to the drainage. Rivers on the steep coastal side cut back faster than those on the gentler inland side, with greater headward erosion. By stage (C) headward retreat has passed the original tectonic divide (marked by the dashed line) and the once-continuous palaeoplain is broken up into smaller individual plateaux. East Victoria (Australia) is a good example*

Eastern Brazil. Workers such as Beaumont *et al.* (2000) think the drainage was initiated on opposite sides of a fault scarp that formed the initial watershed. Other passive margins have the major watershed on the broad swell, inland of the Great Escarpment, as in much of eastern Australia. Such Great Divides may be the crest of the original tectonic marginal swell, with the crest of the swell being the drainage divide. Such divides can migrate, not by a rolling wave of tectonism but by headward erosion of rivers and river capture. With river capture the divide 'leaps' to the far side of the captured stream rather than 'creeps' by steady erosion (Ollier and Pain, 2000).

On some continental margins the major rivers were in existence before continental breakup and can still be traced in the modern landscape, as in Australia and South Africa. It is often found that minor drainage is structurally controlled but major drainage is not, and so was probably initiated on a former erosion surface or sedimentary covering. Drainage may be reversed (indicated by barbed drainage) or modified by river capture (indicated by elbows of capture and wind gaps) or evolve in more complex ways (Ollier and Pain, 2000).

Drainage modification depends on the original direction of drainage in relation to the orientation of the elongated marginal swell. If the ancient drainage was roughly parallel to the swell, headward erosion by coastal rivers will capture plateau rivers. If the original drainage was roughly perpendicular to the swell, headward erosion of coastward flowing rivers will eventually reach the crest of a swell, and then cut beyond it. Isolated plateaux may be left on the old divide. This is the situation in Victoria, Australia, where numerous isolated high plains are preserved south of the watershed known as the Great Divide, and they are higher than the divide (Figure 3.4).

3 Approaches to the study of passive continental margins

With the advent of plate tectonics in the 1960s, the study of continental margins took on new meaning. Several notable papers and books were produced (e.g., Burke and Drake, 1974; Scrutton, 1982; Wilson *et al.*, 2001) but passive margins received less attention than active margins, and geomorphic data were largely neglected. Tectonic geomorphology had a resurgence in the 1980s (e.g., Ollier, 1985; Morisawa and Hack, 1985), in which the basic data are the form of the land. Their interpretation uses geomorphic theory and whatever stratigraphic and geological data can be found to constrain the interpretation. This chapter is largely continuing this tradition. Several other approaches can be used, including the following.

3.1 Fission-track thermochronology

Spontaneous fusion of ^{238}U produces fragments that disrupt the crystal lattice to create tracks, visible under the microscope, which can be counted and measured. The number of tracks is proportional to uranium content, age and thermal history. The method dates the time of cooling below a mineral-dependent rock-annealing temperature. Gleadow and Brown (2000) present a clear account of the principles, techniques and variations within the method. Mineral cooling ages can be combined with assumptions or measurements of the geothermal gradient to estimate at what depth below the surface the samples were located. Fielding (2000: 214) describes some of the problems inherent in the method: high rates of denudation, high local relief or underthrusting of colder rocks can strongly affect and even reverse the normal geothermal gradient, 'and these effects can greatly complicate data interpretation'. There are conflicts between the amount of erosion determined from geological and geomorphic evidence, and the amount of erosion implied by interpretation of fission-track data. As Pillans (1998: 16) wrote: 'Interpretation of fission-track data in terms of denudation history is complex'. Examples will be provided below.

3.2 Numerical surface process models

These are essentially mathematical models in which tectonic and surface processes are quantified, and a model run to see the effect of any given set of assumptions. Tectonic forcing (uplift) may be slow or rapid, and erosion and sedimentation may be quantified in different ways. It is sometimes assumed that the landscape is in a steady state or that there is elastic lithospheric flexure. A discussion is provided by Beaumont et al. (2000), which provides a useful discussion of models in general and a discussion of rifted continental margins and Great Escarpments. They present a model in which the main escarpment was originally coincident with a master fault near the coastline – very different from the downwarp model of Ollier and Pain (1997). There is often conflict between the geomorphic ground truth and model predictions, and some examples will be described below.

3.3 Offshore sediments

A vast amount of offshore data are available, but only occasionally have they been related to geomorphology of neighbouring land areas. Several examples are mentioned below, but clearly much more can be done on this topic.

3.4 Cosmogenic dating

Cosmic ray interactions produce ^{10}Be, ^{36}Cl, ^{26}Al in the atmosphere and lithosphere. Accumulation of these reflects duration of cosmic ray exposure of the upper 1 or 2 m of the Earth's surface. It is highly affected by erosion so can give only a minimum age. Nevertheless it is used (e.g., Cockburn et al., 2000) to determine denudation rates and then draw conclusions on landscape evolution in general, including retreat of Great Escarpments.

4 Examples of mountains on passive margins

The following account is not systematic, because workers in different areas have different data, and ask different questions. The account of Australia comes first and is longer than the others as some of the debates are introduced for the first time. Tectonic mechanisms are discussed after the regional treatment.

4.1 Australia

4.1.1 The Eastern Highlands

Maps of Australia commonly show 'The Great Dividing Range' running inland of the eastern coast. This is a cartographic myth! The Great Divide is real enough, separating coastal from inland drainage, but for most of its length the divide crosses a plateau of low relief, and numerous lakes are located right on it. A much more spectacular landform is the Great Escarpment which, when seen from the coast, looks like a 'range' of mountains (Ollier, 1982).

The Highlands are underlain by varied rocks, eroded to a surface of low relief that Hills (1975) called the Trias-Jura Surface. Most of the palaeoplain is very deeply weathered, and regolith has been dated by palaeomagnetism to at least 60 Ma (Schmidt and Ollier, 1988; Idnurm, 1985). Eocene to Pliocene lava flows on the palaeoplain are preserved in inverted relief, but their very preservation shows vertical lowering of the palaeoplain by erosion is in the order of hundreds of metres, not kilometres (Ollier, 1988).

The relationships of Great Divide, Great Escarpment, coast and continental shelf are shown on Figure 3.5. The original divide was the culmination of a warp of the palaeoplain and, in places, is still in its original position. Elsewhere it has been shifted by headward erosion of the Great Escarpment, river capture, and diversion by tilting, volcanic activity and faulting.

The Eastern Highlands fall into three distinct sectors following tectonic axes. North of Brisbane is the Queensland axis. The morphotectonic lines (divide, escarpment, coast, contental margin) trend north-northwest and are widely spaced, but the region is broken up by numerous N–S faults, complicating the relationships of drainage patterns and escarpments (Ollier and Stevens, 1989). The Great Escarpment is up to 1000 m high, and some continuous stretches are 700 km long. The Great Divide is up to 350 km further inland in some places, but is occasionally coincident with the Great Escarpment: the highest mountain in Queensland, Mt Bartle Frere (1622 m), is right on the Great Escarpment.

The New South Wales (NSW) axis runs between Brisbane and the Victoria–NSW State border, and here the lines run north-northeast and are closely spaced. The NSW divide is the simplest, being a simple warp for much of its length, with the Great Escarpment tens of kilometres east of the Divide, and numerous well-preserved reversed rivers. The highest ground (ignoring volcanoes) is on the Divide itself. The escarpment is frequently embayed by headward erosion of individual valleys.

In the south, the Victoria axis runs E–W. The palaeoplain has been eroded to a number of isolated plateaux known locally as High Plains, and the Great Divide is generally north of, and lower than, the High Plains that mark the original axis of uplift. The Victorian uplift is probably older than that of NSW and Queensland, and geomorphic evolution has progressed further, dissecting a single palaeoplain into many small plateaux.

The Eastern Highlands passive margin is exceptional in having abundant basalt volcanism dating from the Cretaceous to 5000 years ago, giving exceptional opportunity to date the geomorphic evolution of the landscape. There are many old lava flows; large central volcanoes that mark a hotspot trace, and (in the Quaternary) areal volcanism with scores of small scoria cones and maars in northern Queensland and western Victoria. Volcanism roughly follows the main divides (Figure 3.5).

The Great Escarpment cuts across the 19-Ma Ebor volcano, NSW, so here the escarpment is post 19 Ma. Near Innisfail in Queensland a 3-Ma lava flow went over the Great Escarpment, so here the escarpment was in existence at that time, and has not retreated very far since. A 28-Ma basalt flowed down the Clyde River valley, east of the Great Escarpment in southern NSW, indicating that the

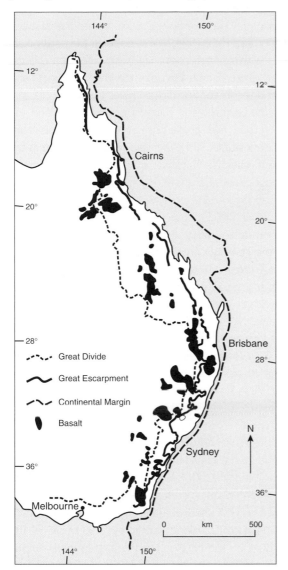

Figure 3.5 *The Eastern Australian passive margin, showing the location of major morphotectonic features*

drainage systems were already established at that time. Some volcanoes erupted after the passage of the escarpment and creation of the coastal plain, such as the 25-Ma Tweed volcano (1204 m) in NSW. Eocene and Palaeocene lavas are found on the coastal plain about 30 km southwest of Brisbane. The escarpment is therefore a diachronic feature. Volcanicity also suggests that tectonic movements occurred over a long period, and not just at the time of creation of continental margins.

Before the modern continental margins were formed, the rivers flowed from land to the east and south of Australia to the Great Artesian Basin (Ollier and Pain, 1994). With breakup the palaeoplain was warped down to the coast, reversing the rivers (with the barbed drainage) and creating the Great Divide on a plateau that was already high. Reversed rivers include the Clarence River (Haworth and Ollier, 1992), the Hunter River (Galloway, 1967) and the lower Shoalhaven River (Ollier, 1978), and many other examples have been claimed, including the Daintree and Tulley Rivers in Queensland and the Genoa and Cann Rivers in Victoria. In western Victoria, river profiles preserved beneath basalt are steeper than profiles of the present-day valleys, indicating uplift in the highlands and subsidence in the plains.

Bishop and Goldrick (2000:235) disagree with these major changes in drainage pattern. They wrote '. . . the valley-filling lavas throughout southeast Australia indicate persistence of drainage directions from the latest Mesozoic or early Cenozoic'. Young and McDougall (1982) and Young and Wray (1999) believe the Eastern Highlands have remained stable, rather than downwarped to the coast, and Bishop and Goldrick (2000) have extended the stability concept to the entire highlands. The single main piece of evidence for the idea of post-Triassic stability, traced through the literature by Brown (2000), is the presence of horizontal Triassic strata on the plateau above the escarpment west of Nowra, which lies south of Sydney on the New South Wales coast. Young and McDougall (1982) take this to indicate that the area has remained stable ever since Triassic times, and that downwarping to the coast was impossible. This alleged long-term stability is then extended to much of eastern Australia. But the argument is not even true in the Nowra area. The structure of the Triassic is revealed on a map (Standard, 1969) and shows a gentle dip to the north complicated by monoclines. The section described by Young and McDougall (1982) is simply a strike section. In reality there is plenty of evidence of mobility, including faulting (Brown, 2000).

An alternative hypothesis to coastal downwarp comes from researchers working on fission-track data in the NSW sector of the highlands (e.g., Moore et al., 1986). In brief, their results suggest late Cretaceous uplift (not downwarp) to the east, so the coastal lowlands must be made by later erosion, removing several kilometres of rock. Bishop and Goldrick (2000) summarize the major issues of current debate. Did escarpment retreat involve dissection of an uplifted plateau or a downwarped rift shoulder? They wrote (p. 234): 'The 100 Ma old lavas and Mesozoic weathering profiles on the NSW coast are inconsistent with deep denudation having occurred along and across the whole of the coastal strip below the escarpment . . .'. South of Bega on the NSW coast there is a superb coastal 'facet' of old downwarped surface (as shown in Figure 3.1) with deep and intense weathering suggesting Mesozoic age (Kubiniok, 1988). On the coast Brown (2000) reports basalts 27–30 Ma old and Oligocene to early Miocene sediments. This is precisely where fission-track work and Apatite (U-Th)/He age constraints (Persano et al., 2002) indicate kilometres of erosion at the coast, which should surely strip these materials.

Similarly in north Queensland, one interpretation of fission-track data is erosion of 0.8–3.0 km of overlying rock (Marshallsea et al., 2000), but the geology of the region 'suggests that an explanation in terms of deeper burial may be untenable. Heating due to hot fluid flow may be a more realistic mechanism' (p. 779).

Early workers such as Andrews (1910) were impressed by the great faults in southern NSW, where lava flows hang high at the top of fault scarps. The Long Plain Fault, for example, still makes a clear fault-scarp with the 22-Ma Kiandra Basalts at the top, and has a throw of 800 m or more. They believed the highlands were formed during the 'Kosciusko Orogeny' in the very late Cenozoic. It now seems possible that there was an early phase of warping and uplift to make the major features of the continental margin, and then a later uplift in some areas, including the plateau of Mt Kosciusko (2229 m), Australia's highest mountain (Ollier and Taylor, 1988).

4.1.2 Southwest Australia

The southwest of Western Australia has only small erosional mountains, but it illustrates some of the classical passive mountain features in a region of very subdued relief. The planation surface is under-lain by ancient rocks and was already flat and deeply weathered by the Cretaceous. Chains of salt lakes mark the course of huge ancient rivers, some of which flowed from the south before Australia and Antarctica separated. There are two passive margins meeting at roughly 90°.

With the separation from Antarctica, the southern coastal region was downwarped, creating a new watershed on the axis 300–400 m high, 60–120 km inland and stretching 650 km. On the seaward side of this axis the rivers were reversed (Clarke, 1994). The oldest sediments found in the palaeo-valleys are Eocene. There is no Great Escarpment.

The western margin is dominated by the Darling Fault which separates the West Australian Plateau, cut across Archaean rocks, from a basin on the west containing over 10 km of Silurian to Cretaceous sediments. East of the Darling Fault is the Darling Uplift, a marginal swell along a N–S axis, 1000 km long, 60–80 km wide, and uplifted 150–200 m. Uplift created a depression on its eastern side, where middle Eocene fluviatile sediments accumulated, thus dating the uplift (Beard, 1999). Near Perth the Darling Fault makes the Darling Range, a west-facing fault line escarpment 250 m high. North and south of Perth, Mesozoic rocks are preserved west of the Darling Fault. These are bounded to the west by low gradient erosional escarpments that separate them from the coastal plain. This compos-ite escarpment is by no means 'great' but may be analogous to Great Escarpments of other continen-tal margins, and has a total length of about 900 km.

4.2 India

All of peninsular India can be envisaged as a palaeoplain sloping from west to east with two main ranges, the Western and Eastern Ghats. The word 'ghat' simply means an ascent, and may refer to any hill or escarpment. The continent is markedly asymmetrical, and the Eastern Ghats are lower than the Western Ghats. The area has a very long history as a land surface and was probably very flat before the breakup of Gondwana. Bauxites on the palaeoplain have a suggested Jurassic age (Valeton, 1994).

4.2.1 Western Ghats

The Western Ghats of India comprise a Great Escarpment that runs parallel to the western coast of India (Ollier and Powar, 1985), backed by a plateau (Figure 3.6). The escarpment extends without a break for over 1500 km and is seldom more than 60 km from the coast. The Western Ghats attain their greatest altitude in the south (2900 m in the Nilgiri and Cardomom Hills) and laterite-capped plateaux prevail. The edge of the escarpment is almost entirely coincident with the divide between eastern- and western-flowing rivers.

In the north the escarpment is cut across the late Cretaceous Deccan Basalts, and in the south across metamorphic and igneous rocks of the Precambrian Shield (Figure 3.7), with no significant change in landform on these very different bedrocks. The length, continuity and similarity of form indi-cate a single, post-Cretaceous process of uplift and scarp recession. The Deccan plateau is built of a

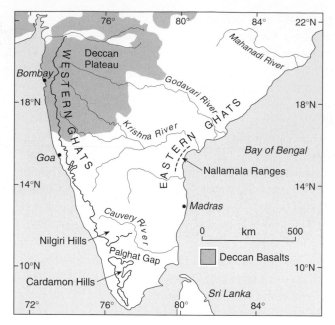

Figure 3.6 *Map of peninsular India showing the scarp of the Western Ghats, which is a clear line, and the location of the Eastern Ghats, which are less distinct*

huge stack of relatively thin lava flows. Widdowson (1997) suggested that some laterites on the plateau mark the original constructional surface, but his map of laterite shows a dendritic pattern that suggests a former river, and appears to be a perfect example of inversion of relief. The time of extrusion of the Deccan Traps (Cretaceous–Tertiary boundary) is the same as the formation of a new continental margin with the opening of the Arabian Sea, which in turn led to the creation of the Western Ghats (Widdowson and Gunnell, 1999).

A monocline called the Panvel flexure runs parallel to the coast in the vicinity of Bombay where the horizontal Deccan Traps of the inland are bent down to angles of 8° or more. It can be traced for at least 120 km and the axis of the flexure is several tens of kilometres west of the Great Escarpment. Offshore there is a horst and graben complex, and the flexure has been related to faulting in various ways (Dessai and Bertrand, 1995; Sheth, 1998).

Most east-flowing streams near the crest of the Western Ghats have narrow valleys that only become broad and flat when they enter the eastern plains but some, such as the Godavari and Krishna (Kistna) Rivers, have broad valleys even in their source region. The divide between the east-flowing Indrayani River and the west-flowing Ulhas River, near Bombay (Mumbai), is very inconspicuous, and a broad flat valley is continuous from one side of the divide to the other. After flowing across this valley on the plateau, the Ulhas River descends the western scarp by a precipitous waterfall, in front of which is a narrow gorge 9 km long and 300 m deep. The broad valley of the Indrayani–Ulhas headwaters is hard to explain unless an original broad valley was formed by a single major river and then warped (Ollier and Powar, 1985).

Widdowson and Gunnell (1999) showed several phases of laterite formation on the coastal plain (developed on basalts in the north and Archaean rocks to the south). The elevation of the coastal laterite (60–200 m), together with associated development of an entrenched drainage, indicates that widespread uplift affected the margin during late Tertiary times. Cambering of the palaeosurface is

Figure 3.7 *The Western Ghats in southern India, near Kodaikanal. Here the Great Escarpment is cut across Precambrian metamorphic rocks and granites, yet the general form is the same as that where it is cut across the basalts of the Deccan plateau*
Photo: C.D. Ollier.

consistent with seaward flexure in response to denudational unloading onshore and sedimentary loading offshore.

Figure 3.6 shows a triangular area south of the Palghat Gap, including the Cardamom Hills, which is bounded on all sides by an escarpment. Presumably scarps retreated on both west and east sides of India from the ocean, but the northern boundary is hard to explain, as indeed is the Palghat Gap itself. This would be a good area for further research, offering unique constraints.

4.2.2 Eastern Ghats

The Eastern Ghats lie behind a coastal plain 50–70 km broad, without the grandeur or massive escarpment of the Western Ghats. The major northeast-axis of uplift of the Eastern Ghats is parallel to a system of extensional rift basins that contain Cretaceous to Miocene deltaic deposits (Nash *et al.*, 1999). The Godavari River divides the Eastern Ghats into north and south sections. The southern part is further divided into two: south of the latitude of Madras is the edge of the Mysore Plateau, geomorphically similar to the Western Ghats; the northern part consists of structural ridges making the Nallamala Ranges. In the northern Eastern Ghats, between the Godavari and Mahanadi Rivers, no major rivers cross and minor rivers have been diverted along the strike to form tributaries to the main rivers. The Great Escarpment reappears in the north, on both sides of the Mahanadi River. Behind the escarpment are isolated highland areas capped by planation surface remnants with duricrust, including bauxite, at elevations locally over 900 m (Nash *et al.*, 1999). It is significant that the escarpments of the Western Ghats developed on local upwarps and fault blocks, and could not have retreated from a massive fault scarp near the east coast.

The drainage pattern of peninsular India is essentially simple, from a watershed along the Western Ghats escarpment. The courses of most of the major east-flowing rivers are deflected by the Eastern Ghats, but the Mahanadi River drainage system crosses the uplift in a major gorge, presumably antecedent. The uplift of the Eastern Ghats is therefore younger than the drainage system, and so younger than the uplift of the Western Ghats.

4.3 Southern Africa

Much of southern Africa is a plateau, bounded by marginal swells and a Great Escarpment (Ollier and Marker, 1985; Partridge and Maud, 1987; Partridge, 1998). The plateau is generally over 1200 m in elevation, and over 3000 m in Lesotho. The Great Escarpment makes a huge arc all around southern Africa from Angola, through Namibia, around South Africa and to the Limpopo River (Figure 3.8). It is highest in the Drakensberg (3299 m), where the rocks are generally horizontal, and the upper part consists of 1500 m of Triassic basalt (Figure 3.9). The marginal swells bound the central depression of the Kalahari Basin.

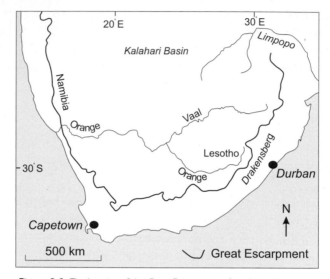

Figure 3.8 *The location of the Great Escarpment of southern Africa*

Figure 3.9 *View of the Drakensberg, showing the Great Escarpment, plateau remnants and outlying hills*
Photo: Björn Björnsen. Reproduced with the permission of Art Publishers (Pty) Ltd, Republic of South Africa.

Partridge (1998) reviewed the morphotectonics of southern Africa and concluded:

1 much of Africa possessed high elevation prior to the rifting that made new continental boundaries;
2 the ground surface was lowered by 1–3 km during the Cretaceous;
3 neogene uplift re-established high elevations, especially in the eastern half of the subcontinent.

The suggestion that southern Africa was already high when new continental margins appeared seems reasonable. If the Western Rift valley of Uganda were to widen, the plateaux of Uganda and Congo would be high from the start. The Cretaceous African Surface is recognized by widespread deep weathering and massive ferricretes and silcretes, indicating a long period of stability after planation. The suggestion of 3 km of erosion seems hardly consistent with preservation of such deep regolith. Unfortunately there are no lava flows to date the surface, as in Australia. De Wit (1999) describes gravels on both the high plateau surface and on the coastal plain that are of probable late Cretaceous age, confirming a Cretaceous age for the blocking out of the major features of southern Africa.

Partridge (1998: 167) wrote: 'The evidence for large-scale Neogene uplift is now beyond question … The largest movements post-date the Miocene and have contributed both to the anomalous elevation of the eastern hinterland and to the strong east–west climatic gradient across southern Africa.' This late and dramatic uplift that elevated southern Africa (and much of interior Africa up to the Red Sea) is indicated by a wide variety of evidence. River-long profiles are convex up; early Pliocene marine deposits have been uplifted 400 m; remnants of the African Surface have been warped from 3 m km^{-1} to 40 m km^{-1}. These and other data show total uplifts of 700–900 m along the axis of warping within the last 5 Ma (Partridge, 1998).

As early as 1933 Du Toit argued that major river divides within the interior of southern Africa were controlled by axes of epeirogenic flexure. The model was extended by King (1963) who proposed that the divide between coastal rivers and interior rivers is a line of flexure which he termed the Escarpment axis. Flexuring along the Escarpment axis is generally ascribed to rift flank uplift associated with the disruption of Gondwanaland (e.g., Partridge and Maud, 1987). Partridge (1998) argues that there is little firm evidence for inland migration of an isostatically triggered flexural bulge. However, both King (1963) and Gilchrist and Summerfield (1991) took the view that erosion subsequent to uplift would have resulted in the coastal divide migrating inland, and that the original line of flexure may have been located closer to the continental margin.

Moore (1999) presents a new figure of axes of epeirogenic uplift (his figure 8) and recognizes three axes: the Escarpment axis; the Etosha–Griqualand–Transvaal; and the Ovamboland–Kalahari–Zimbabwe axes. These roughly concentric lines of crustal flexure are broadly parallel to the edge of the subcontinent. This requires a far more elaborate tectonic explanation than simple rise to the rift as Gondwanaland broke up. Moore agrees with others that the initial uplift along the Escarpment axis was associated with the late Jurassic–early Cretaceous rifting that initiated the disruption of Gondwana, and believes the other axes are younger. For details of suggested ages and the reasoning behind them see Moore (1999).

The major west-flowing drainage system of southern Africa is the Orange River and its main tributary the Vaal River, which rise in the mountains of the east and flow to the Atlantic Ocean, described in detail by De Wit (1999). The major east-flowing rivers are the Zambezi, Save and Limpopo, described in detail by Moore and Larkin (2001). According to De Wit (1999), the Orange River system had established itself on a well-planed surface, more or less as it is today, by the late Cretaceous. The lower Orange River is an antecedent river that flowed to the Atlantic, close to its present course, for at least 60 Ma. Above the escarpment gravels on various rivers, including the proto-Orange terraces, up to 40 m above the present river, have Miocene fossils. In the middle Orange River the proto-Orange terrace is 100 m above the present river level. One river terrace is dated by mammalian fossils at 17–19 Ma.

Moore and Larkin (2001) trace the evolution of the east-flowing rivers since the disruption of Gondwana, using evidence of geomorphology, mineralogy of old alluvium, continental sediments, offshore sediments and the relationships of freshwater fish. They suggest that the Limpopo provided a conduit linking the major inland drainage to the Indian Ocean immediately following disruption of Gondwana. Its course may have been along an aulcogen (a plate tectonics term: the failed arm of a triple junction).

The occurrence of common fish species in different rivers provides compelling evidence for earlier links between them. The number of species common to two rivers divided by their total number of species provides a similarity index. Using this technique Skelton (1994) showed southern Africa can be divided into two major icthyofaunal provinces, designated the Southern and Zambezi provinces. The high similarity index between fish of the Okavango River (east flowing) and the Cunene River (Atlantic flowing) points to a former link between the two, attributed to river capture by Moore and Larkin (2001). Capture also explains the occurrence of common fish species in the Congo River system and the mid–lower Zambezi River.

Evidence from marine sediments is available for both east and west coasts. East coast offshore sediments range from Jurassic to Quaternary (Dingle et al., 1983). Data are summarized by Moore and Larkin (2001), with interpretation of geomorphic significance. From this, they trace changes in palaeodrainage from Jurassic to Pleistocene. Opening of the Atlantic Ocean was also associated with the early development of major depocentres on the western continental margin (Dingle et al., 1983). Sediment was supplied to these basins by the Orange River (Dingle and Hendy, 1984) and the Trans-Karoo (De Wit, 1999), which both drained the interior of southern Africa. Brown et al. (1995) report that a depocentre developed after 103 Ma near the mouth of the Orange River, as a result of loading by the Orange River delta systems.

4.3.1 Internal drainage hypothesis

The fluvial and marine data outlined above indicate persistent drainage to the sea from the earliest period of continental breakup, which is in sharp contrast to the internal drainage hypothesis of others. Thomas and Shaw (1991) envisage that marginal warping of the continent associated with continental breakup of Gondwana initiated an endoreic (internal) drainage system that only recently broke through the girdle of high ground behind the Great Escarpment. Gilchrist et al. (1994) performed numerical experiments that were related to southwestern Africa. They write (p. 12 211): '. . . the primary effect of rifting in the subsequent landscape evolution is that it generates two distinct drainage regions. A marginal upwarp, or rift flank uplift, separates rejuvenated rivers that drain into the subsiding rift from rivers in the continental interior that are deflected but not rejuvenated. The two catchments evolve independently . . .'. Since the Great Escarpment of southern Africa makes a complete arc, this might suggest that the inner part may be an area of internal drainage, supporting the view of Thomas and Shaw (1991). They maintain that the Orange River assumed its present course after a 'recent' capture. Geomorphic evidence tends to suggest otherwise.

Brown et al. (2000) provide an overview of fission-track data from southern Africa, and present data on fission-track age, palaeotemperature and denudation. Along the Atlantic margin their data show 3–4 km of erosion, generally increasing towards the coast, between 118 Ma and present. Some other approaches to the African continental margin are summarized in Section 5 below.

4.4 The Appalachians

Eastern North America is a passive margin that has been intensely studied since the birth of modern geomorphology, with an enormous literature and many controversies (Thornbury, 1965; Mills et al., 1987). Folded Palaeozoic rocks underlie the area. In the Triassic all older structures were planed off. From west to east the region can be simplified into five provinces.

4.4.1 Appalachian Plateaux

This includes the Allegheny (N) and Cumberland (S) Plateaux. The unglaciated parts of the plateaux is lowest in the west and highest in the east, up to 1300 m in West Virginia. The eastern part of the Allegheny Plateau is the Allegheny Mountain section, more dissected than in the west. The Allegheny Front along the eastern margin of the plateau is an imposing escarpment rising 300 m or more above the valley floors of the adjacent Ridge and Valley Province.

4.4.2 Ridge and Valley Province

This is a dissected palaeoplain cut across highly folded Palaeozoic rocks. Bevelled cuestas and accordant ridges on the harder rocks, though small in area, indicate a formerly continuous planation surface of very low relief. The ridge tops are at about 650 m, and relief is around 300 m.

4.4.3 The Blue Ridge Province

The term 'ridge' is a misnomer: it is not a ridge, with slopes on either side, but an escarpment on the eastern side of a plateau, with many outliers, especially in the south. Summits of Blue Ridge rise over 1000 m and the Blue Ridge scarp looms 800 m above the Piedmont. Pazzaglia and Gardner (2000) argue that parts of the Blue Ridge may be equivalent to a Great Escarpment.

4.4.4 The Piedmont Province

The Piedmont plateau, cut across metamorphic rocks, lies 250 m asl, with generally low relief (<20 m) although a few gorges have cut down 180 m. The Piedmont is deeply weathered, with saprolite commonly 10–20 m deep, which was formerly thought to indicate great age (Mills et al., 1987), although some recent interpretations suggest it is Quaternary (Stanford et al., 2001). The boundary between resistant rocks of the Piedmont and soft rocks of the Coastal Plain is expressed as a seaward-facing escarpment with 50–150 m relief called the Fall Zone.

4.4.5 Coastal Plains

The Coastal Plains province has a wedge of sediment stretching from an inland margin to the continental self, and the oldest sediments are Cretaceous. The basal unconformity is known as the Fall Zone peneplain.

4.4.6 General theory of evolution

In the Appalachians, planation surfaces are indicated by local 'plateau' names as well as geomorphic evidence of many bevelled cuestas and accordant strike ridges. At the simplest, the situation could be regarded as that in Figure 3.1, with one palaeoplain warped down beneath the coastal plains. Indeed Ashley (1935) held that there is only one peneplain and other surfaces result from local differential erosion, and Wright (1942) thought the main upland surface (the Schooley peneplain) was 'possibly equivalent to the Fall Zone peneplain' (Thornbury, 1962: 240). The situation is complicated by the recognition of many peneplains by some (detailed in Thornbury, 1965: 79), and the denial of their existence by others. Poag and Sevon (1989) used the continental margin sedimentary record to identify periods of quiescence (when planation occurred) and uplift, and attribute a late Cretaceous age to the upland (Schooley) surface.

The modern drainage divide of the Appalachians is not parallel to the provinces described earlier. In the north it lies far to the west, so long rivers rise on the Allegheny Plateau and drain to the Atlantic. The divide then swings across the Ridge and Valley Province, and in the south is the Blue Ridge. Here, streams that descend the eastern side to the Piedmont are short and steep compared with those that rise on the western flank, and it is here that the Great Escarpment of the Appalachians may be located. The divide that crosses the Appalachians diagonally might be close to an original axis of warping, or it might result from complex drainage-pattern evolution.

Such long-term landscape evolution might be regarded as the classical view: Thornbury (1965) and Mills *et al.* (1987) review early ideas. It must be remembered that early workers lived at a time when continental drift was not acceptable, especially in America, so the very concept of a passive margin was unavailable. Pazzaglia and Gardner (2000) write that modern work does not support simple evolution of features inherited from Cretaceous precursors, and present evidence of much younger movements. Terraces of the lower Susquehanna River converged at the river mouth, diverged through the Piedmont and re-converged north of the Piedmont, suggesting isostatic uplift in the middle. This swell is largely confined to the Piedmont. They use time lines based on a wide range of evidence to constrain flexural deformation of the margin in response to offshore deposition and continental denudation. Total uplift of the central Appalachian Piedmont is only about 90 m in the past 15 Ma. They also found that to account for sediment in offshore basins requires no less than 1.1 km of erosion since the Miocene, if spread evenly. The concept of the eroded slab is supported by fission-track data, which suggest removal of 1.5 km in the last 20 Ma.

Stanford *et al.* (2001) studied the Schooley erosion surface in the type area in the Piedmont of New Jersey and, in great contrast to earlier workers, concluded that it is no older than late Miocene and subsequent landscape evolution is related to glacioeustatic events in the Plio-Pleistocene.

In summary, there is a broad marginal swell in Atlantic North America comparable with that on other passive margins. This may be of considerable antiquity, as in the classical views, but it has undoubtedly been modified by younger uplift and erosion, especially in the Piedmont Province.

4.5 Brazil and South America

In South America the only significant area of passive margin mountains is in southern Brazil, where the Brazilian Plateau is bounded by two Great Escarpments. One is the Serra do Mar Escarpment, which extends 800 km with a maximum height of 2245 m (Maak, 1969). It separates a coastal strip from the interior plateau, and all major streams drain westwards on the gentle inland slope to the Parana. The other is the Escarpment of the Planalto Plateau, underlain by the Serra Geral volcanics of Cretaceous age. A small part of this has been described by Martins *et al.* (in press). The escarpment reaches 1600 m, and may be up to 370 km inland. The contrast between the topography, soils and processes of the Planalto and the area draining to the southeast is clear and great. The Planalto surface is regarded as part of the Sul Americana surface, of Cretaceous–Pliocene age.

Martins *et al.* (in press) believe that the relief of southeastern Brazil is related to the far field effects of the Tertiary Andean uplift. In the offshore Palotas Basin the most widespread unconformity is Miocene, 10–16 Ma. In the Pelotas Basin widespread décollement in Miocene shales implies post-depositional seaward tilt caused by onshore uplift.

Gallagher *et al.* (1994) analysed fission-track data of southeast Brazil. There is a regional trend of young fission-track ages close to the coast becoming older with distance inland. Gallagher *et al.* concluded that 3 km has been eroded from the coastal plain, and 1 km inland. Maps of fission-track age, palaeotemperature and denudation are presented in Brown *et al.* (2000).

There are also undocumented passive margin mountains in northern Brazil (Jean-Pierre Peulvast, personal communication, 2003). Furthermore, northern South America has a series of highlands and mountains running from Venezuela through Surinam and Guyana. These do not have all the classical features of passive continental margins but have interesting relationships, described by Zonneveld (1954). There is evidence for coastal flexure, and attempts have been made to correlate erosion surfaces with offshore deposition. Interestingly, the highest summits are not on the major divide. The escarpments are huge, and have the world's highest waterfalls: Angel Falls in Venezuela has a single drop of 986 m.

4.6 Scandinavia

Scandinavia is commonly described as having Caledonian mountains, but although the rocks and structures may be Palaeozoic (Caledonian orogeny) the present mountains are much younger (Reusch, 1901). It is also generally thought that glaciation was dominant in forming the landscape, but in reality many palaeoforms are preserved. Holtedahl (1953) looked upon Scandinavia as part of a rifted continent (a bold view long before plate tectonics) and recognized what we now see as passive margin landform assemblages on the opposed coasts of Norway and Greenland, and Greenland and Baffin Island.

A broad surface of low relief (generally called the Palaeic surface in Scandinavia) extends over the divide between the Atlantic and the Baltic. It was identified as early as 1901 by Reusch, who realized that it had existed before a late upheaval of the land. Many other surfaces have been named – the envelope surface, the Tuipal surface, the preglacial landscape, and others – which are described, explained and referenced in Gjessing (1967) and Lidmar-Bergström *et al.* (2000).

Schipull (1974) made a thorough study of a part of the surface in southwestern Norway. He found remains of old meandering valleys and small intact parts of an exhumed sub-Cambrian surface, and concluded that the Quaternary was a period of conservation of pre-existing relief on the plateaux. In some places the land suddenly becomes steep at a Great Escarpment. The Great Escarpment is much degraded by glaciation in northern Scandinavia but it is still evident in southern Norway. Glaciation deepened the valleys and steepened the slopes, but the Great Escarpment was in existence before ice ever appeared. A landscape with gentle slopes, the Palaeic surface, extends over the divide in southern Norway from about 1000 m asl to summit plateaux at 2000 m.

The uplift of the palaeoplain may be related to offshore sedimentation. Doré (1992) regarded the envelope surface as equivalent to the base-Tertiary surface offshore, and thought the surface evolved by late Mesozoic base-levelling. Jensen and Schmidt (1992) paid particular attention to the base-Quaternary surface offshore, which cuts across older strata, and interpreted the uplift of Norway as a mainly Neogene event with a magnitude of 1500–2000 m caused by broad regional warping without faulting. Riis (1996) suggested the envelope surface was of Jurassic age, but concluded from an overview of the offshore sediments that there were two main uplift phases – Palaeogene and Neogene. The early one had a maximum in northern Scandinavia, the later one in southern Norway.

4.7 Greenland

Weidick (1976) provides a clear account of the Greenland mountains. The present coastal strip contains alpine landscapes and elevated plateaux, with one high summit level and some minor ones. The highest and oldest planation surface cuts across the Tertiary basalt as well as Precambrian bedrock and so was formed since the deposition of the basalts, that is in Miocene times or later. Several kilometres of the basalt flows were eroded and uplift therefore took place in the late Miocene, Pliocene or Pleistocene. Since uplift, the margin has been deeply eroded, to give the appearance from the coast of a steep rugged mountain range. However, inland of the mountains there is a plateau dissected only by large glacial valleys.

The Miocene climate of Greenland was similar to that of southern Europe today, so glaciation, if any existed, would have been limited then to small local glaciers on the highest areas. Glaciation appears to have developed rapidly about 3 Ma. The Great Escarpment of Greenland may therefore be pre-glacial, as it is in Norway. Glaciation straightened and steepened the valleys, but major landscape features are older. Peulvast (1988) pointed out that, as they are parts of the same Caledonian orogen broken apart by the seafloor spreading, the Lofoten–Vesteralen area of Norway and the Scoresby Sund area of Greenland should have comparable morphotectonics and histories.

Nielsen and Brooks (1981) provide a tectonic description of the continental flexure. The warp is gentle, over a coastal strip 50 km wide with a vertical displacement of about 8 km, and the maximum dip is 20°. A dyke swarm parallel to the coast adds to the structural complexity. The flexure is tensional, associated with a large number of rotated fault blocks.

Dam et al. (1998) and Clift et al. (1998) attribute uplift of Greenland to a mantle plume, starting at 63 Ma. Clift et al. (1998) interpret apatite fission-track data as showing 2.5 km of erosion over much of the Greenland margin, with 4–6 km of erosion in part of east Greenland.

4.8 Antarctic

The passive margins of Antarctica should be unique because of different geomorphic processes: the work of running water has been suppressed for millions of years. Studies of marine sediments indicate that Cenozoic Antarctic ice sheet activity dates back to 45 Ma, and continental scale glaciation to about 40–36 Ma (Hambrey et al., 1989; Cooper et al., 1991) (this long glacial period contrasts with the otherwise similar Greenland). Arguments continue about the role of ice and water. Is the landscape essentially glacial in origin, or is glaciation superimposed on a pre-glacial water-eroded landscape?

Eastern Antarctica is a fragment of Gondwana. Plateaux (palaeoplains) have been observed all around the continent at the margin of the ice sheet. Escarpments are also common. Kerr et al. (2000) and Näslund (2001) review earlier work.

The Transantarctic Mountains of eastern Antarctica run along the edge of a plateau that rises gradually from the interior, forming as a dramatic, major escarpment (Kerr et al., 2000). The high plateau is about 50 km wide near the McMurdo Dry Valley but up to 200 km across in the Queen Alexandra Mountains. In some locations, such as the Royal Society Range, the escarpment forms a more or less continuous topographic feature with a vertical relief of 2000–2800 m. The highest section of the plateau flanks the Ross Sea embayment and peaks reach 4000–4500 m.

A major debate concerns whether the Transantarctic Mountains experienced uplift of about 1 km Ma^{-1} since the early or middle Pleistocene (Behrendt and Cooper, 1991), or whether the mountains have remained at their present level since the Miocene (favoured by Kerr et al., 2000).

In Queen Maud Land (Dronning Maud Land to Scandinavians) a flat, high-elevation plateau has been eroded into isolated remnants. The surface existed in the Early Permian. Näslund (2001) wrote that the morphology of Dronning Maud Land supports the idea that the escarpment formed by scarp retreat, following the rifting and separation of east Antarctica and southern Africa. Näslund's interpretation of landscape evolution may be summarized as follows:

1 southern Africa and Queen Maud Land separated in the late Jurassic;
2 Mesozoic rifting was associated with uplift and formation of flood basalts, and an escarpment formed at the new passive continental margin;
3 prior to the Oligocene, mountain glaciers eroded valleys on the palaeoplain and escarpment;
4 formation of alpine landforms in the mountain ranges ceased when cold-based Cenozoic ice sheets formed in the Miocene, inhibiting glacial erosion and preserving existing landforms.

The Prince Charles Mountains consist of large flat-topped massifs with accordant summits that are the remnants of a pre-glacial surface of low relief. An Eocene lava flow on the surface suggests that erosion of the surface must have been younger than Eocene (Tingey, 1985). Fission track data suggest about 5 km of denudation in the Dry Valley area and denudation of up to 10 km near the coast (Fitzgerald, 1994).

5 Tectonic relationships and geomorphic evolution

5.1 Classical geomorphology

This term is used, sometimes pejoratively, to refer to the work of geomorphologists such as Davis (1899), Cotton (1918) and King (1953). Although they differed substantially in detail, these workers saw geomorphology as a history of landscape, based on a model of erosion leading to planation, followed by new erosion phases after tectonic uplift. They argued about the processes involved in forming plains, the nature of slope retreat and the age of landscape. Some modern workers seem to be at pains to separate themselves from these ideas, and King in particular is attacked for his landscape models. Brown et al. (2000: 274), for instance, wrote: 'Our assessment is that the approach advocated by King and adopted by other researchers is founded on unverified assumptions and therefore does not provide a viable basis for assessing the relationships between tectonics and macroscale landscape development.' Another example is provided by Fleming et al. (1999), who did cosmogenic dating in the Drakensberg region, and claimed that summit lowering is sufficient to prevent the long-term intact survival of erosion cycle surfaces formed in the Mesozoic that were previously inferred for this region by King, Partridge and others. Sometimes it seems these writers are not really familiar with King's views and achievements. In fact King described the Escarpment axis, he presented a cross-section of Africa similar to that in Figure 3.10, he described the evolution of the Natal Monocline in detail and he discussed the complications of isostatic adjustment (King, 1955).

Nevertheless, landscapes do consist of plains, separated at times by steeper country. Anyone working on Great Escarpments must admit that the escarpment is steep, compared with the surface inland from it, which is usually a plain of some sort with local names such as Tableland, Plateau, Highveld, etc. The origin of plains and escarpments is still a valid topic for research, today as in the 'classical' past.

5.2 Symmetry of continents and asymmetry of passive margins

Cross-section diagrams of several continents suggest a rough symmetry (Figure 3.10). Thus, the cross-section of southern Africa shows in the west the coastal plain of Namibia is backed by the Great Escarpment (here the local name). The Khomas Plateau is the palaeoplain, which gradually sinks to the Kalahari Basin and then rises again to become the Highveld. This ends abruptly at the Great Escarpment of the Drakensberg, and then comes the coastal plain of Natal.

Australia has the high plateaux of western Australia and eastern Australia on either side of the central lowlands, which actually go below sea level at Lake Eyre. South America has the swell of the Andes, the central depression of the Parana lowlands and the Brazilian Plateau, with its Great

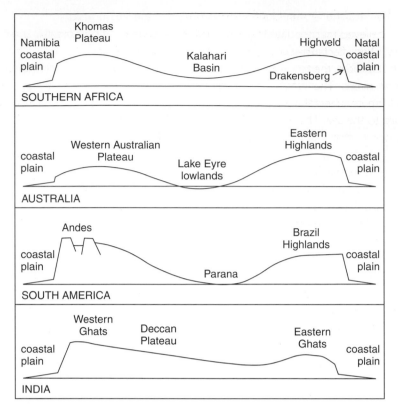

Figure 3.10 *The rough symmetry of marginal swells and Great Escarpments displayed by several continents*

Escarpment, the Serro do Mar. The symmetry is incomplete because the western side is an active margin. The same situation is found in North America, with the Central Plains between high land on west and east, but only the east is a passive margin with plateaux and a partial Great Escarpment. India has the Western Ghats and Eastern Ghats with a lowland between the two.

Antarctica and Greenland both have huge ice caps in the centre and mountains on the rim of marginal swells. It is generally thought that the weight of ice depresses the crust (e.g., Kearey and Vine, 1990; Brown and Mussett, 1993), which is no doubt correct, but since many other continents also have central depressions this might not be the complete answer. Such basins would provide ideal conditions for the collection of ice in the right climate.

5.3 Absence of mountains on some passive continental margins

The general pattern of passive margins, with a marginal swell and Great Escarpment, is not present on all passive margins, and the first big problem is – why not? Australia has a superb example along the eastern edge of the continent, but why is there no marginal swell on the south coast that matches Antarctica so well? Instead thick Tertiary rocks have remained virtually horizontal around the Great Bight and Nullarbor Plain. In Africa the marginal swell is beautifully simple in the south, but is missing elsewhere. The passive margin geomorphology of southern Brazil is absent in Argentina. Clearly, whatever explanation might account for the features typical of so many passive margins should also account for their absence in others.

5.4 Rift valleys and the breakup of continents

It is generally believed that the first stage in continental breakup is formation of a rift valley, so the nature of the rifting affects the subsequent landscape evolution (Davis, 1984). In simple rifting, such as in the Red Sea today, the top of the rift fault is the watershed between coastal-draining rivers and inland-draining rivers. The rift slope is steeper and so erodes faster, giving rise to a Great Escarpment. The continental margins of Brazil and western India might have started like this, with no downwarp to the sea. The Western Rift Valley in Uganda has a different style. There is a rise to a warp or shoulder that is about 50 km from the rift fault, which affects pre-fault drainage patterns (Ollier, 1990). The Australian margin in NSW appears to be of this type, and perhaps the Scandinavian passive margin.

Aulacogens (failed rifts) may also be part of the story, as in the submarine rift of the Queensland coast of Australia. Triple junctions, where three rifts join at angles of 120°, are a popular feature of mega-tectonics. Lester King (personal communication, 1976) believed that where two marginal swells meet, an area of extra uplift is produced. In Australia the north-northeast monocline of New South Wales meets the E–W monocline of the Victorian Highlands, with Mt Kosciusko in the corner – the highest mountain in Australia. In South Africa the Natal monocline meets the Cape Province monocline at the Kompassberg, the highest point in Cape Province. Likewise the East Brazil monocline intersects the Paraiba monocline at Pico da Bandiera, the second highest point in Brazil. All three examples have the same orientation, and the monoclines meet at about 120°.

The breakup of Pangea and its daughters, Gondwanaland and Laurasia, occurred at different times in different places (e.g., Kearey and Vine, 1990), so passive margins are of varying age, and have had different lengths of time for later geomorphic evolution.

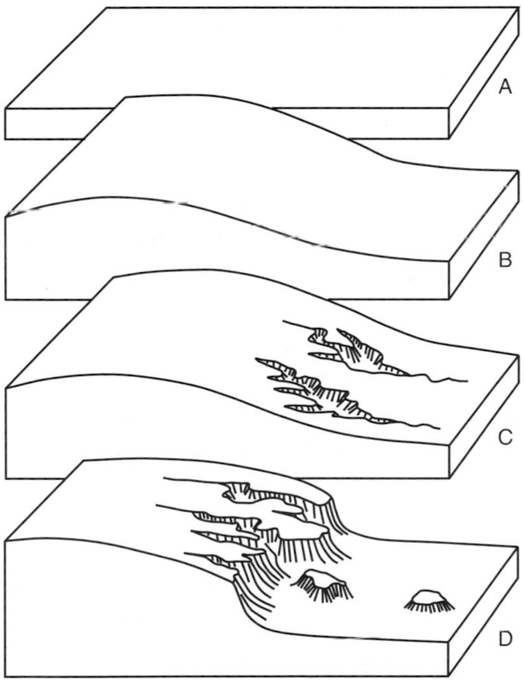

Figure 3.11 *The origin of the Great Escarpment by coalescence of valley heads. (A) The initial palaeoplain; (B) the warped palaeo-plain; (C) valley incision starts on the steeper, coast-facing slope; (D) coalescence of steep valley heads to form a continuous escarpment, with occasional outliers*

5.5 The origin of the Great Escarpment

Two very different models have been proposed for the origin of Great Escarpments. One, typified by Ollier (1982), suggests that the original continental margin was an asymmetrical swell. On the steeper, coastal side erosion was sufficiently rapid to create first steep gorges and valleys, and these eventually coalesced to create the main escarpment and isolated plateaux (Figure 3.11). The scarp did *not* originate as a major feature at the coast that migrated inland.

Beaumont *et al.* (2000) exemplify another model. They propose formation of an initial tectonic escarpment along a master fault located parallel to a coastline. On their figure it is shown close to the coastline, but they say it could equally be farther offshore. The fault represents one flank of a rift system. The fault scarp is then eroded and retreats. In their own words: 'The main escarpment was originally coincident with the master fault but has retreated into the uplands . . .' (Beaumont *et al.*, 2000: 42). It should be possible to locate such huge faults. This model is tacitly assumed by fission-track workers such as Gilchrist *et al.* (1994) and Brown *et al.* (2000) who depict increasing amounts of vertical erosion towards the coast.

5.6 'Pinning' of the Great Escarpment

In recent years several authors have advocated a model of Great Escarpments that involves the escarpment being 'pinned' to a drainage divide. The original theoretical model for this was given in Kooi and Beaumont (1996). They wrote (p. 3375): 'An essential condition for model escarpments to retreat in a uniform substrate is that the top of the escarpment be maintained as a drainage divide, separating the plateau drainage basin from the drainage system on and below the escarpment, so that retreat, drainage capture, and divide migration occur in concert.'

Other workers, using different kinds of physical evidence such as fission-tracks and cosmogenic isotopes to calculate denudation rates, have adopted this model. Cockburn *et al.* (2000) used cosmogenic isotopes and fission-track thermochronology data to determine denudation rates in Namibia, and argued that the west coast escarpment is fixed at a drainage divide: 'The escarpment may have originated only a few kilometers [*sic*] oceanward of its present location and its subsequent slow rate of retreat has been controlled by pinning at the drainage divide, possibly enhanced by flexural isostatic rebound (p. 434).' Van der Wateren and Dunai (2001) worked on denudation rates based on cosmogenic isotope measurements in Namibia, and state: 'The location of the Escarpment is to a high degree controlled by a pre-existing drainage divide and has only shifted a few kilometres inland' (p. 273). Fleming *et al.* (1999) use cosmogenic dating to work out rates of erosion in the Drakensberg, and claim (p. 209): 'The present escarpment would have originally grown vertically through differential denudation as a feature pinned at the seaward flank of the drainage divide, with subsequent inland retreat of only a few kilometres.'

Of course, the escarpment cannot be pinned to the divide in places such as eastern Australia, where the top of the escarpment is not a divide, and where the Great Divide and Great Escarpment are often tens or even hundreds of kilometres apart.

5.7 Uplift or downwarp at the margin

There are two main models for passive margin evolution (Figure 3.12). With a simple rise to the rift there has to be massive faulting offshore. Slope retreat moves from the initial faulted margin to the present position of the Great Escarpment, as in the model of Beaumont *et al.* (2000), described above. Huge volumes of rock would be eroded and no old regolith would survive at the coast. The alternative is warping of the palaeoplain to below sea level. Valleys eroding the steeper, coastal side coalesce to form a Great Escarpment, which then retreats, as in the model of Ollier (1982) described above. This simultaneously accounts for the plateau between the Great Divide and the Great Escarpment, the breakup unconformity which is simply the downwarped palaeoplain, and the preservation of a belt of old land surface features and regolith near the coast. This model equates the shoulder of present-day rift valleys with the marginal swell of passive continental margins, which in turn implies that the marginal swell dates back to the earliest days of continental breakup.

5.8 Time of uplift and downwarp

In the first place it may not be necessary to explain uplift. Some margins may have been high originally, such as the high plateaux bounding many present-day rift valleys. We then want to know the time of downwarp, to make the breakup unconformity beneath marine sediments offshore. If marginal fea-

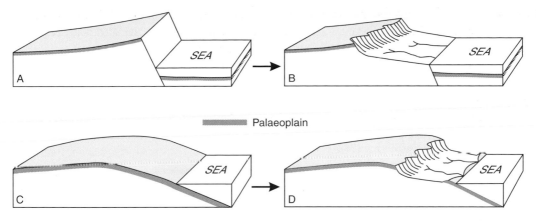

Palaeoplain

Figure 3.12 *Two possible relationships of a regolith-covered planation surface to tectonics at a passive continental margin. Top: with continuous rise of the planation surface to the continental margin (A) any continuation of the surface has to be faulted down below sea level. When erosion produces a coastal zone backed by a Great Escarpment (B) no trace of the old regolith will be preserved on the coastal lowland. Bottom: with downwarp of the planation surface from the marginal swell to below sea level the regolith-covered plain crops out at the coast, and some palaeo-landforms may be preserved close to the coast (D), though not at the foot of the Great Escarpment*
After Ollier and Pain (2000).

tures relate to a rift valley phase, they should be older than seafloor spreading, which follows the rift period. If they display younger uplift, either they were not related to initial formation of continental margins, or there has been a secondary movement at a later stage. Some margins have been traced back to the Mesozoic, such as the Australian Trias-Jura surface of Hills (1975). Elsewhere the landscape evolution is thought to be Miocene and younger, as in the northern Piedmont of the USA (Stanford *et al.*, 2001).

There may be more than one period of movement. Partridge (1998) has Mesozoic precursors for the landscape, and also Pliocene uplift of the South African high plains of up to 800 m. In the Appalachians the palaeoplain may date back to the Cretaceous but there is also evidence of Miocene or younger uplift in the Piedmont province. In Australia, the Eastern Highlands were once thought to be formed by the Plio-Pleistocene 'Kosciusko orogeny' (Andrews, 1910). This idea was replaced by general belief in early Cenozoic uplift, but some movements of up to 1 km may have occurred in the Pleistocene (Ollier and Taylor, 1988).

5.9 Causes of marginal swells

Table 3.1 shows some possible mechanisms of formation of marginal swells. These are mostly mechanisms for uplift although, as explained above, it is possible that the continent was already high when new continental margins appeared. It is also necessary to discriminate between primary mechanisms and secondary or feedback mechanisms. Mechanisms fall into three main groups: those that account for uplift; those that account for downwarp; and those depending on isostatic adjustment. The mechanism favoured depends on whether there is presumed uplift, downwarp or both. For most margins a wide range of proposals have been made, as indicated by the following selection.

For eastern Australia, Smith (1982) and Karner and Weissel (1984) regarded the uplift as due to the passage of Australia over a hotspot. Wellman (1987) suggested underplating, while Ollier (1978) applied the rift-valley model, especially to the NSW sector. Lambeck and Stephenson (1986) proposed that the present elevation of the Eastern Highlands is a result of isostatic rebound resulting

Table 3.1 – Suggested causes of marginal swells

Uplift mechanisms

1   Passage of the continental margin over a zone of anomalous high heat flow, such as a hotspot or old spreading site (e.g., Smith, 1982; Karner and Weissel, 1984)

2   Thermal expansion of a mantle plume beneath the continental margin (Cox, 1989)

3   Thermal events in the geological past causing uplift (Dumitru *et al.*, 1991)

4   Thermal buoyancy (Partridge, 1998)

5   Underplating of the continental margin by lighter rock, itself of unknown source (e.g., Wellman, 1987)

6   Intrusion of large amounts of igneous rock (Pitcher, 1982)

Downward mechanisms

7   Downwarp of the edges of a rift valley (Ollier, 1982)

8   Thinning of crust ('necking') at the continental margin (Sheth, 1998)

9   Intrusion of dyke swarms at the continental margin (only Greenland, Western Ghats) (Nielsen and Brooks, 1981)

Isostatic response

10 Isostatic response to erosion of the continental margin, especially by scarp retreat (denudation rebound) (King, 1955)

11 Uplift to compensate for subsidence offshore, caused in turn by loading of deposited sediments (Gilchrist and Summerfield, 1990)

12 Isostatic compensation for the weight of an ice cap (only Antarctica, Greenland) (Kearey and Vine, 1990)

Other mechanisms

13 Delayed response to erosion of much older orogenic belts (Lambeck and Stephenson, 1986)

from erosion since at least 150 Ma. This is still supported by mathematical modellers of passive margins, who leave out such basic features as the Great Escarpment (van der Beek *et al.*, 1999).

Gunnell and Fleitout (2000) review models of Western Ghats formation. They do not entirely reject the possibility that shoulder uplift could be achieved through tectonic processes, such as underplating, but suggest that the rim bulge could also have been obtained during the Cenozoic by passive denudational rebound alone. For southern Africa, Partridge (1998) favours a buoyancy force originating from a massive low density anomaly in the Earth's mantle. Smith (1982) suggested migration of the continent over a source of heat parallel to the east coast. King (1955) noted secondary isostatic uplift, and Brown *et al.* (2000) propose thermal uplift followed by 3–4 km of erosion. Martins *et al.* (in press) think the relief of southeastern Brazil is related to the far field effects of the Tertiary Andean uplift. Models of uplift in Antarctica reviewed by Kerr *et al.* (2000) include glacial loading, thermal buoyancy (perhaps related to separation from Australia at 90 Ma but initiation of uplift at 60 Ma), flexural uplift as a result of lithospheric necking and differential denudation across the margin. Some models can only apply to particular cases, such as glacial loading in Antarctica and Greenland, or loading by a dyke-swarm west of the Western Ghats. Isostatic feedback mechanisms, whereby erosion of the margin or deposition offshore leads to uplift of the land and subsidence offshore, is thought to be a major process in the Appalachians (Pazzaglia and Gardner, 2000) and the Western Ghats

(Widdowson and Gunnell, 1999). Mathematical models of marginal swells or Great Escarpments, such as those of Beaumont *et al.* (2000) and Gilchrist *et al.* (1994) build in various other feedback mechanisms.

6 Conclusions

Mountains on passive continental margins exist as high points on escarpments, erosional mountains carved out of plateaux, and occasionally as volcanoes. They are a class of mountains that do not fit into the simple plate tectonic story in which mountains are formed at collisional sites. Work on mountains has been very subordinate to work on the broader geomorphology of the continental margins, especially the swells and the Great Escarpments. Many techniques have been used in the study of continental margins, from conventional geology to application of geophysics and measurement of physical properties that may be surrogates for palaeotemperatures or depth of cover, and can be interpreted in geomorphic terms such as amount, age or rates of erosion.

What is clear from this review is that there are many basic differences in ideas of how passive margins and their mountains originate. There is great disagreement on how the margins are uplifted or downwarped and what the tectonic mechanisms involved are, and also on geomorphic processes of river incision and escarpment retreat, and secondary mechanisms such as isostatic response to erosion and offshore sedimentation. We must try to resolve the great conflicts that seem to recur repeatedly between experimental data such as apatite fission track (AFT) and evidence from regolith studies and basic geological and geomorphic mapping. At present our imagination and models seem to be running ahead of accepted basic information.

It is unlikely we shall ever agree on the cause of mountain building on passive margins before we have assembled the facts more systematically. At the least we need detailed maps of escarpments, divides (where different from the top of the escarpment) and present and palaeodrainage. Regolith and surficial sediments (and volcanic deposits where present) provide important constraints. It is important to relate proposed onshore models to the record provided by offshore sediments. Since many proposed mechanisms require major coastal or offshore faults, a search for such faults is required. It is also important to base any general explanation on a world survey, and not on the limited geology and landform information provided by a single area. This chapter offers no ultimate solutions, but highlights the gaps in present understanding of mountains on passive continental margins.

References

Andrews, E.C., 1910. Geographical unity of Eastern Australia in Late and post Tertiary time. *Journal of the Proceedings of the Royal Society of New South Wales*, **44**: 420–80.

Ashley, G.H., 1935. Studies in Appalachian mountain structure. *Geological Society of America Bulletin*, **46**: 1395–436.

Beard, J.S., 1999. Evolution of the river systems of the south-west drainage division, Western Australia. *Journal of the Royal Society of Western Australia*, **82**: 147–64.

Beaumont, C., Kooi, H. and Willett, S., 2000. Coupled tectonic–surface process models with applications to rifted margins and collisional orogens. In Summerfield, M.A. (ed.), *Geomorphology and global tectonics*. Chichester: Wiley, 29–55.

Behrendt, J.C. and Cooper, A., 1991. Evidence of rapid Cenozoic uplift of the shoulder escarpment of the Cenozoic West Antarctic rift system and a speculation on possible climatic forcing. *Geology*, **19**: 315–19.

Bishop, P. and Goldrick, G., 2000. Geomorphological evolution of the East Australian continental margin. In Summerfield, M.A. (ed.), *Geomorphology and global tectonics*. Chichester: Wiley, 225–54.

Bremer, H., 1985. Randschwellen: a link between plate tectonics and climatic geomorphology. *Zeitschrift für Geomorphologie, Supplementband,* **54**: 11–21.

Brown, G.C. and Mussett, A.E., 1993. *The inaccessible Earth.* London: Chapman and Hall.

Brown, L.F., Jr, Benson, J.M., Brink, G.J., Doherty, S., Jollands, A., Jungslager, E.H.A., Keenan, J.H.G., Murtingh, A. and van Dwyk, N.J.S., 1995. *Sequence stratigraphy in offshore South African divergent basins, an atlas on exploration for Cretacous lowstand traps by Soeker (Pty) Ltd.* American Association of Petroleum Geologists, Studies in Geology No. 41.

Brown, M.C., 2000. Cenozoic tectonics and landform evolution of the coast and adjacent highlands of south-east New South Wales. *Australian Journal of Earth Sciences,* **47**: 245–57.

Brown, R.W., Gallagher, K., Gleadow, A.J.W. and Summerfield, M.A., 2000. Morphotectonic evolution of the South Atlantic margins of Africa and South America. In Summerfield, M.A. (ed.), *Geomorphology and global tectonics.* Chichester: Wiley, 255–81.

Burke, C.A. and Drake, C.L. (eds), 1974. *The geology of continental margins.* Berlin: Springer-Verlag.

Clarke, J.D.A., 1994. Evolution of the Lefroy and Cowan palaeodrainage channels. *Australian Journal of Earth Sciences,* **41**: 55–68.

Clift, P.D., Carter, A. and Hurford, A.J., 1998. The erosional and uplift history of NE Atlantic passive margins: constraints on a passing plume. *Journal of the Geological Society of London,* **155**: 787–800.

Cockburn, H.A.P., Brown, R.W., Summerfield, M.A. and Seidl, M.A., 2000. Quantifying passive margin denudation and landscape development using a combined fission-track thermochronology and cosmogenic isotope analysis approach. *Earth and Planetary Science Letters,* **179**: 429–35.

Cooper, A.K., Barrett, P.J., Hinz, K., Traube, V., Leitchenkov, G. and Stagg, J.M.J., 1991. Cainozoic prograding sequences of the Antarctic continental margin: a record of glacio-eustatic and tectonic events. *Marine Geology,* **102**: 175–213.

Cotton, C.A., 1918. Mountains. *New Zealand Journal of Science and Technology* **1**: 280–85. Reprinted in Cotton, C.A., 1955. *New Zealand geomorphology.* Wellington: New Zealand University Press.

Cox, K.G., 1989. The role of mantle plumes in the development of continental drainage patterns. *Nature,* **342**: 873–76.

Dam, G.A., Larsen, M. and Sønderholm, M., 1998. Sedimentary response to mantle plumes: implications from Paleocene onshore successions, West and East Greenland. *Geology,* **26**: 207–10.

Davis, G.H., 1984. *Structural geology of rocks and regions.* New York: Wiley.

Davis, W.M., 1899. The geographical cycle. *Geographical Journal,* **14**: 481–504.

Dessai, A.G. and Bertrand, H., 1995. The 'Panvel Flexure' along the Western India continental margin: an extensional fault structure related to Deccan magmatism. *Tectonophysics,* **241**: 165–78.

De Wit, M., 1999. Post-Gondwana drainage and the development of diamond placers in western South Africa. *Economic Geology,* **94**: 721–40.

Dingle, R.V. and Hendy, Q.B., 1984. Late Mesozoic and Tertiary sediment supply to the eastern Cape Basin (SE Atlantic) and palaeo-drainage systems in south-western Africa. *Marine Geology,* **56**: 13–26.

Dingle, R.V., Siesser, W.G. and Newton, A.R., 1983. *Mesozoic and Tertiary geology of Southern Africa.* Rotterdam: A.A. Balkema.

Doré, A.G., 1992. The base Tertiary surface of southern Norway and the northern North Sea. *Norsk Geologisk Tidsskrift,* **72**: 259–65.

Dumitru, T.A., Hill, K.C., Coyle, D.A., Duddy, I.R., Foster, D.A., Gleadow, A.J.W., Green, P.F., Laslett, G.M., Kohn, B.P. and O'Sullivan, A.B., 1991. Fission track thermo-chronology: applications to continental rifting of south-eastern Australia. *APEA Journal,* **31**: 131–42.

Du Toit, A.L., 1933. Crustal movements as a factor in the geographical evolution of South Africa. *South African Geographical Journal*, **24**: 88–101.

Fielding, E.J., 2000. Morphotectonic evolution of the Himalayas and Tibetan Plateau. In Summerfield, M.A. (ed.), *Geomorphology and global tectonics*. Chichester: Wiley, 201–22.

Fitzgerald, P.G., 1994. Thermochronologic constraints on post-Paleozoic tectonic evolution of the central Transantarctic Mountains. *Tectonics*, **13**: 818–36.

Fleming, A., Summerfield, M.A., Stone, J.O., Fifield, L.K. and Cresswell, R.G., 1999. Denudation rates for the southern Drakensberg escarpment, SE Africa, derived from in-situ produced cosmogenic ^{36}Cl: initial results. *Journal of the Geological Society of London*, **156**: 209–12.

Gallagher, K., Hawkesworth, C. and Mantovani, M., 1994. The denudation history of the onshore continental margin of S.E. Brazil inferred from fission track data. *Journal of Geophysical Research*, **99**: 18117–45.

Galloway, R.W., 1967. Pre-basalt, sub-basalt, and post-basalt surfaces of the Hunter Valley, New South Wales. In Jennings, J.N. and Mabbutt, J.A. (eds), *Landform studies from Australia and New Guinea*. Canberra: ANU Press, 293–314.

Gilchrist, A.R. and Summerfield, M.A., 1990. Differential denudation and flexural isostasy in formation of rifted-margin upwarps. *Nature*, **346**: 739–42.

Gilchrist, A.R. and Summerfield, M.A., 1991. Denudation, isostasy and landscape evolution. *Earth Surface Processes and Landforms*, **16**: 555–62.

Gilchrist, A.R., Kooi, H. and Beaumont, C., 1994. Post-Gondwana geomorphic evolution of southwestern Africa: implications for the controls on landscape development from observations and numerical experiments. *Journal of Geophysical Research*, **99**(B6): 12211–28.

Gjessing, J., 1967. Norway's paleic surface. *Norsk Geografisk Tidsskrift*, **20**: 273–99.

Gleadow, A.J.W. and Brown, R.W., 2000. Fission-track thermochronology and the long-term denudational response to tectonics. In Summerfield, M.A. (ed.), *Geomorphology and global tectonics*. Chichester: Wiley, 57–75.

Godard, A. (ed.), 1982. Les bourrelets marginaux des hautes latitudes. *Bulletin of the Association of Geographers of France, Paris*, **489**: 239–69.

Gunnell, Y. and Fleitout, L., 2000. Morphotectonic evolution of the Western Ghats, India. In Summerfield, M.A. (ed.), *Geomorphology and global tectonics*. Chichester: Wiley, 321–38.

Hambrey, M.J., Larsen, B., Ehrmann, W.U. and Ocean Drilling Program Leg 119 Shipboard Party, 1989. Forty million years of Antarctic glacial history yielded by leg 119 of the Ocean Drilling Program. *Polar Record*, **25**: 99–106.

Haworth, R.J. and Ollier C.D., 1992. Continental rifting and drainage reversal: the Clarence River of eastern Australia. *Earth Surface Processes and Landforms*, **17**: 387–97.

Hills, E.S., 1975. *The physiography of Victoria*. Melbourne: Whitcombe and Tombs.

Holtedahl, O., 1953. On the oblique uplift of some northern lands. *Norsk Geografisk Tidsskrift*, **14**: 132–39.

Idnurm, M., 1985. Late Mesozoic and Cenozoic palaeomagnetism of Australia I. A redetermined apparent polar wander path. *Geophysical Journal*, **83**: 399–418.

Jensen, L.N. and Schmidt, B.J., 1992. Late Tertiary uplift and erosion in the Skagerack area: magnitude and consequences. *Norsk Geologisk Tidsskrift*, **72**: 275–79.

Jessen, O., 1943. *Die Ranschwellen der Kontinente*. Gotha: Justus Perthes.

Karner, G.D. and Weissel, J.K., 1984. Thermally induced uplift and lithospheric flexural readjustment of the eastern Australian highlands. *Geological Society of Australia Abstracts*, **12**: 293–94.

Kearey, P. and Vine, F.J., 1990. *Global tectonics*. Oxford: Blackwell Scientific.

Kerr, A., Sugden, D.E. and Summerfield, M.A., 2000. Linking tectonics and landscape development in a passive margin setting: the Transantarctic Mountains. In Summerfield, M.A. (ed.), *Geomorphology and global tectonics*. Chichester: Wiley, 303–19.

King, L.C., 1953. Canons of landscape evolution. *Bulletin of the Geological Society of America*, **64**: 721–51.

King, L.C., 1955. Pediplanation and isostasy: an example from South Africa. *Quarterly Journal of the Geological Society of London*, **111**: 353–54.

King, L.C., 1963. *South African scenery*. Edinburgh: Oliver and Boyd.

Kooi, H. and Beaumont, C., 1996. Large-scale geomorphology: classical concepts reconciled and integrated with contemporary ideas via a surface processes model. *Journal of Geophysical Research*, **101**: 3361–86.

Kubiniok, J., 1988. *Kristallinvergrusung an Beispielen aus Südostaustralien und Deutschen Mittelgebirgen*. Kölner Geographische Arbeiten 48. University of Cologne.

Lambeck, K. and Stephenson, R., 1986. The post-Palaeozoic uplift history of south-eastern Australia. *Australian Journal of Earth Sciences*, **13**: 253–70.

Lidmar-Bergström, K., Ollier, C.D. and Sulebak, J.R., 2000. Landforms and uplift history of southern Norway. *Global and Planetary Change*, **24**: 211–31.

Maak, R., 1969. Die Serra do Mar im Staate Parana. *Die Erde*, **100**: 327–47.

Marshallsea, S.J., Green, P.F. and Webb, J., 2000. Thermal history of the Hodgkinson Province and Laura Basin, Queensland: multiple cooling episodes identified from apatite fission track analysis and vitrinite reflectance data. *Australian Journal of Earth Science*, **47**: 779–97.

Martins, D.P., Verdum, R. and Potter, P.E., in press. Geomorphic studies of Tres Forquilhas Valley, RS: significance for the origin of the Planato Escarpment.

Mills, H.H., Brackenridge, G.R., Jacobson, R.B., Newell, W.L., Pavich, M.J. and Pomeroy, J.S., 1987. Appalachian mountains and plateaus. In Graf, W.L. (ed.), *Geomorphic systems of North America*. The Geology of North America, Special Centennial Volume, 2. Boulder, CO: Geological Society of America, 5–50.

Moore, A.E., 1999. A reappraisal of epeirogenic flexure axes in southern Africa. *South African Journal of Geology*, **102**: 363–76.

Moore, A.E. and Larkin, P.A., 2001. Drainage evolution in south-central Africa since the breakup of Gondwana. *South African Journal of Geology*, **104**: 47–68.

Moore, M.E., Gleadow, A.J.W. and Lovering, J.F., 1986: Thermal evolution of rifted continental margins; new evidence from fission tracks in basement apatites from southeastern Australia. *Earth and Planetary Science Letters*, **78**: 255–70.

Morisawa, M. and Hack, J.T. (eds), 1985. *Tectonic geomorphology*. Boston, MA: Allen & Unwin, 3–25.

Nash, C., Chakravartula, P. and Crowe, W., 1999. Landforms of Northern Orissa, India from analysis of satellite imagery and aircraft digital terrain data. In Rao, A.T., Rao, D. and Yoshida, M. (eds), *Eastern Ghats granulites*. Gondwana Research Group memoir no. 5. Osaka: Field Science Publishers, 1–13.

Näslund, J.O., 2001. Landscape development in western and central Dronning Maud Land, East Antarctica. *Antarctic Science*, **13**: 302–11.

Nielsen, T.F.D. and Brooks, C.K., 1981. The E. Greenland rifted continental margin: an examination of the coastal flexure. *Journal of the Geological Society of London*, **138**: 559–68.

Ollier, C.D., 1978. Tectonics and geomorphology of the Eastern Highlands. In Jennings, J.N. and Mabbutt, J.A. (eds), *Landform studies from Australia and New Guinea*. Canberra: ANU Press, 5–47.

Ollier, C.D., 1982. The Great Escarpment of eastern Australia: tectonic and geomorphic significance. *Journal of the Geological Society of Australia*, **29**: 13–23.

Ollier, C.D. (ed.), 1985. Morphotectonics of passive continental margins. *Zeitschrift für Geomorphologie, Supplementband,* **54**: 1–9.

Ollier, C.D., 1988. *Volcanoes.* Oxford: Blackwell.

Ollier, C.D., 1990. Morphotectonics of the Lake Albert Rift Valley and its significance for continental margins. *Journal of Geodynamics,* **11**: 345–55.

Ollier, C.D. and Marker, M.E., 1985. The Great Escarpment of southern Africa. *Zeitschrift für Geomorphologie, Supplementband,* **54**: 37–56.

Ollier, C.D. and Pain, C.F., 1994. Landscape evolution and tectonics in southeastern Australia. *AGSO Journal of Australian Geology and Geophysics,* **15**: 335–45.

Ollier, C.D. and Pain, C.F., 1997. Equating the basal unconformity with the palaeoplain: a model for passive margins. *Geomorphology,* **19**: 1–15.

Ollier, C.D. and Pain, C.F., 2000. *The origin of mountains.* London: Routledge.

Ollier, C.D. and Powar, K.B., 1985. The Western Ghats and the morphotectonics of Peninsular India. *Zeitschrift für Geomorphologie, Supplementband,* **54**: 57–69.

Ollier, C.D. and Stevens, N.C., 1989. The Great Escarpment in Queensland. In Le Maitre, R.W. (ed.), *Pathways in geology: essays in honour of Edwin Sherbon Hills.* Blackwell: Melbourne, 140–52.

Ollier, C.D. and Taylor, D., 1988. Major geomorphic features of the Kosciusko–Bega region. *BMR Journal of Australian Geology and Geophysics,* **10**: 357–62.

Partridge, T.C., 1998. Of diamonds, dinosaurs and diastrophism: 150 million years of landscape evolution in southern Africa. *South African Journal of Geology,* **101**: 167–84.

Partridge, T.C. and Maud, R.R., 1987. Geomorphic evolution of southern Africa since the Mesozoic. *South African Journal of Geology,* **90**: 179–208.

Pazzaglia, F.J. and Gardner, T.W., 2000. Late Cenozoic landscape evolution of the US Atlantic passive margin: insights into a North American Great Escarpment. In Summerfield, M.A. (ed.), *Geomorphology and global tectonics.* Chichester: Wiley, 223–302.

Persano, C., Finlay, M.S., Bishop, P. and Barfod, D.N., 2002. Apatite (U-Th)/He age constraints on the development of the Great Escarpment on the southern Australian passive margin. *Earth and Planetary Science Letters,* **200**: 79–90.

Peulvast, J.P., 1988. Pre-glacial landform evolution in two coastal high latitude mountains: Lofoten – Vesteralen (Norway) and Scoresby Sund area (Greenland). *Geografiska Annaler,* **70A**: 351–60.

Pillans, B., 1998. *Regolith dating methods. A guide to numerical dating techniques.* Perth: Cooperative Research Centre for Landscape Evolution and Mineral Exploration.

Pitcher, W.S., 1982. Granite type and tectonic environment. In Hsu, K.J. (ed.), *Mountain building processes.* London: Academic Press, 19–40.

Poag, C.W. and Sevon, W.D., 1989. A record of Appalachian denudation in post-rift Mesozoic and Cenozoic sedimentary deposits of the U.S. middle Atlantic continental margin. *Geomorphology,* **2**: 119–57.

Reusch, H., 1901. Nogle bidrag till fortaaelsen af Hvorledes Norges dale og fjelde er blevne til. *Norges Geologiske Undersoekelse,* **32**: 124–263, Aarbog 1900.

Riis, F., 1996. Quantification of Cenozoic vertical movements of Scandinavia by correlation of morphological surfaces with offshore data. *Global and Planetary Change,* **12**: 331–57.

Schipull, K., 1974. Geomorphologische Studien in zentralen Sudnorwegen mit Beitragen uber Regelung- und Steurungssysteme in der Geomorphologi. *Hamburger Geographische Studien,* **31**, 91 pp.

Schmidt, P.W. and Ollier, C.D., 1988 Palaeomagnetic dating of Late Cretaceous to Early Tertiary weathering in New England, N.S.W., Australia. *Earth-Science Reviews,* **25**: 363–71.

Scrutton, R. (ed.), 1982. *Dynamics of passive margins*. Geodynamics Series, Volume 6. Boston, MA: Geological Society of America.

Sheth, H.C., 1998. A reappraisal of the coastal Panvel flexure, Deccan Traps, as a listric-fault-controlled reverse drag structure. *Tectonophysics*, **294**: 143–49.

Skelton, P.H., 1994. Diversity and distribution of freshwater fishes in east and southern Africa. *Annales Musee Royal de l'Afrique Centrale. Serie in Quarto-Zoologie*, **275**: 95–113.

Smith, A.G., 1982. Late Cenozoic uplift of stable continents in a reference frame fixed to South America. *Nature*, **296**: 400–404.

Standard, J.C., 1969. Hawkesbury Sandstone. In Packham, G.H. (ed.), The geology of New South Wales. *Journal of the Geological Society of Australia*, **16**: 407–16.

Stanford, S.D., Ashley, G.M. and Bremner, G.J., 2001. Late Cenozoic fluvial stratigraphy of the New Jersey Piedmont: a record of glacioeustasy, planation, and incision on a low-relief passive margin. *Journal of Geology*, **109**: 265–76.

Thomas, D.S.G. and Shaw, P.A., 1991. *The Kalahari environment*. Cambridge: Cambridge University Press.

Thornbury, W.D., 1962. *Principles of geomorphology*. New York: Wiley.

Thornbury, W.D., 1965. *Regional geomorphology of the United States*. New York: Wiley.

Tingey, R.J., 1985. Uplift in Antarctica. *Zeitschrift für Geomorphologie, Supplementband*, **85**: 85–99.

Valeton, I., 1994. Element concentration and formation of ore deposits by weathering. *Catena*, **21**: 99–129.

Van der Beek, P.A., Braun, J. and Lambeck, K., 1999. Post-Palaeozoic uplift history of southeastern Australia revisited: results from a process-based model of landscape evolution. *Australian Journal of Earth Sciences*, **46**: 157–72.

Van der Wateren, F.M. and Dunai, T.J., 2001. Late Neogene passive margin denudation history – cosmogenic isotope measurements from the central Namib desert. *Global and Planetary Change*, **30**: 271–307.

Weidick, A., 1976. Glaciation and the Quaternary of Greenland. In Escher, A. and Warr, W.S. (eds), *Geology of Greenland*. Odense: Geological Survey of Greenland, 431–58.

Wellman, P., 1987. Eastern Highlands of Australia: their uplift and erosion. *BMR Journal of Australian Geology and Geophysics*, **10**: 277–86.

Widdowson, M., 1997. Tertiary palaeosurfaces of the SW Deccan, Western India. Implications for passive margin uplift. In Widdowson, M. (ed.), *Palaeosurfaces: recognition, reconstruction and palaeoenvironmental interpretation*. Geological Society of London Special Publication, No. 120. London: Geological Society of London 221–48.

Widdowson, M. and Gunnel, Y., 1999. Lateritization, geomorphology and geodynamics of a passive continental margin: the Konkan and Kanara coastal lowlands of western peninsular India. *Special Publication of the International Association of Sedimentology*, **27**: 245–74.

Wilson, R.C.L., Whitmarsh, R.B., Taylor, B. and Froizheim, N. (eds), 2001. *Non-volcanic rifting of continental margins: a comparison of evidence from land and sea*. Geological Society Special Publication 187. London: Geological Society of London.

Wright, F.J., 1942. Erosional history of the southern Appalachians. *Journal of Geomorphology*, **5**: 151–61.

Young, R.W. and McDougall, I., 1982. Basalts and silcretes on the coast near Ulladulla, southern New South Wales. *Journal of the Geological Society of Australia*, **29**: 425–30.

Young, R.W. and Wray, R.A.L., 1999. The longterm development of river valleys: evidence from the passive margin of southeastern Australia. *Transactions of the Japanese Geomorphological Union*, **20**: 1–19.

Zonneveld, J.I.S., 1954. Geomorphological notes on the Continental Border in the Guyanas, N. South America. *Zeitschrift für Geomorphologie, Supplementband*, **54**: 71–83.

4

The evolution of the mountains of New Zealand

Paul W. Williams

1 Introduction: geological origins

The New Zealand microcontinent straddles the boundary of the Pacific and Australian plates. Much of the continental crust is beneath sea level (Figure 4.1), but the land rising above it forms mountainous chains that run through the central axis of present-day New Zealand. The Lord Howe Rise to the northwest of New Zealand lies 500–2000 m below sea level and comprises continental crust on the Australian plate, whereas the Campbell Plateau and Chatham Rise are similar marine plateaux to the southeast of New Zealand, but constitute part of the Pacific plate.

The oldest known rocks in New Zealand are about 680 Ma and were once part of the supercontinent of Gondwana. Sediments eroded from that landmass accumulated in marine basins and were subsequently uplifted around 400 Ma in the Tuhua orogeny. The mountains developed at that time were themselves eroded, their sediments accumulating in a marine trench known as the New Zealand Geosyncline together with the eroded fragments of nearby volcanoes (Suggate *et al.*, 1978). This occurred between 320–140 Ma, and the sediments hardened into the greywacke rock that is so characteristic of New Zealand mountains today. Sediment accumulation ceased in the New Zealand Geosyncline because of the commencement of a new mountain-building phase known as the Rangitata orogeny. This culminated 140–120 Ma and coincided with the commencement of breakup of the Gondwana supercontinent. Gee and Kirkpatrick (1997) provide a depiction of the tectonic movement associated with the breakup of Gondwana and the formation of New Zealand (Figure 4.2).

The new landmass created by the Rangitata orogeny extended over about 30° of latitude and was about half the size of modern Australia, but it was located polewards of the present position of New Zealand. It finally broke away from Gondwana in the late Cretaceous, about 80 Ma. The opening of the Southern Ocean between Australia and Antarctica also commenced at about this time, with the pace of separation accelerating some 55 Ma. By 70 Ma New Zealand was entirely surrounded by ocean; consequently, whereas the Rangitata landmass had originally been on the eastern margins of a large continent, by the early Tertiary it was located to the east of the spreading Tasman Sea. It was in an environment that was becoming progressively more oceanic and was exposed to mid-latitude westerly winds; a situation that has persisted to this day. By 60 Ma the Tasman Sea had reached its full width and the central North Island was located about 55°S 170°W, whereas today the equivalent point is at 39°S and 176°E. The Rangitata landmass had been reduced to an archipelago of swampy islands near sea level and the rifted eastern margin of the Tasman Sea had accumulated considerable amounts of land-derived sediments eroded from this landmass, as well as thick carbonate marine sediments from an Oligocene sea. Remnants of the denuded low relief surface of 60 Ma are still evident in the New Zealand landscape, where it is now termed the Late Cretaceous peneplain.

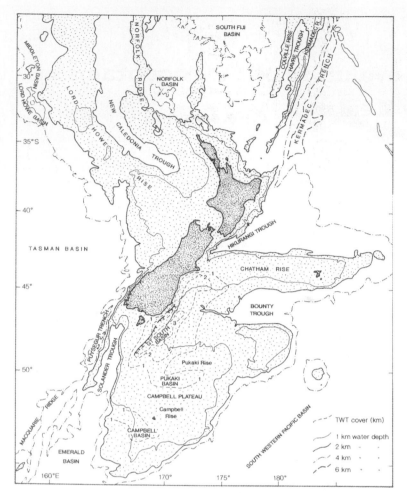

Figure 4.1 *The microcontinent of New Zealand. Continental crust is stippled and extends to about 2 km water depth. Important features are the marine plateaux to the northwest and southeast, and the trench and ridge systems to the northeast and southwest of the country that mark the continuation across the ocean floor of the boundary between the Pacific and Australian plates*
Source: *from Kamp (1992). Reproduced with the permission of Pearson Education (NZ).*

About 30 Ma, a convergence zone between the Pacific and Australian plates developed through the country, with the result that a new phase of mountain building was initiated, although significant mountain building did not get under way until the late Cenozoic. This is known as the Kaikoura orogeny, and the mountain building associated with it appears to be accelerating. Migration of the plate boundary has resulted in a zone of active volcanism moving southeast from the Northland Peninsula in the Mio-Pliocene to the Taupo Volcanic Zone (TVZ) in the Pleistocene. King (2000) provides a reconstruction of New Zealand's configuration in 13 steps over the last 40 Ma.

Uplift started at different times across the country. In parts of eastern and southern North Island, for example, marine sediments still accumulated until the Plio-Pleistocene. In places, lower Pliocene sediments rest unconformably on Mesozoic greywacke basement rocks, which had been denuded to low relief in the late Cretaceous–early Tertiary, but were trimmed again by marine abrasion prior to the deposition of the Pliocene sediments. Quaternary uplift and erosion has since removed much of this sediment to reveal the underlying erosion surface cut across the greywacke. The upland surface

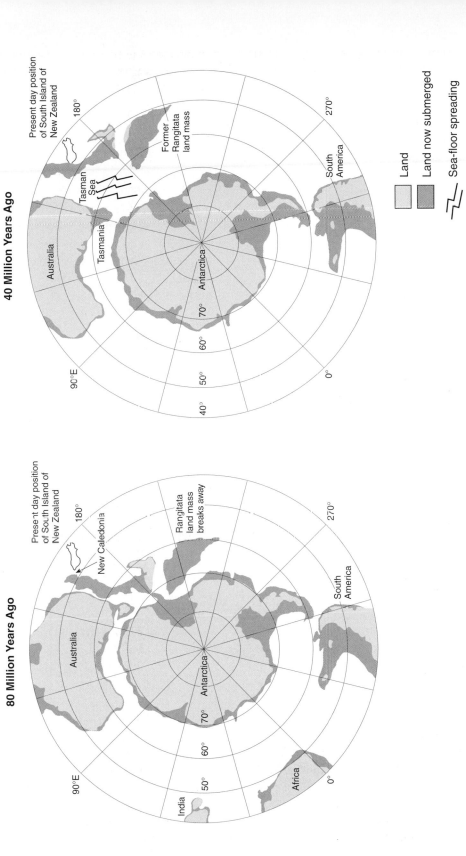

80 Million Years Ago

Present day position of South Island of New Zealand

New Caledonia

Australia

Rangitata land mass breaks away

90°E

180°

50°

60°

70°

India

Antarctica

Africa

South America

0°

270°

40 Million Years Ago

Present day position of South Island of New Zealand

Australia

Tasmania

Tasman Sea

180°

Former Rangitata land mass

90°E

40°

50°

60°

70°

Antarctica

South America

0°

270°

Land

Land now submerged

Sea-floor spreading

Figure 4.2 *The breakup of Gondwana and the formation of New Zealand*
Source: *simplified from Gee and Kirkpatrick (1997).*

91

around Wellington (sometimes referred to as the K Surface) can therefore be considered an abrasion surface of lower Pliocene age (Ota *et al.*, 1981). The surface varies in height up to 460 m and is folded with axes trending N–S. Individual fault blocks are tilted towards the west.

2 The plate boundary through New Zealand

The inception of the Australian–Pacific plate boundary through New Zealand was in two phases. Seafloor spreading to the south caused continental rupturing in western New Zealand by the end of the Eocene and, in the second phase, changing subduction kinematics caused the strike-slip to convergent plate boundary (Alpine Fault) to form in the early Miocene (King, 2000). There has been about 800 km of relative motion between the Australian and Pacific plates since the middle Eocene.

The present-day plate boundary is defined to the north of New Zealand by the SSW–NNE striking Hikurangi–Kermadec Trench, which attains more than 10 000 m in depth, and by the associated volcanic arcs of the Kermadec and Colville Ridges. The plate boundary to the south of New Zealand is marked by the SW–NE striking Puysegur Trench and Macquarie Ridge. The Colville–Kermadec and Macquarie Ridges are seismically active, whereas the Kermadec Ridge is also volcanically active with underwater vents. These are aligned with the volcanic White Island in the Bay of Plenty and the TVZ in the North Island. However, volcanism is now extinct on the Macquarie Ridge. The Hikurangi and Puysegur trenches are connected by a system of faults that extend obliquely across the South Island, with associated faults also traversing the east of the North Island (Figure 4.3). The principal dislocation that marks the plate boundary across the South Island is known as the Alpine Fault. The similarity of rocks in Fiordland to those in NW Nelson, on opposite sides of the plate boundary, indicates that about 460 km of dextral offset has occurred on-land along the Alpine Fault. Lateral offset on the fault was initiated around the start of the Miocene c.23–22 Ma (King, 2000), although most of the movement has probably taken place in the last 10 Ma.

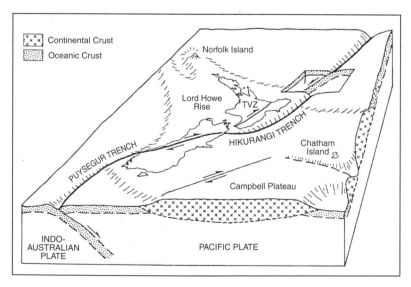

Figure 4.3 *A simplified three-dimensional depiction of the plate boundary and associated fault system running across New Zealand. The Alpine Fault provides the terrestrial link between the Hikurangi Trench to the northeast and the Puysegur Trench to the southwest. An important feature of the situation is the opposite polarity of subduction in the northeast compared with the southwest*
Adapted from Williams (1991), reproduced with the permission of Catena Verlag GMGH.

The Pacific and Indo-Australian plates are both moving north at a rate of about 30 mm a^{-1}, but slight differences in the directions and rates result in relative movement between the two. A particularly important feature of the plate boundary is the different polarity of subduction beneath the eastern North Island compared with beneath the southwestern South Island (Figure 4.3). Plate convergence at the Hikurangi margin is obliquely to the southwest, with the Pacific plate plunging beneath the North Island, whereas in the Puysegur Trench it is to the southeast, with the Indo-Australian plate plunging beneath Fiordland. In both cases the plate convergence and subduction is accompanied by compression, uplift and translational strike faulting. These two subduction zones of opposite polarity are linked by an intracontinental region of oblique convergence across the Southern Alps, where the mantle lithosphere has shortened horizontally, approximately in concert with the overlying crust (Stern *et al.*, 2000).

In the northeast, the denser oceanic crust of the Pacific plate converges against and is drawn under the more buoyant continental material of the North Island at a rate of about 50 mm a^{-1}. The plate dips at 12°–15° until it is 250 km west of the Hikurangi Trench, after which the dip increases. Melting of oceanic crust occurs below about 85 km depth (Figure 4.4) and plumes of molten rock rise towards the surface. Back-arc tension is opening the Bay of Plenty region at a rate of about 7 mm a^{-1} and thinning and fracturing the continental crust. The triangular TVZ is the result. In the central part of the zone, basement rocks have subsided progressively to 2–3 km below sea level (Healy, 1992). The rifting has permitted magma plumes to break through and generate volcanic activity, that first became active in the TVZ about 2 Ma. Because the Pacific plate is migrating south relative to the Australian plate, the apex of the TVZ and its associated volcanism is also likely to move southwards.

In southwest South Island, thinned continental crust of the Australian plate is being subducted beneath Fiordland. Since the early Eocene, this plate boundary has experienced a progressive change of displacement between the two plates from extension to dextral strike-slip to very oblique convergence. At present, the northeast-trending Fiordland margin is obliquely underthrust by the Australian plate towards the northeast (Figure 4.5). The angle of plate convergence is about 65° from orthogonal and

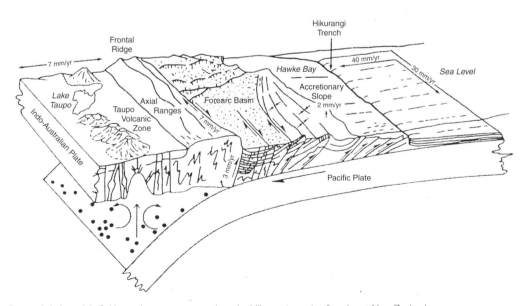

Figure 4.4 *A model of oblique plate convergence along the Hikurangi margin of northeast New Zealand*
Source: *from Williams (1991), modified from Cole and Lewis (1981), reproduced with the permission of Catena Verlag GMGH.*

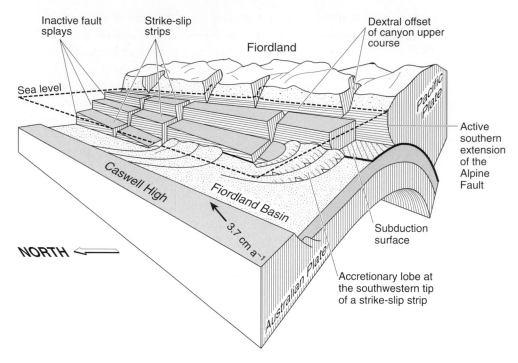

Inactive fault
splays

Strike-slip
strips

Fiordland

Dextral offset
of canyon upper
course

Sea level

Pacific Plate

Active
southern
extension
of the
Alpine
Fault

Caswell High

Fiordland Basin

3.7 cm a⁻¹

NORTH

Subduction
surface

Australian Plate

Accretionary lobe at
the southwestern tip
of a strike-slip strip

Figure 4.5 *A model of oblique plate convergence along the Fiordland margin of southwest New Zealand*
Source: *from Delteil* et al. *(1996).*

the subduction surface is estimated to dip landward at about 10° (Delteil *et al.*, 1996). The rate of relative plate convergence is about 37 mm a⁻¹. The main branch of the Alpine Fault is thought to lie along the Fiordland coast, and closely spaced long rectilinear scarps on the steep continental slope are interpreted as splays of the Alpine Fault system. The heads of submarine canyons of presumed Pleistocene age appear collectively to be well offset to the northeast from the mouths of fjords, suggesting that major Quaternary strike-slip has occurred on the continental shelf, probably along the Alpine Fault (Delteil *et al.*, 1996). The Fiordland region is still seismically very active, which is consistent with on-going convergence.

3 The axial mountain ranges of New Zealand

3.1 North Island

The mountain ranges that traverse the axis of New Zealand are shown on Figure 4.6 as land rising above 1000 m. In the North Island these mountains have been uplifted to over 1700 m. They extend from East Cape to Cook Strait and include individual ranges such as the Raukumara, Huiarau, Kaweka, Kaimanawa, Ruahine and Tararua Ranges. All are composed dominantly of Mesozoic greywacke, argillite and schist. Most have an essentially NE–SW orientation, roughly parallel to the East Coast. These mountains are mainly horst blocks and collectively form a frontal shoulder or ridge, seawards of which is an accretionary borderland of folded and imbricately strike-faulted Neogene sediments (Kamp, 1988; Williams, 1991). The upland surfaces of the ranges are generally plateau-like, but with deeply incised valleys. Each range constitutes a dissected fragment of the late Cretaceous peneplain, with the low relief surface often having been trimmed further by marine abrasion in the mid-Tertiary. Upland surfaces are especially evident in the concordant summits of the Raukumara Range near East

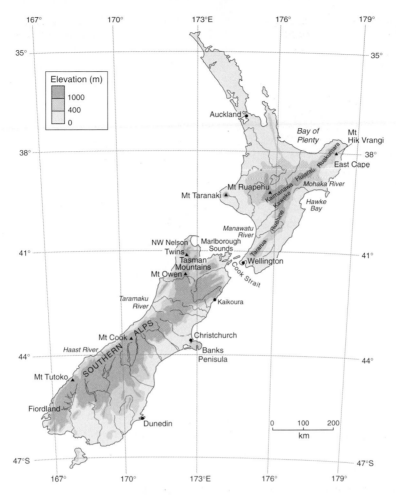

Figure 4.6 *The axial ranges of New Zealand extend from East Cape in the northeast of the North Island to Fiordland in southwestern South Island. The highest point on these mountains in the North Island is Mt Hikurangi (1752 m) in the Raukumara Range and in the South Island is Aoraki/Mt Cook (3764 m). The Main Divide along the Southern Alps is usually about 30–60 km inland from the coast. It lies to the east of the Alpine Fault (see Figures 4.8 and 4.12) and passes through Mt Cook*

Cape, in the Kaimanawa Mountains to the southeast of Lake Taupo, the Ruahine Range southeast of Hawke Bay and in the hills around Wellington. However, the ranges were generally not emergent until the early Pleistocene (Beu *et al.*, 1980; Ota *et al.*, 1981) and, at the time of uplift, were veneered with Tertiary sediments, most of which have since been stripped away by erosion, thereby exhuming the underlying erosion surface which has itself been incised.

During glacial phases of the Pleistocene, areas above 1500 m may have experienced permanent snow and ice cover. Small glacial troughs have been identified in the highest valleys of the Tararua Range, but no terminal moraines have been recognized.

3.2 South Island

The axial ranges continue to the south of Cook Strait and traverse the South Island obliquely from northeast to southwest (Figure 4.6). They also become significantly higher. As in the North Island, they are dominantly composed of Jurassic and Triassic greywackes, argillites and schists. The Hikurangi plate boundary meets the northeast coast of the South Island just north of Kaikoura, where the Hikurangi Trench brings deep water close inshore (Figure 4.3). The associated subduction is reflected inland by a series of NE–SW striking faults that are part of the Alpine Fault system. Individual mountain ranges in the northeast of the South Island, such as the Seaward Kaikoura (to 2610 m) and Inland Kaikoura (to 2885 m), are separated by these faults, which have been exploited by erosion to form deep valleys.

The axial ranges reach their greatest height in Aoraki/Mt Cook (3764 m) in the Southern Alps of the central South Island. In this region the plate boundary is no longer marked by a trench, but is defined by the Alpine Fault, a transpressive dextral system. The greywacke sediments become progressively metamorphosed into greenschist, biotite schist and amphibolite schist as the Alpine Fault is approached. The highest mountains of the Main Divide lie to the east of the Alpine Fault, but roughly parallel to it. Summit erosion surfaces such as found in the axial ranges of the North Island are completely absent in the central Southern Alps, because the Tertiary cover beds and the underlying erosion surface on the greywacke have been totally and deeply denuded and dissected to yield a mountain range of sharp peaks incised by deep valleys. Glaciation in the Quaternary contributed to the dissection of the Southern Alps and the extension of local relief, unlike in the greywacke ranges of the North Island where only a few small cirque glaciers were supported.

In southwest New Zealand the axial ranges acquire a different character. In this region the Alpine Fault runs off-shore and merges with the Puysegur Trench. On-shore there is a major change in rock type. The Fiordland block is composed mainly of Palaeozoic gneissic metasediments and granitic gneisses, which are considerably more resistant to erosion than the Mesozoic schists, greywackes and argillites of the ranges further north. The mountains in this region have a well-defined summit surface (Figure 4.7) that rises from about 1500 m in the southwest to over 2000 m in the northeast (Augustinus, 1992). The highest peak is Mt Tutoko (2746 m). The surface is deeply dissected by glacial valleys and fjords, the walls of which commonly rise precipitously more than 1000 m asl.

Because of the dextral slip along the Alpine Fault, rocks similar to those found in Fiordland are also found in the northwest of the South Island, in NW Nelson, where peaks rise to 1700–1900 m. Mt Owen (1875 m) and The Twins (1826 m) are two of the highest and are composed of Ordovician marble. Remnant patches of Tertiary sediments indicate that they were originally much more extensive in NW Nelson, but they are being stripped to reveal an exhumed late Cretaceous erosion surface on the underlying Palaeozoic metasediments. This gives rise to distinctly plateau-like skylines in parts of the region, especially recognizable in the accordant summits of the Tasman Mountains; thus echoing the similar-aged summit surface of Fiordland. In NW Nelson there were valley glaciers in the Pleistocene, but none was sufficiently extensive enough to reach the sea; so fiords are not found in NW Nelson.

4 Mountain-building processes

4.1 Uplift at the plate boundary

Mountains grow when the rate of rock uplift exceeds the rate of denudation, thus permitting the surface to be uplifted. Where the rates balance, a steady state is achieved, and the gross landform geometry does not change, except in detail. Uplift at an active plate boundary is a consequence of crustal compression. This is often considered to be related to subduction, where a slab of relatively dense oceanic crust is forced beneath a slab of less dense continental crust. Yet, in the case of continental crust to continental crust convergence, the mechanism is less clear and subduction may not be involved. New Zealand provides examples of these contrasting plate boundary conditions, i.e., the

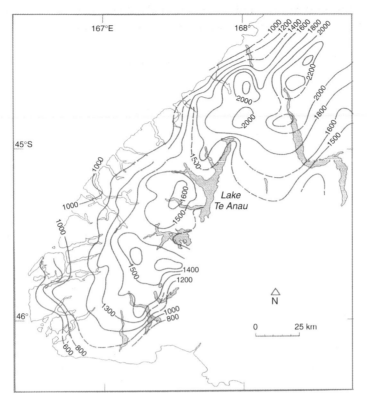

Figure 4.7 *The summit surface of mountains in Fiordland. The surface is assumed to approximate an old erosion surface. Remnant patches of inclined Eocene marine sediments resting unconformably on truncated Palaeozoic basement rocks at over 1200 m indicate that the surface was trimmed by marine erosion in the early Tertiary, buried by sediments, warped, uplifted and then exhumed by denudation. The shaded areas represent major lakes*
Source: *reprinted from Geomorphology, 4, Augustinus, P.C., Outlet glacier trough size–drainage area relationships, Fiordland, New Zealand, pp. 347–61, © 1992, with permission from Elsevier.*

east coast of the North Island in the case of uplift induced by subduction and the central Southern Alps in the case of uplift induced by continental crust-to-crust convergence.

The relative movement between neighbouring plates can change over time. Consequently, the zones of active convergence, extension and transcurrent movement also change. As a result, the rates of rock uplift and denudation also change over the long time periods during which mountains are built. So a snapshot evaluation of these rates over given intervals may not yield a representative view of the long-term situation or permit a realistic assessment to be made of whether or not the mountains are in a steady state. With that caveat in mind, late Quaternary uplift rates in New Zealand are shown in Figure 4.8 and may be compared with modern erosion rates in Figure 4.9.

The uplift and erosion data for the North Island suggest that there is a general lack of balance between present regional uplift rates and denudation rates. Furthermore, a major zone of strongly negative gravity anomalies runs across the island from East Cape to Cook Strait, indicating an upward hydrostatic pressure and hence a tendency to rise. In a comparison of uplift rate and erosion rate evidence for the North Island, Adams (1979) noted that in the Tararua Range the erosion rate is far less than the uplift rate; so those mountains at least must be growing.

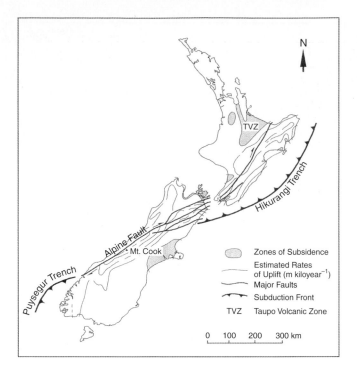

Figure 4.8 *A late Quaternary uplift map for New Zealand*
Source: *from Williams (1991) with modifications after Williams (1988), Pillans (1986), Wellman (1979) and others, reproduced with the permission of Catena Verlag GMGH.*

Off the east coast of the North Island seafloor spreading results in convergence at a rate of about 58 mm a^{-1} near East Cape. This reduces to 49 mm a^{-1} near Cook Strait (see Figure 4.6 for location). The oblique convergence leads to crustal compression that is taken up by uplift that attains up to 3 mm a^{-1} in the axial ranges, folding and shearing. Kamp (1988) described the tectonic geomorphology along this margin and noted that north of Hawke Bay all typical forearc, arc and back-arc elements occur, whereas to the south there are no arc and back-arc features. The lack of these elements in the south is considered by Kamp (1988) to reflect the dominance of crustal compression in the south compared with crustal extension in the north.

Since the early Pleistocene, an uplift of about 2 km is considered to have occurred along the frontal ridge inland of Hawke Bay (Berryman, 1988). This has been associated with a rate of strike-slip on the faults of about 8 km in the same interval. Maximum uplift rates in this region in the late Quaternary attain about 3 m ka^{-1} (Figure 4.8), although the back-arc region is one of subsidence or only slight uplift, tension rather than compression, and much volcanism. In the southern part of the margin, the subduction complex is wider, and a broadly downwarped basin with compressive folds extends up to 350 km from the trench. It is notable that the highest parts of the axial ranges are not fold-mountains, but are composed of uplifted, tilted and warped basement blocks. Folds are generally more evident in the Tertiary sediments of the accretionary prism, although on-going folding is occurring in the greywacke rocks of the Wellington region (Ota *et al.*, 1981). In the southern North Island, uplift rates range from 0.75 to 4 mm a^{-1} for the growing anticlines, and from 0.5 to 2.2 mm a^{-1} for the growing synclines (Ghani, 1978).

To the south of Cook Strait, plate convergence in the northeast of the South Island occurs at a rate

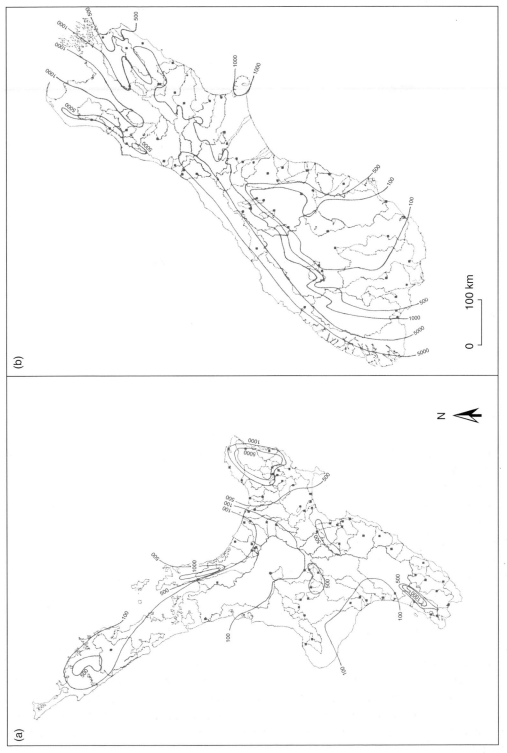

Figure 4.9 *Specific mean annual sediment yield (t km⁻² a⁻¹) in (a) the North Island of New Zealand and (b) the South Island of New Zealand Source: from Mosley and Duncan (1992). Reproduced with the permission of Pearson Education (NZ).*

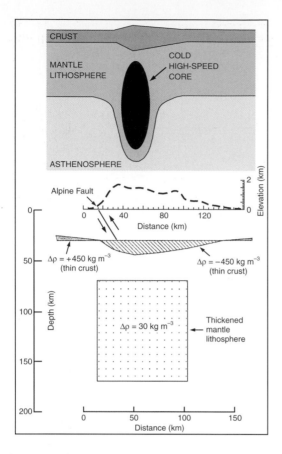

Figure 4.10 *A schematic representation of homogeneous shortening and thickening of the New Zealand lithosphere across the Southern Alps of the central South Island. The upper diagram depicts the probable mode of intracontinental convergence in the mantle beneath a region of crustal shortening such as the Southern Alps. The lower diagram provides field evidence that supports the model depicted above. It shows a generalized topographic profile across the central Southern Alps and modelled rock density conditions beneath it. $\Delta\rho$ refers to the density contrast between rocks. Between the mantle (3350 kg m^{-3}) and oceanic crust (2900 kg m^{-3}) $\Delta\rho = 450$ kg m^{-3}. Measured and calculated Bouger gravity anomalies along the W–E transect reveal a 30 mgal (1 mgalileo $= 0.01$ mm s^{-2}) difference that can be accounted for by a dense mass beneath the thickest crust. Modelling suggests that it could be provided by a two-dimensional body (with $\Delta\rho = 30$ kg m^{-3}) centred at a depth of 120 km that is 90 km wide and has a depth extent of 100 km Adapted from Stern* et al. *(2000).*

of about 33 mm a^{-1}. This also leads to compression, strike-slip and uplift, which in the late Quaternary has attained up to 5 m ka^{-1}. But there are also zones of subsidence. This is best exemplified in the Marlborough Sounds, where block subsidence at the coast has led to the development of a superb ria coastline. However, coastal valleys such as the Pelorus also have low level terraces, indicating recent relative uplift of the land. The interplay of tectonic and eustatic movement is complex.

The central South Island is a zone of intracontinental convergence and has the highest peaks of the Southern Alps. The Alpine Fault, which runs through the region, acts as an oblique transform connecting the two subduction zones of opposing polarity. Plate reconstructions indicate about 90 km of convergence across the central portion of the Alpine Fault in the past 7 Ma plus about 25 km of convergence in the 12 Ma prior to 7 Ma BP (Walcott, 1998). Strike-slip occurs at a present-day rate of about 38 mm a^{-1} and convergence at 10–12 mm a^{-1}. Evidence provided by Stern et al. (2000) appears to show that shortening of the overlying crust is diffuse and that intracontinental subduction of one slab beneath another does not occur, but instead mantle lithosphere thickens penetratively beneath the thickened crust (Figure 4.10). An excess mass in the mantle is also required by the observed gravity anomalies. Their work indicates a relatively cool laterally symmetrical core of 80–100 km width within the lithospheric root, centred at a depth of about 120 km and with a depth extent of 100 km. This is consistent with the mantle lithosphere behaving more as a deformable continuum than as a lithospheric plate, as generally understood in plate tectonics. Stern et al. (2000) conclude that the high-density mantle anomaly provides sufficient force to maintain the crustal root, and that it is approximately twice as thick as that necessary to support the topography. The maximum crustal thickness is about 44 km and is located just east of the highest peaks. This reduces to about 27 km near both the east and west coastlines.

The topographic manifestation of crustal compression in this region is the construction of the Southern Alps, which in gross form constitute an elongate asymmetric dome parallel to the plate boundary. Tippett and Kamp (1993a,b, 1995a,b) have made a major contribution to our understanding of the evolution of these mountains. They determined from zircon fission track dating that Cenozoic rock uplift commenced between 8 and 5 Ma, depending on location in the Southern Alps. Rock uplift began about 8 Ma in the southern parts of the Southern Alps, compared with about 5 Ma in northern parts, and the eastern margin of the Alps began to be uplifted about 3 Ma. The initiation of surface uplift propagated southeastwards from the Alpine Fault at a rate of 30 km Ma^{-1}. Rock uplift reached a maximum of about 19 km just east of the Alpine Fault in the Mt Cook region with lesser amounts at the northern (c.10 km) and southern (c.8 km) margins of the Southern Alps. Rock uplift decreases almost exponentially to values of 3 km with distance to the southeast of the fault. Whereas rock uplift started on average about 6–7 Ma, surface uplift commenced later, about 4–5 Ma. Almost 2 km of erosion occurred before significant mean surface uplift began, and erosion has continued during further uplift. In the first 5 Ma of rock uplift Tertiary cover rocks were removed progressively eastwards, thereby exposing the greywacke basement. Throughout the region, mountain elevation and relief have changed with respect to time and the amount of rock uplift.

Three domains of uplift are recognized by Tippett and Kamp (1995b) based on long-term rates of rock uplift: domain 1 is an area of no uplift that lies to the east of the Southern Alps; domain 3 lies adjacent to the Alpine Fault and has undergone the most uplift; and domain 2 lies between the two. The boundary between domains 2 and 3 is located at 1 mm a^{-1}, either side of which the long-term rock uplift rate changes dramatically. Domain 3 has undergone long-term (since about 7 Ma) rock uplift rates of 1.0–2.7 mm a^{-1}, with shorter-term (since 1.3 Ma) rates of 6–9 mm a^{-1}, and modern rates of 8–10 mm a^{-1}. By about 1.5 Ma the present mountain range front began to form and Quaternary gravels began to accumulate to the southeast of the Southern Alps, forming the Canterbury Plains. Tippett and Kamp (1995b) conclude that in domain 2 to the east of the Main Divide the mountains are not stable equilibrium forms, but have continued to increase in elevation and evolve in form since the start of uplift. By contrast, in domain 3 (between the Alpine Fault and the Main Divide) they consider mountain elevations to be in stable equilibrium with respect to rock uplift rate. The mean surface elevation is greatest (and reaches almost 2000 m) along a zone 20–30 km to the southeast of and parallel to the Alpine Fault, but decreases to about 750 m both to the north and south. To the west of the Main Divide surface elevation decreases more rapidly than it does to the east and therefore the mountain elevations of the Southern Alps define an elongate asymmetric dome. The variables that contribute to the gross geomorphological characteristics of the central Southern Alps are summarized in Figure 4.11.

4.2 Volcanism

The plate tectonic activity in New Zealand has also given rise to large volcanic mountains. The best example, and also the highest mountain in the North Island, is the frequently active Mt Ruapehu (2797 m) at the southern apex of the TVZ. Ruapehu is a stratovolcano made of andesitic lavas and tuff. It is about 230000 years old and it is situated immediately to the south of an older volcano, Mt Tongariro (1968 m), the summit of which has been riven by past explosive activity. Its lavas have been dated to 260000 years BP (J. Stipp, cited in Healy, 1992). Between the two is the actively growing cone of Mt Ngauruhoe (2291m), which is only 5000 years old. Its eruptive deposits occupy glacial troughs from the Last Glacial Maximum, lava flows in some cases being contained by glacial moraines, the termini of which descend to about 1100 m. Deposition of volcanic debris by radially draining rivers and lahars has constructed great ring plains around these volcanoes, extending to a radius of up to 20 km from the volcanic centre. The ring plain deposits of Mt Ruapehu have banked against the faulted margin of the Kaimanawa Range and have diverted the Tongariro River north into Lake Taupo.

The andesitic volcanism of Mts Tongariro, Ruapehu and Ngauruhoe is uncharacteristic of the TVZ,

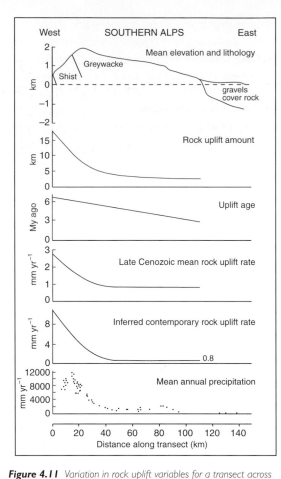

Figure 4.11 *Variation in rock uplift variables for a transect across the central Southern Alps*
Source: *from Tippett, J.H. and Kamp, P.J.J., Geomorphic evolution of the Southern Alps, New Zealand. Earth Surface Processes and Landforms, 20, 177–92, 1995. © John Wiley & Sons Limited. Reproduced with permission.*

most of which is associated with large calderas and ignimbrite sheets resulting from explosive rhyolitic volcanism and collapse over the last 750 000 years and especially in the last 300 000 years. In addition there are numerous smaller rhyolitic domes and scoria cones. The largest of the calderas contains Lake Taupo (358 m asl), the largest lake in New Zealand. The lake is 40 km long and 28 km wide and covers an area of 620 km², but occupies only the bottom part of the caldera. Taupo Volcano last erupted about 1800 years ago, when it ejected more than 105 km³ of debris, but this was a small eruption compared with many that had preceded it (Wilson *et al.*, 1986). The TVZ has a faulted eastern boundary against the axial ranges of the North Island (Figure 4.4).

Another large andesitic stratovolcano is Mt Taranaki/Egmont (2518 m), which is located beyond the limits of the TVZ on the west coast of the central North Island. It is also a young volcano (~16 500 years old), that developed on an older volcanic centre that first commenced activity over 575 000 years BP. Its ring plain has a radius of about 25 km. Mt Taranaki has not erupted since AD 1755 (Neall, 1992).

Basaltic eruptions in the Miocene of Lyttleton and Akaroa volcanoes on Banks Peninsula near Christchurch produced mountains that today have a relief of more than 900 m.

5 Interaction of mountain building and erosion

New Zealand is oriented approximately NE–SW athwart the mid-latitude westerly wind belt. Consequently, as the mountains have grown, orographic precipitation effects have increased and windward–leeward rainfall patterns have developed. The greatest precipitation occurs parallel to the axis of the Southern Alps, but displaced to the west of the highest peaks. In those areas annual precipitation commonly exceeds 6 m and can exceed 10 m. By contrast, average annual precipitation in rainshadow areas to the east of the Southern Alps is typically less than 1000 mm and can be as low as 400 mm. Precipitation commonly falls as snow in winter and sustains many permanent glaciers. Windward–leeward effects also occur across the mountains of the North Island, but the contrast is not so marked as in the South Island because the mountains are lower. Snow is common in winter in the central North Island, and Mt Ruapehu supports a few small glaciers. The treeline rises from about 1000 m in Fiordland to 1400 m in the central North Island and in the alpine zone frost shattering becomes a prominent weathering process.

As precipitation increases so does runoff and erosion. Figure 4.9 provides estimates of modern

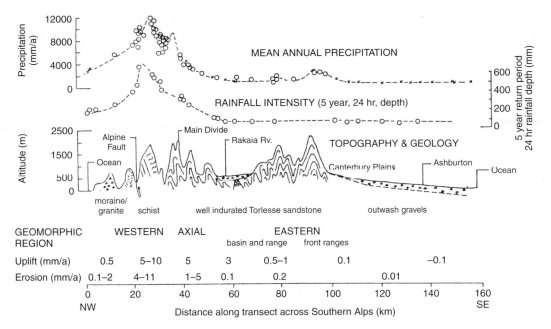

Figure 4.12 *Cross-section through the central Southern Alps of New Zealand, showing the relationship between topography, precipitation and erosion rates*
Source: *from Whitehouse (1988). Reproduced with the permission of Gebrüder Borntraeger Verlagsbuchhandlung, Stuttgart.*

mean annual sediment yields of rivers in the North and South Islands. Figure 4.12 illustrates a transect across the Southern Alps in which a comparison is made between altitude and precipitation, and rates of uplift and erosion. Evaluation of this kind of information, plus evidence from sedimentation rates and its comparison with late Quaternary uplift rates led Adams (1979, 1985) to suggest that a dynamic balance may have been achieved in the Southern Alps between mountain building and denudation. However, more recent work by Tippett and Kamp (1993a,b, 1995a,b) described above has shown the situation to be far more complex. The uplift rates shown in Figure 4.8 apply only to the late Quaternary, and so do not provide a representative picture of long-term uplift that might be compared with long-term rates of denudation. For example, while the maximum uplift rate beside the Alpine Fault is at present about 8–10 mm a^{-1}, for most of the period over which uplift has occurred in the Southern Alps the rate has averaged only 2.7 mm a^{-1}. Nevertheless, while a simple dynamic balance may not have been achieved, the inter-relationship between rates of uplift, precipitation and erosion is undeniable and supports the view that erosion rates adjust to high rates of tectonically driven rock uplift (Montgomery and Brandon, 2002).

It is noticeable that the fastest rates of uplift are in the regions where the mountains are highest, and this in turn is in the zone of vigorous intracontinental convergence, where there is a deep lithospheric root produced by thickening of the mantle beneath the mountains (Figure 4.10). Subduction is not involved. Tippett and Kamp (1995b) note that there are strong and consistent relationships between the rock uplift parameters (amount, age, rate) and elevation/relief parameters in the Southern Alps and, therefore, the pattern of rock uplift is the primary influence on the landscape development in the region. Factors other than uplift, especially those that affect the magnitude and style of erosion, also contribute to the nature of landscape development, as implied by Figure 4.12. Elevation and relief are greatest where there is a combination of rapid uplift rate and vigorous erosion (which is promoted by high rainfall and high potential energy).

The generally high rainfall of New Zealand has given rise to large rivers that have maintained their courses as uplift has occurred. Consequently, river valleys within the mountains have interesting and very long histories. On the accretionary slope of the Hikurangi margin in the North Island, for example, the valleys have extended eastwards as accretion and uplift have occurred, and so their downstream reaches are progressively younger as the coast is approached. Such river systems have also adjusted to strike-slip faulting across their paths. Fluvial incision has accompanied uplift and has exploited fault zones. Rivers such as the Mohaka, Ngaruroro and Tukituki that flow into Hawke Bay (Figure 4.6) have consequently adjusted to differential rates of uplift, differential rates of displacement across fault traces along their path, and to the growth of fold axes, especially in their lower reaches. The maintenance of a river's course in the face of uplift is especially well illustrated by the Manawatu River in the North Island (Figure 4.6), which flows NE–SW right across the spine of the axial ranges. In the South Island, several rivers (notably the Haast and Taramakau) flow west from the Main Divide across the Alpine Fault to the west coast. As uplift of the Southern Alps has occurred, the Main Divide has migrated progressively eastwards (Tippett and Kamp, 1995b) and so the valleys of these rivers become progressively younger with distance towards their headwaters. When the Kaikoura orogeny started in New Zealand in the late Cenozoic, most of the land emerging from the sea was veneered with 1–2 km of Tertiary sediments. The rivers that flowed across these cover beds eventually stripped most of them away, often maintaining their courses as uplift and block faulting occurred across their path, and so became superimposed upon the underlying basement rocks.

6 Conclusion

There is clear evidence in New Zealand that mountains (other than volcanoes) have resulted from the uplift of a previously eroded landmass veneered by young Tertiary sediments. Vestiges of the old land-mass, which was planed to near sea level in the late Cretaceous and in some places trimmed again in the mid-Tertiary, are often evident in the landscape as exhumed surfaces revealed by the stripping of Cenozoic cover beds (Figure 4.7). In these places the mountain ranges can be seen to be uplifted fault blocks, commonly tilted, sometimes warped, and always deeply dissected by valleys that have developed in the late Tertiary and Quaternary. Where uplift rates have been particularly rapid and sustained, the high mountains produced have interacted with the westerly circulation and led to greater precipitation and enhanced erosion. This has resulted in the remnants of old erosion surfaces being completely destroyed, and so concordant plateau tops have been replaced by intersecting valleyside slopes, sharp ridges and peaks. Along the highest part of the Main Divide of the Southern Alps, average elevation attains almost 2 km, but rock uplift since the initiation of mountain building has been about 19 km. Therefore, considerable denudation has clearly accompanied rock uplift.

The relationship between uplift and erosion is complex throughout New Zealand and data on rates are spatially and temporally uneven. Consequently, it is not easy to evaluate whether mountains in different parts of the country are growing, declining or are in a steady state. Nevertheless, in the central part of the Southern Alps against the Alpine Fault, in domain 1 of Tippett and Kamp (1995b), the mountains appear to have attained a balance with respect to rock uplift rate. By contrast in domain 2, to the east of the Main Divide, both elevation and relief are still growing. In the axial ranges of the North Island, late Quaternary uplift rates are relatively high (approximately equal to the average long-term rate along the central part of the Alpine Fault) and appear greater than modern erosion rates; so mountains are likely to be growing in that region. It is less clear what is happening in Fiordland for, although the late Quaternary uplift rate is relatively low, the Palaeozoic rocks are particularly resistant, so denudation may also be relatively low, as implied by the preservation of the extensive upland surface (Figure 4.7).

Although warping and the growth of folds can sometimes be shown to be associated with uplift, as in the southern North Island (Ghani, 1978; Ota *et al.*, 1981), New Zealand mountains are not fold mountains. In most cases the axial ranges are composed of uplifted and tilted faulted blocks. This is particularly clear along the Hikurangi and Fiordland margins, where subduction appears to be the main mechanism leading to compression and uplift. However, it is less clear in the central Southern Alps, because uplift associated with diffuse crustal shortening has resulted in up to 19 km of rock being denuded. Consequently, the erosion surface across the Mesozoic basement rocks has long since disappeared.

References

Adams, J.E., 1979. Late Cenozoic erosion in New Zealand. Unpublished PhD Thesis, Victoria: University of Wellington.

Adams, J.E., 1985. Large-scale tectonic geomorphology of the Southern Alps, New Zealand. In Morisawa, M. and Hack, J.T. (eds), *Tectonic geomorphology*. Boston, MA: Allen & Unwin, 105–28.

Augustinus, P.C., 1992. Outlet glacier trough size–drainage area relationships, Fiordland, New Zealand. *Geomorphology*, **4**: 347–61.

Berryman, K., 1988. Tectonic geomorphology at a plate boundary: a transect across Hawke Bay, New Zealand. *Zeitschrift für Geomorphologie, Supplementband*, **69**: 69–86.

Beu, A.G., Grant-Taylor, T.L. and Hornibrook, N. De B., 1980. *The Te Aute Limestone facies: Poverty Bay to northern Wairarapa 1:250 000*. NZ Geological Survey Miscellaneous Map Series 13. Wellington: Department of Scientific and Industrial Research.

Cole, J.W. and Lewis, K.B., 1981. Evolution of the Taupo-Hikurangi subduction system. *Tectonophysics*, **72**: 1–21.

Delteil, J., Herzer, R.H., Sosson, M., Lebrun, J.-F., Collot, J.-Y. and Wood, R., 1996. Influence of pre-existing back-stop structure on oblique tectonic accretion: the Fiordland margin (southwestern New Zealand). *Geology*, **24**: 1045–48.

Gee, M. and Kirkpatrick, R., 1997. Gondwanaland: tectonic movement and the formation of the New Zealand land mass. In McKinnon, M. (ed.), *New Zealand historical atlas*. Auckland: Bateman, plate 4.

Ghani, M.A., 1978. Late Cenozoic vertical crustal movements in the southern North Island, New Zealand. *New Zealand Journal of Geology and Geophysics*, **21**: 117–25.

Healy, J., 1992. Central volcanic region. In Soons, J.M. and Selby, M.J. (eds), *Landforms of New Zealand*. Auckland: Longman Paul, 256–86.

Kamp, P.J.J., 1988. Tectonic geomorphology of the Hikurangi margin: surface manifestations of different modes of subduction. *Zeitschrift für Geomorphologie, Supplementband*, **69**: 55–67.

Kamp, P.J.J., 1992. Tectonic architecture of New Zealand. In Soons, J.M. and Selby, M.J. (eds), *Landforms of New Zealand*. Auckland: Longman Paul, 1–30.

King, P.R., 2000. Tectonic reconstructions of New Zealand: 40 Ma to the Present. *New Zealand Journal of Geology and Geophysics*, **43**: 611–38.

Montgomery, D.R. and Brandon, M.T., 2002. Topographic controls on erosion rates in tectonically active mountain ranges. *Earth and Planetary Science Letters*, **201**: 481–89.

Mosley, M.P. and Duncan, M.J., 1992. Rivers. In Soons, J.M. and Selby, M.J. (eds), *Landforms of New Zealand*. Auckland: Longman Paul, 91–106.

Neall, V.E., 1992. Landforms of Taranaki and the Wanganui Lowlands. In Soons, J.M. and Selby, M.J. (eds), *Landforms of New Zealand*. Auckland: Longman Paul, 287–307.

Ota, Y., Williams, D.N. and Berryman, K.R., 1981. *Parts sheets Q27, R27 & R28 – Wellington Late Quaternary tectonic map of New Zealand 1:50 000. With notes* (1st edition). Wellington: DSIR.

Pillans, B.J., 1986. A late Quaternary uplift map for the North Island, New Zealand. *Royal Society of New Zealand Bulletin*, **24**: 409–17.

Stern, T., Molnar, P., Okaya, D. and Eberhart-Phillips, D., 2000. Teleseismic P wave delays and modes of shortening the mantle lithosphere beneath South Island, New Zealand. *Journal of Geophysical Research*, **105** (B9): 21615–31.

Suggate, R.P., Stevens, G.R. and Te Punga, M.T. (eds), 1978. *The geology of New Zealand.* Wellington: Government Printer.

Tippett, J.M. and Kamp, P.J.J., 1993a. Fission track analysis of the late Cenozoic vertical kinematics of continental Pacific crust, South Island, New Zealand. *Journal of Geophysical Research*, **98**: 16119–48.

Tippett, J.M. and Kamp, P.J.J., 1993b. The role of faulting in rock uplift in the Southern Alps, New Zealand. *New Zealand Journal of Geology and Geophysics*, **36**: 497–504.

Tippett, J.M. and Kamp, P.J.J., 1995a. Quantitative relationships between uplift and relief parameters for the Southern Alps, New Zealand, as determined by fission track analysis. *Earth Surface Processes and Landforms*, **20**: 153–75.

Tippett, J.M. and Kamp, P.J.J., 1995b. Geomorphic evolution of the Southern Alps, New Zealand. *Earth Surface Processes and Landforms*, **20**: 177–92.

Walcott, R.I., 1998. Modes of oblique compression: Late Cenozoic tectonics of the South Island of New Zealand. *Reviews of Geophysics*, **36**: 1–26.

Wellman, H.W., 1979. An uplift map for the South Island of New Zealand, and a model for the uplift of the Southern Alps. In Walcott, R.I. and Cresswell, M.M. (eds), The origin of the Southern Alps. *Royal Society of New Zealand Bulletin*, **18**: 13–20.

Whitehouse, I.E., 1988. Geomorphology of the central Southern Alps, New Zealand: the interaction of plate collision and atmospheric circulation. *Zeitschrift für Geomorphologie, Supplementband*, **69**: 105–16.

Williams, P.W., 1988. Introduction. In Williams, P.W. (ed.), The geomorphology of plate boundaries and active continental margins. *Zeitschrift für Geomorphologie, Supplementband*, **69**: v–ix.

Williams, P.W., 1991. Tectonic geomorphology, uplift rates and geomorphic response in New Zealand. *Catena*, **18**: 439–52.

Wilson, C.J.N., Houghton, B.F. and Lloyd, E.F., 1986. Volcanic history and evolution of the Maroa-Taupo area, Central North Island. In Smith, I.E.M. (ed.), Late Cenozoic volcanism in New Zealand. *Royal Society of New Zealand Bulletin*, **23**: 194–223.

PART 3 Functional Mountain Geomorphology

Proglacial zone, Mittivakkat Glacier, Greenland
Photo: P.N. Owens

5
Processes, rates and patterns of mountain-belt erosion

Niels Hovius, Dimitri Lague and Simon Dadson

1 Introduction

The tectonic evolution of active plate boundaries is controlled not only by how the crust and mantle parts of the lithosphere deform in compression, but also by how material is removed by erosion (Koons, 1989; Beaumont et al., 1991). Climate exerts a fundamental control on the tectonic development of orogens by driving selective erosion of crustal mass (Molnar and England, 1990; Willett, 1999). In turn, climate is moderated by the impact of mountain belts on atmospheric circulation patterns (Kutzbach et al., 1989), and the draw-down of carbon dioxide by rapid weathering of fresh silicate crust exposed in active orogens (Raymo and Ruddiman, 1992) and the burial of organic carbon in nearby basins (Pacala et al., 2001; Lyons et al., 2002). Thus, erosion provides a first-order, two-way link between tectonic and atmospheric processes. This link is most effective in active, compressional orogens that source most of the clastic sediment eroded from the present-day continents (Milliman and Syvitski, 1992) and a significant component of the chemical and organic fluxes to the oceans (Gaillardet et al., 1999; West et al., 2002).

Consider, for example, the India–Asia collision. The elevation of the Tibetan Plateau, as a result of this collision, may have caused the onset or intensification of the southeast Asia monsoon (Prell and Kutzbach, 1992). Precipitation-driven erosion along the margins of the plateau may have promoted rapid weathering of silicate crust, which in turn may have contributed to the marked global cooling since the early Eocene (Raymo and Ruddiman, 1992). In addition, erosion may have moderated the evolution of the India–Asia collision zone, especially in the Himalaya (Fielding et al., 1994; Beaumont et al., 2001). There, the intensification of erosion associated with the onset of Northern Hemisphere glaciation (Zhang et al., 2001) may have reactivated major structures within the mountain range that had become extinct during less erosive times.

In active orogens, a simple set of processes is responsible for erosion and landscape evolution. Fluvial and/or glacial processes drive down local base level against a backdrop of continuous or episodic rock uplift (Burbank et al., 1996). On hillslopes, mobilization of sediment is generally limited by weathering of the intact rock mass (Heimsath et al., 1997), but, where rock uplift and base-level lowering outpace weathering, interfluves steepen and become subject to frequent bedrock landslides (Hovius et al., 1997). Feedbacks exist between processes operating in valleys and on hillslopes, first because valley lowering drives hillslope erosion (Densmore et al., 1997), and second because the products of hillslope erosion add to the sediment load of the river and become tools for valley-lowering processes (Hovius et al., 2000; Hartshorn et al., 2002).

In this chapter, the major processes of mountain-belt erosion will be reviewed, and their rates and patterns substantiated, with a focus on equilibrium conditions of active uplift and active denudation.

The linkages of different processes across the upland landscape will be demonstrated and, finally, the role of erosion in the topographic and tectonic evolution of orogens and concomitant sedimentary fill of nearby basins will be explored. An emphasis will be placed on fluvial landscapes. We acknowledge the vigour (Hallet et al., 1996) and importance (Brozovic et al., 1997) of glacial processes, but have elected to ignore them here for the sake of simplicity. Information on the role of glacial processes in controlling erosion rates is presented in Caine, Björnsson and Hewitt (all this volume), and a specific regional model is given in Slaymaker and Owens (this volume). The role of storage in sediment routing from mountain-belts is not explored in detail in this chapter; for further information see Owens and Slaymaker, and Caine (this volume).

2 Basics

Local base level can be defined as the lowest topographic point within a geomorphologically continuous landscape. Such points line up along channel thalwegs or glacial troughs. Most elements of the surrounding landscape will respond dynamically to any change of base level. Erosional landscape evolution is, therefore, intimately controlled by fluvial or glacial base-level lowering against advecting bedrock. A river has a natural tendency towards equilibrium, adapting its local geometry in order to incise at a rate equal and opposite to the ambient rock uplift rate, and evacuate all material supplied from upstream, making use of the available runoff. Hence, rock uplift rate is a first-order control on the long-term incision rate of rivers, while runoff and runoff variability, and the mechanical properties of the exposed rock mass, are additional controls on fluvial wear. In most catchments (outside the arid and semi-arid zones, but see also Mackin, 1963, for a cautionary note), water discharge and sediment load increase systematically in a downstream direction. Channel slope decreases downstream and channel width increases downstream in order to accommodate increased fluxes and maintain constant rates of channel bed lowering. In concert with this, the dominant channel process changes downstream as does the degree of channel bed cover. Starting from the top and going downstream, one might find the following sequence of process domains in a river catchment: (bedrock) hillslope, debris flow channel, bedrock river channel, mixed bedrock–alluvial channel, coarse bed alluvial river channel, and fine bed alluvial river channel. In any given drainage basin these systematic downstream patterns may be obscured by the influence of local tectonic features such as active faults, variability of rock mass properties and the stochastic nature of sediment transfer from hillslope to channel.

Regional surveys of topographic attributes of mountain ranges have served to highlight basic downstream patterns. Montgomery (2001), for example, has quantified the relationship between upstream drainage area and local slope in the Olympic Mountains in the Pacific Northwest of the USA. Using a 10-m grid digital elevation model (DEM), he found an order of magnitude variation in the grid cell slope for a given drainage area. This scatter reflects the variance of climate, rock uplift rate and rock mass properties across the mountain range, as well as the range of erosional processes captured, and errors in the source data (cf. Dietrich et al., 1992; Montgomery and Foufoula-Georgiou, 1993). Underlying the scatter is a simple trend of decreasing local slope with increasing drainage area above a cutoff length scale (Figure 5.1). Mean slope values increased from low values at the drainage area of a single grid cell (100 m^2) to high values at drainage areas of ~1000 m^2. This high coincided with the transition from hillslopes to runoff-dominated topography in the Olympic Mountains. At larger drainage areas, several domains could be discerned, each with a different, negative power function scaling of area and slope. In a downstream order, these domains were interpreted to reflect the dominance of unchannelized runoff, colluvial processes, fluvial bedrock incision and channel alluviation.

Numerous studies have considered the prevalence of Earth surface processes within the context of local slope and upstream area, where the latter is a proxy for water discharge (Montgomery and

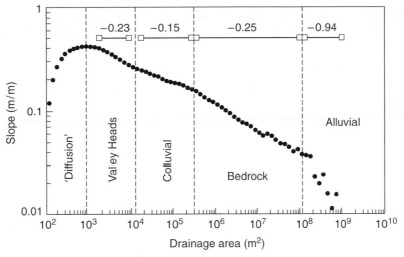

Figure 5.1 *Plot of log-bin averaged drainage area versus slope derived from a 10-m grid DEM of the Olympic Mountains, Pacific Northwest, USA. The plot shows the mean slope of the individual DEM grid cells for each 0.1 log interval in drainage area. Numbers show the exponent for a power function regression of values in the segments of the plots indicated by bars connecting open squares. Dashed vertical lines divide the plot into areas considered to reflect different geomorphic zones of the landscape*
Source: Montgomery, D.R., 2001. *Slope distributions, threshold hillslopes, and steady-state topography*. American Journal of Science, 301: 432–54. Reprinted by permission of the American Journal of Science.

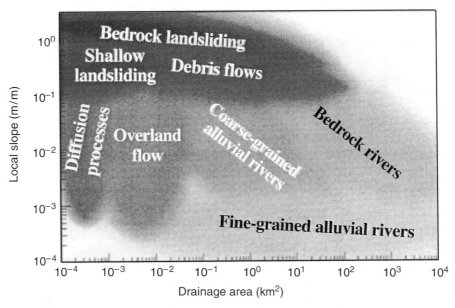

Figure 5.2 *Plot of upstream drainage area versus local slope showing the characteristic domains of key processes in erosional landscapes. Within these domains, the exact nature of the area–slope relationship for a given process depends on local conditions, and specifically on the rock uplift rate, mean precipitation and precipitation variability, and rock mass properties*

Foufoula-Georgiou, 1993; Ijjasz-Vasquez and Bras, 1995; Montgomery *et al.*, 1996; Sklar and Dietrich, 1998; Tucker and Bras, 1998; Lague and Davy, 2003). A compilation of results is given in Figure 5.2. With the exception of glacial and periglacial processes, all physical processes operating on a mountain belt can be accommodated in this diagram.

3 Fluvial bedrock incision

Of all erosion processes, fluvial bedrock incision is crucial because it lowers base level and drives mass wasting of adjacent hillslopes. Fluvial incision occurs by a combination of mechanisms, including impact abrasion by sediment grains moving along the channel bed or in suspension, block removal due to hydraulic forces, cavitation of large air bubbles in turbulent flow and solution (Tinkler and Wohl, 1998; Whipple *et al.*, 2000a). All these processes are aided by progressive weathering of the rock mass exposed in the river channel. The dominant type of wear may vary along a river channel, because of both changing hydraulic conditions and variations in substrate properties. For example, massive rocks without joints are eroded mainly by abrasion, and possibly by cavitation, whereas heavily jointed rock surfaces are lowered primarily by block removal. There have, so far, been few systematic, observational studies of fluvial bedrock wear (Sklar and Dietrich, 2001; Hartshorn *et al.*, 2002), but results to date have identified sediment availability and calibre, and rock mass properties as key controls on fluvial bedrock wear.

Sediment supplied to a river influences bedrock incision in two opposing ways: it promotes abrasion of exposed bedrock and limits the extent of bedrock exposure (cf. Gilbert, 1877; Howard *et al.*, 1994). Thus, the maximum incision rate should occur when sediment supply to a stream is moderate with respect to its transport capacity (Sklar and Dietrich, 1998). This was demonstrated in an abrasion mill study by Sklar and Dietrich (2001). At low supply rates, erosion rates increased rapidly as the mass of abrasive sediment was increased (Figure 5.3). But, at higher supply rates, erosion rates peaked and then declined, following the formation of an immobile sediment layer over the substrate. Erosion continued at very slow rates even when the rock surface was fully covered. Measured erosion rates peaked near the coarse end of the range of grain sizes tested (fine gravel) and declined rapidly with decreasing grain size, becoming negligible for grain sizes smaller than coarse sand. In this experiment, gravel moved as bed load whereas sand moved primarily in suspension, rarely impacting the rock

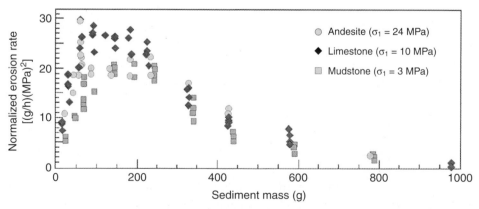

Figure 5.3 *Variation in erosion rate with sediment mass in an abrasion mill. The mill consisted of a rock-floored cylinder in which circular water flow was driven by a propeller. Abrasion of the rock disc occurred primarily through impact of saltating grains of uniform size (6 mm diameter in this experiment). Data for three rock types have been collapsed by multiplying erosion rate (g/h) by the square of the tensile strength (MPa)*
After Sklar and Dietrich (2001); reproduced with the permission of the Geological Society of America.

surface. These results have prompted Sklar and Dietrich (2001) to propose that, in rivers, the minimum channel slope required for incision to occur is set by the threshold of motion for the coarsest grain sizes present in sufficient quantities to bury the channel bed in the absence of transport and particle breakdown. Sklar and Dietrich (2001) also found that rock resistance to fluvial abrasion scaled with the square of rock tensile strength. However, they had limited their study to bedrock that was not jointed or fractured, and therefore did not capture the process of block removal by particle impact and/or hydraulic lifting. Such limitations are hard to overcome in laboratory experiments and instead require study in natural channels.

A field study of fluvial bedrock wear in the Central Range of Taiwan has yielded insights into the mechanisms, rates and patterns of fluvial bedrock wear in a natural river channel (Hartshorn et al., 2002). The rocks at the site of this study comprised schists and a prominent quartzite bed exposed in a channel with a simple, parabolic cross-section and an upstream drainage area of 435 km². The densely foliated schists had a tensile strength of roughly half that of the coarsely jointed quartzite and, in an abrasion mill test, the former was found to abrade approximately four times faster than the latter. However, detailed measurements of changes in channel bed topography over a 22-month interval between February 2000 and December 2001 revealed that natural wear rates were higher in quartzite than in schist: median estimates of erosion across the active channel were 8.5 mm for the quartzite and 6 mm for the schist. Interestingly, these rates compare well with independent estimates of rock uplift and exhumation in the catchment of 3–6 km Ma^{-1}. However, important spatial and temporal patterns emerged during the course of the study, with maximum local erosion of 182 mm in quartzite and 69 mm in schist over the 22-month period. High erosion rates in quartzite reflect both its current, relatively exposed aspect above the schist and the effect of controls on erosion resistance, such as spacing of joints, which are not captured by usual measures of rock strength. Joint block removal dominated erosion of quartzite, whereas uniform surface lowering by abrasion occurred in schists (Figure 5.4). Observed lithological differences in wear rate are consistent with the notion that, where and when active, the removal of blocks is a more effective style of fluvial bedrock erosion than abrasion (Hancock et al., 1998; Whipple et al., 2000a,b).

Erosion of the study channel was greater by up to an order of magnitude during the wet season of 2000 than during the subsequent dry and wet seasons of 2001 combined (Figure 5.4). This wear-rate peak was largely due to the impact of supertyphoon Bilis, an event with a 20-year return interval. Bilis caused a 12-m-deep flood at the study site, with water discharges of about 65 times the long-term average. During this flood, erosion was greatest at intermediate elevations within the active channel. Subsequent erosion mostly occurred low in the channel, with wear rates increasing toward the low flow line. For peak flow conditions at the study site, it can be calculated that a metre-sized boulder would have bounced up to several metres above the bed. But most of the sediment in the channel is much finer and would travel in turbulent suspension. Only particles <2 mm in diameter would have travelled in significant numbers at elevations of >0.3 times the flow depth, or 4 m during peak flow. This implies that maximal wear rates at mid levels of peak flow were due to rare but significant impacts of large, saltating boulders, and to continuous abrasion by very coarse sand and finer material in suspension transport.

Ultimately, however, base-level lowering occurs in the thalweg of a river channel, and it was there that erosion due to typhoon Bilis was relatively subdued. Prorated for the 20-year return period of the flood caused by this typhoon, wear rates at the base of the channel were only 1.7 mm a^{-1} for quartzite and 0.3 mm a^{-1} for schist. In contrast, erosion of the channel base in both rock types approached values of 2–6 mm a^{-1} for the interval December 2000 to December 2001, during which no major floods occurred. This suggests that valley lowering at the Taiwanese study site is driven by

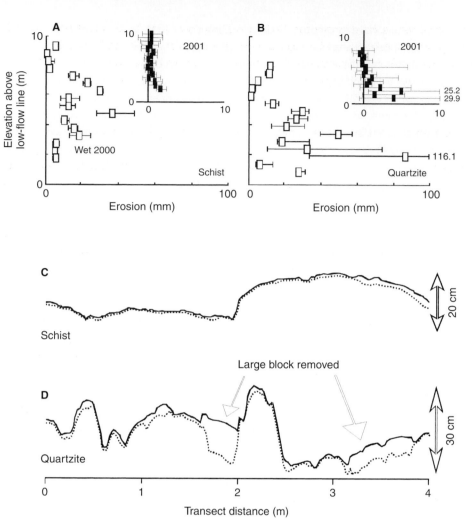

Figure 5.4 *Erosion (mm) of the Liwu (Taiwan) channel for schists (A) and quartzite (B) between February 2000 and December 2001. Boxes represent median values of erosion as a function of elevation above the low-flow line, and whiskers represent the 25th – and 75th – percentile values. The elevation range has been divided into 50-cm intervals, and all measurements within an interval combine to produce the median values for that elevation. Open boxes represent the combined median values from the wet season of 2000 (including a 20-year flood), and inset black boxes represent combined median values from subsequent seasons. Inset graphs have the same axis labels as the larger graphs. Erosion represents vertical surface lowering at a point. In schists, erosion was dominated by impact abrasion, and resulted in parallel surface retreat (C). In quartzites, removal of joint blocks caused extreme variations in local wear rates and a high roughness of the channel bed (D). In (C) and (D), solid and dashed lines represent the channel bed topography in February 2000 and December 2001, respectively*
Source: *Hartshorn, K., Hovius, N., Dade, W.B. and Slingerland, R.L. 2002. Climate-driven bedrock incision in an active mountain belt. Science, 297, 2036–38. Reproduced with the permission of the American Association for the Advancement of Science.*

relatively frequent flows of low to moderate intensity (cf. Wolman and Miller, 1960), and that rare, large floods are more important in widening the bedrock channel than they are in driving down the base level. However, such floods help transmit the effect of accumulated thalweg lowering to adjacent hillslopes.

Having probed the details of fluvial bedrock incision, the challenge is now to bridge the gap with incision laws that are relevant to the long-term erosional evolution of orogens (Whipple et al., 2000a). In this respect, it is critical to understand how rivers adapt their local geometry in order to keep up with rock uplift. For a given drainage basin with spatially uniform substrate, precipitation and rock uplift rate, two geometric relationships are crucial. First, the local slope, S, of bedrock channels decreases with increasing upstream drainage area, A, according to a simple power function

$$S = k_s A^{-\theta}, \qquad (1)$$

where k_s is a steepness index, and θ is a slope–area exponent (Snyder et al., 2000; Kirby and Whipple, 2001). For most studied rivers, θ is restricted to values between 0.3 and 0.6, while k_s appears to increase systematically with the mean rock uplift rate of the catchment. Second, channel width, w, increases with upstream drainage area following a power function

$$w = cA^b, \qquad (2)$$

where b is a scaling exponent, ranging between 0.3 and 0.5 (Snyder, 2001; Montgomery and Gran, 2001) and c is a constant. These relationships gloss over some important complications. For example, Lavé and Avouac (2001) found that some river channels in the Siwaliks foothills of the Himalayan chain have responded to locally increased rock uplift rates by narrowing their channels, rather than steepening their slopes (cf. Harbor, 1998; Montgomery and Gran, 2001). Similarly, a river might achieve the increase of bed shear stress required to cope with an increase of sediment supply to the channel either by an increase of channel slope or a decrease of channel width (Howard et al., 1994). At present it is unknown what controls the partitioning between the lateral and longitudinal response of a river channel to changes in substrate properties, uplift rate, discharge, and sediment volume and calibre.

Notwithstanding these complications, some considerable progress has been made in the theory of the long-term evolution of river channels. An early formulation of fluvial bedrock incision was proposed by Howard and Kerby (1983). It states that the vertical incision rate, I, is a function of the flow intensity, ψ

$$I = -K_b(\psi - \psi_c)^{\xi}, \qquad (3)$$

where K_b is an erosion efficiency factor, ψ_c is the critical flow intensity above which incision is possible, and ξ is a positive exponent. The flow intensity is taken as either the basal shear stress, τ, or the unit stream power, $\omega = \tau V$, where V is the flow speed. It is useful to cast eq. (3) in terms of topographic observables such as channel bed slope and drainage area. This is done assuming that the geometric evolution of the channel is robustly captured by eqs (1) and (2). Neglecting the critical flow intensity for channel wear, eq. (3) can be rewritten as

$$I = KA^m S^n, \qquad (4)$$

where K is an erosion efficiency factor depending on substrate properties, basin hydrology, flow geometry and climate, A is the upstream drainage area, S is the local channel gradient, and m and n are positive exponents. This incision model, in either its original or simplified form, has been used in numerous studies (e.g., Seidl and Dietrich, 1992; Howard, 1994, 1998; Howard et al., 1994; Seidl et al., 1994; Rinaldo et al., 1995; Tucker and Slingerland, 1996; Sklar and Dietrich, 1998; Stock and Montgomery, 1999; Whipple and Tucker, 1999; Whipple et al., 1999, 2000a; Tucker and Bras, 2000; Snyder et al., 2000; Kirby and Whipple, 2001; Snyder, 2001; Whipple, 2001; Tucker and Whipple, 2003). Theoretically, the slope exponent n should be $\sim$2/3 for the shear stress model, and 1 for the unit stream power model (Howard et al., 1994; Whipple and Tucker, 1999), but Whipple et al. (2000a) have argued that for hydraulic block removal $n = 2/3 - 1$, and for suspended load abrasion $n = 5/3$. Using data from natural channels to calibrate the incision model, a wide range of exponent values is found (Seidl and Dietrich, 1992; Stock and Montgomery, 1999; Snyder et al., 2000; Whipple et al., 2000a; Kirby and Whipple, 2001): n ranges between 0 and 2, and m ranges between 0.1 and 0.5. These ranges reflect not only the variations in substrate, uplift rate and basin hydrology, and sediment volume and calibre, but also the fact that the threshold of incision, ψ_c, cannot always be neglected (Snyder, 2001; Lavé and Avouac, 2001).

It should be noted that, at topographic steady state, the right-hand term of eq. (4) is equal to the uplift rate, U. This leads to the following relationship between slope and drainage area (cf. Howard, 1980)

$$S = \left(\frac{U}{K}\right)^{1/n} A^{m/n}. \qquad (5)$$

This equation provides a physical basis to the observed slope–area relationship (eq. (1)) in bedrock channels, and allows quantification of the parameters of the incision model (eq. (4)), given knowledge of channel geometry and uplift rate (Snyder et al., 2000). Inversely, eq. (5) permits the estimation of rock uplift rates from measured channel slopes and drainage areas (Lague et al., 2000; Kirby and Whipple, 2001).

4 Hillslope mass wasting

River channels occupy only a minor part of upland landscapes. The bulk of their sediment load is derived from interfluves where bedrock is exposed to physical and chemical weathering processes, driven by climate and modulated by vegetation. These processes cause disintegration of coherent bedrock and the formation of regolith, and may selectively remove or modify some mineral components. Given sufficient topographic energy, weathering products are eroded from the interfluves by hillslope mass wasting processes, whose rate is thought to depend on the local surface gradient as well as the probability distributions of their triggers. Eventually, the eroded material is transferred onto the valley floor, where its removal is a function of the transport capacity of the fluvial system (Hovius et al., 2000).

Hillslope mass wasting is often represented as a diffusion process, in which the hillslope sediment transport rate, Q_s, is proportional to the local topographic slope, and its spatial variation is proportional to the vertical erosion or aggradation rate of the substrate, such that

$$\frac{\partial z}{\partial t} = \kappa \frac{\partial^2 z}{\partial x^2}, \qquad (6)$$

where x is distance from the divide, z is elevation, t is time, and κ is a diffusion coefficient. This expression implies that the steady-state profile of hillslopes dominated by diffusion processes, and underlain by a homogeneous substrate, is parabolic. Thus, the topographic fingerprint of diffusion is a positive correlation between local gradient and upslope area (Figure 5.1). Convex-up hillslopes are common in upland landscapes with low erosion rates, where mass wastage occurs by splash, wash and creep. In tectonically active mountain belts, they tend to be limited to sections of drainage divides not recently affected by slope failure.

Splash, wash and creep are limited by the rate of production of regolith by weathering of intact rock mass. This is a slow process, limited by the kinetics of the chemical reactions involved. Where the rate of valley floor lowering is greater than the rate of regolith production, weathering-limited mass

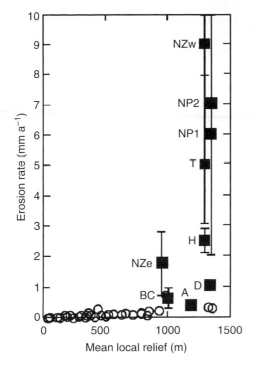

Figure 5.5 *Plot of erosion rate versus mean local relief (measured over 10 km) from mostly tectonically inactive areas (open circles) and tectonically active, convergent areas (solid squares): A is European Alps; BC is British Columbia, Canada; D is Denali portion of the Alaska Range; H is central Himalaya; NP is Nanga Parbat region, western Himalaya; NZ is Southern Alps, New Zealand; and T is Taiwan*
Source: *reprinted from* Earth and Planetary Science Letters, 201, *Montgomery, D.R. and Brandon, M.T., Topographic controls on erosion rates in tectonically active mountain ranges, pp. 481–89.* © 2002, *with permission from Elsevier.*

wasting cannot keep pace with local base-level lowering, and valley sides become progressively undercut. Eventually, this will give rise to landslides involving not only weathered material, but also unweathered rock mass. Bedrock landslides dominate hillslope mass wasting in tectonically active mountain belts. The following is evidence for this scenario.

A power-law relation exists between the rates of erosion and silicate weathering across a range of climates and catchment sizes (Millot et al., 2002), implying that the two are intimately linked, both through weathering-limited mass wasting and the erosional refreshing of the weathering front. But, in a number of active mountain belts, erosion rates are up to an order of magnitude higher than would be expected from measured silicate weathering rates (J. West, personal communication, 2003). There, weathering rates may be at the kinetic limit for a given substrate and climate, a limit that is subdued by the absence of continuous, organic-rich soils, because of relentless mass wasting: active orogens have bedrock landscapes.

These landscapes yield sediment at a rate that is independent of local relief (Montgomery and Brandon, 2002) (Figure 5.5). A well-defined, linear relation exists between erosion rates and local relief, calculated over 10 km, for catchments outside areas of active mountain building (cf. Ahnert, 1970). Erosion rates in active orogens vary by an order of magnitude whereas mean local relief over 10 km is fairly constant, between 1.0 km and 1.5 km. This implies that topographic slope is not a first-order control on the rate of hillslope mass wasting in active mountain belts, that erosion rates are set, instead, by some external forcing, and that there is a limit to local relief. This limit is imposed by bedrock landslides.

The fundamental relation for the initiation of failure of an infinite slope with angle α is

$$\tan \alpha = \left(\frac{C}{\rho g y \cos^2 \alpha}\right) + \tan \phi, \qquad (7)$$

where C is the cohesion of the rock mass and ρ its density, ϕ is the angle of internal friction, g is the gravitational acceleration and y is the depth of the failure plain. This relation demonstrates how a slope might become unstable if there were a change in any of the parameters involved. For example, the slope may increase through undercutting by river erosion at its base. Similarly, the frictional or cohesive strengths may decrease by weathering of material, seismic shaking or wetting of the rock mass, which also increases the weight of the slide block.

Rock mass strength decreases with increasing spatial scale because of the influence of spatially distributed discontinuities. The mountain-scale strength of the rock mass limits relief in bedrock land-scapes, and landslide-dominated landscapes are close to this limit (Schmidt and Montgomery, 1995). In such landscapes the maximum hillslope height is determined by the spacing between higher-order streams and the bulk mass strength of the interfluves. Given effective fluvial bedrock incision, it may therefore be expected that dry mountain belts have greater relief than their wetter equivalents.

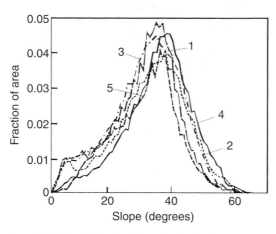

Figure 5.6 *Slope distributions from subregions in the northwestern Himalaya. Slopes were calculated as best-fit planes to a 4 × 4 grid cell matrix in a ~90 m DEM. Areas 1–3 have apatite fission track ages of 0–1 Ma; areas 4 and 5 have apatite fission track ages of 1–6 Ma. Regardless of the ten-fold contrast in denudation rates, implied by these fission track ages, there are few significant differences in slope statistics among them*
Source: *Burbank, D.W., Leland, J., Fielding, E., Anderson, R.S., Brozovic, N., Reid, M.R. and Duncan, C., Bedrock incision, rock uplift and threshold hillslopes in the northwestern Himalayas. © 1996, Nature, Volume 379, pp. 505–10. Reproduced with permission.*

The rock mass control on topographic development was illustrated in a study of the topography of the northwestern Himalaya (Burbank et al., 1996) (Figure 5.6). There, the frequency distributions of slopes were found to be essentially indistinguishable among different mountain regions, despite differences in denudation rates of up to an order of magnitude. In each region, most slopes fell between 20° and 45°, the mean slope was 32 ± 2°, and the modal slope was only marginally steeper. This similarity of slope distributions suggests homogenous topographic characteristics in this landscape, largely independent of denudation variations, and set by rock mass strength. The rapid decrease in the frequency of slopes steeper than 35° implies that such slopes are unstable and prone to collapse. They do not, in general, survive for geomorphologically significant amounts of time. Interestingly, this cutoff value is only slightly higher than the maximum stable slope in loose, dry sand, implying that the rock mass strength in the northwest Himalaya, and probably most other mountain belts, is determined by through-going discontinuities rather than the properties of the intact rocks: to first order, mountains are built of cohesionless material.

If landslides dominate the erosion of active mountain belts, it is important to quantify their long-term impact. Extrapolating short-term geomorphic observations to timescales pertinent to landscape evolution and orogen dynamics requires an understanding of the scaling behaviour of the processes involved, in particular the magnitude and frequency with which they occur (Wolman and Miller, 1960). Landslide size distributions generally exhibit power-law scaling over a limited scale range: the number of landslides is a negative power function of the landslide size (Figure 5.7). This holds true whether the landslide size is defined as the scar area (Hovius et al., 2000) or the total area disturbed (Pelletier et al., 1997), and whether landslides are triggered over a long period of time (Hovius et al., 1997) or almost instantaneously (Harp and Jibson, 1996). For an idealized landslide size distribution to be power-law distributed across the size range, ($x \in [c, \infty)$, the size probability density is defined as

$$p(x) \equiv \beta c^{\beta} x^{-\beta-1}, \quad c > 0, \beta > 0 \qquad (8)$$

where β is the power-law scaling exponent (Stark and Hovius, 2001). The scaling exponent explicitly determines the impact of large versus small landslides on integrated measures such as the total area

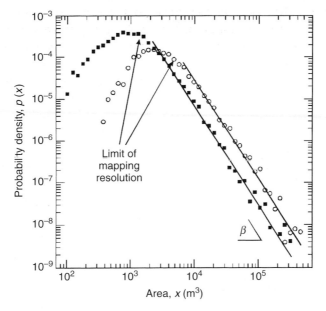

Figure 5.7 *Examples of landslide size distributions from the western Southern Alps, New Zealand, plotted as a probability density function p(x) plotted in log [p(x)] versus log(x) form. Solid squares show the probability density of landslides in the Whataroa catchment, mapped at 1:25 000, N = 3986; open circles show the probability density of landslides in a larger part of the western Southern Alps, mapped at 1:50 000, N = 5086. The datasets show similar scaling of landslide magnitude and frequency. Above a cutoff size, related to the resolution of the mapping and/or a break in the failure mechanism, the data scale as a power-law. This portion of the data is the tail end of the distribution and represents about a quarter of the observed landslides. β is the slope of the best fit power-law, and values are almost identical at β = 1.45 for both datasets*
Source: *reproduced from Stark and Hovius (2001).*

disturbed or the volume of material yielded. In the western Southern Alps of New Zealand, it was found that $\beta \approx 1.44$–1.48; in the Central Mountains of Taiwan $\beta \approx 1.11$ (Stark and Hovius, 2001). These estimates suggest that the total area of landslide disturbance is set by the number of smallest landslides. If it is assumed that mean landslide thickness, y, scales linearly with landslide size, such that $y(x) = \epsilon x$, then the total volume, V, of material yielded by landslides is given by

$$V = 2\beta\epsilon\chi \int_{L_0}^{L_1} x^{2-2\beta}\, dx, \qquad (9)$$

where χ is a rate coefficient, and L_0 and L_1 are the minimum and maximum possible width of a landslide in the region, respectively. From this it follows that when $\beta < 1.5$, and $L_1 \gg L_0$, denudation due to landsliding is dominated by the largest events.

Using parameter values derived from the observed scaling behaviour of landslides in the western Southern Alps, Hovius *et al.* (1997) obtained an estimate for the regional denudation rate because of landsliding of 9 ± 4 mm a^{-1}. This compares well with estimates of erosion by river gauging (Griffiths, 1979), implying that erosion of the western Southern Alps occurs principally through landsliding. The slope of the observed landslide size–frequency distribution demonstrates that the erosional work is done in equal measures by large and small events.

A scenario for the erosion of mountain belts has now emerged in which the rate of bedrock uplift is matched by the rate of valley lowering (steady-state longitudinal river profiles) but surpasses the rate of weathering. Then interfluves grow until topographic elements become unstable and collapse, producing bedrock landslides. Given sufficient transport capacity of the rivers, this type of landscape yields sediment at a rate that is solely determined by the rate of rock uplift. It is characteristic of most active compressional mountain belts and responsible for the production of >80% of all clastic material that is eroded from the present-day continents.

5 Sediment production and routing

The output of sediment from active mountain belts is not uniform in space and time. On geological timescales, changes in tectonic and climatic boundary conditions give rise to changes in sediment delivery at the mountain front (e.g., Zhang et al., 2001; Clift and Gaedicke, 2002). On shorter timescales, the variability of catchment conditions causes variability of sediment production and supply (e.g., Trustrum et al., 1999; Hovius et al., 2000; Gomez et al., 2002). This variability can be treated as a fluctuation of instantaneous erosion rates about a long-term mean. It arises from the stochastic nature of geomorphic systems (Benda and Dunne, 1997). Sediment influx to channel networks is driven by rainstorms, earthquakes and fires of varying intensity and size. These stochasts, with their characteristic probability distributions, occur on landscapes over which there is considerable variability in topography and soil properties, which also vary through time, recording the history of earlier erosion events. Similarly, the onward transport of sediment in rivers occurs as a complex series of pulses, circumscribed in space and time, but with spatial and temporal characteristics that can be quantitatively related to stochastic factors. We illustrate this with examples from Taiwan and the Oregon Coast Range in the USA.

In the eastern Central Range of Taiwan, landslides of all sizes disturb the surface in approximately equal measures (Stark and Hovius, 2001). This size–frequency distribution is not preserved in the downstream sediment loads of rivers draining the mountain belt (Hovius et al., 2000). These loads are strongly dominated by high-magnitude, low-frequency events, but the largest measured events are an order of magnitude smaller than the largest recent landslides in the eastern Central Range. A loosely defined power-law relation exists between flow rate and sediment transport in Taiwanese rivers. Sediment transport rates are highest during episodes of intense precipitation, when hillslope mass wasting rates are high and the transport capacity of the recipient river network is optimal. However, there is a range of water discharges for which sediment transport is possible but not always observed. Rivers in the Central Mountains of Taiwan transport sediment when it is supplied from the hillsides. Supply occurs principally during earthquakes and rainstorms, and its magnitude scales with the magnitude of seismic and climatic triggers of mass wasting. Once in the channel, removal of hillslope debris occurs at a pace set by the transport capacity of the stream at the site of supply. Most sediment is removed within weeks from the trigger event, but the largest landslide deposits may persist for longer (Hovius et al., 2000).

Evidence is now emerging of the importance of the concatenation of erosion triggers. Stream gauging records on major Taiwanese rivers (Water Resources Agency (WRA), 1970–2002) show that the sediment yield from the Taiwan orogen has doubled since a M_w (moment magnitude) 7.6 earthquake struck central west Taiwan in 1999. This earthquake triggered an estimated 22 000 landslides, and average sediment concentrations in river water have increased by up to 8000 ppm in the epicentral area. Notably, erosion rates were lower than average during the first year after the earthquake, but have since risen dramatically because of the remobilization of landslide debris during typhoons. The amount of sediment shifted during these storms has been much larger than expected from prior relations of water discharge and sediment load, indicating that co-seismically generated debris has remained in the landscape. The sediment cascade from valley side to mountain front may have many steps, and sediment production and transfer together determine the output at the mountain front.

Benda and Dunne (1997) have outlined a similar cascade of sediment for the Oregon Coast Range where sediment is introduced to channel networks by landsliding and debris flows. Colluvium mantles planar portions of the hillslopes and migrates downhill into stream channels or bedrock hollows. In these hollows, colluvium, stabilized by tree roots, accumulates over millennia until root strength is no longer capable of retaining it when pore pressure is elevated by large rainstorms. Wedges of colluvium may become mobilized as shallow landslides, the average frequency of which is controlled by the rate of colluvium production. Erosion resets the soil and a new cycle of thickening begins. In the Coast Range, wildfires kill a portion of the forest in a basin, resetting the forest age, and promoting slope instability. Clustering of erosive debris flows and landslides occurs in the affected area for a period of years following a wildfire. The number of potential landslide sites, the probability of storms intersecting recent fires, and hence the frequency of landsliding, increase with increasing basin area. If the basin area is smaller than or equal to the size of most fires, then the frequency of landslide episodes is similar to the frequency of fires. If the basin is larger than commonly occurring fires, then landslide frequency will increase at the rate at which the occurrence of vegetation disturbance events increases with basin area (Figure 5.8). This last point is important and can be phrased more generally: variability of sediment delivery at the mountain front is maximal for catchments with a size similar to the footprint of the largest possible perturbation, in larger catchments the impact of all perturbations is diluted by downstream admixture of water with relatively low sediment concentrations. In general, the short-term and inter-annual variability of sediment load decreases with increasing catchment size, and the emphasis shifts from high-magnitude, low-frequency sediment discharge events to intermediate magnitude and frequency events (cf. Trustrum et al., 1999).

6 Erosion of mountain belts

Direct measurements of erosion on the scale of an orogen are rare. They are essential to the evaluation of the erosional control on the topographic, structural and thermal evolution of orogens, the exposure pattern of deeply deformed rocks, and the composition of oceans and atmosphere. Such measurements include river gauging data (Fuller et al., 2003), sediment accumulation rates (Métivier et al., 1999; Hall and Nichols, 2002), geomorphic marker surfaces (Abbott et al., 1997), detrital and in situ cosmogenic nuclide concentrations (Vance et al., 2003), and ditto fission track statistics (Brandon and Calderwood, 1990; Kamp and Tippett, 1993; Bernet et al., 2001; see also Caine, this volume). Together these methods cover a range of timescales from 10^0 years to 10^6 years. All are labour-intensive and some require expensive equipment and/or sampling. To cover an entire orogen in sufficient detail is a nontrivial exercise, and the global erosion and exhumation dataset is far from complete.

An alternative approach assumes that channels and hillslopes are intimately coupled such that the rate of surface lowering everywhere is equal to the rate of fluvial bedrock incision, and calculates the latter as a function of stream power (eq. (4)). At the orogen scale it is unrealistic to model erosion rates directly considering the lack of data on the spatial distribution of K values, which vary with channel characteristics, rock type, structure and sediment transport rates. Instead, these local determinants of K have sometimes been folded into an erosion index, $I_E = I/K$, which serves as an incision rate proxy. By varying the exponents m and n in eq. (4), different incision models can be applied. Thus, erosion intensity can be mapped across a mountain belt based on the product of local slope and discharge, determined by summing the annual precipitation over the upslope area in a DEM, and holding K constant. This was done, for example, for the Andes (Montgomery et al., 2001) and the Himalaya (Finlayson et al., 2002).

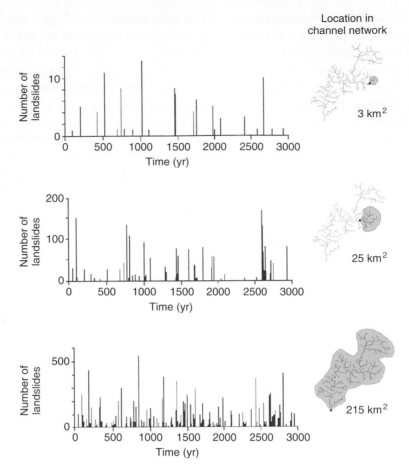

Figure 5.8 *Timing and number of landslides predicted to occur in three model basins of different size, perturbed by a series of triggers as described in the text. Frequency and magnitude (number of failures) of landsliding episodes grow with increasing basin size because of the larger number of landslide source areas and the increasing probability of a landslide trigger*
Source: *Benda, L. and Dunne, T., Stochastic forcing of sediment supply to channel networks from landsliding and debris flow.* Water Resources Research, *vol. 33, pp. 2849–63. © 1997 American Geophysical Union. Reproduced with permission of American Geophysical Union.*

In the Andes, the pattern of I_E values shows that the zone of maximum predicted erosion is on the east side of the range in the northern Andes and on the west side in the southern Andes (Montgomery et al., 2001). Few, localized areas of high I_E were predicted in the central Andes. This 'erosion' pattern matches latitudinal variations in moisture delivery resulting from prevailing wind directions, and implies that the trade winds and polar jet stream affect the orogen-scale morphology of the Andes. Where precipitation rates and associated erosion indices are high, the mountain belt is narrow, and exposure of crystalline rocks reflects deep exhumation. The preserved sedimentary and volcanic cover of the central Andes indicates that exhumation there has been minimal, and the development of a broad, elevated plateau, the Altiplano, implies that tectonic thickening has outpaced erosional mass removal on geological timescales. Its position in the southern arid belt suggests that this dominance of tectonic shortening was due at least in part to climatic limits to erosion.

Similarly, the spatial pattern of the erosion index in the Himalaya is closely tied to the asymmetric distribution of precipitation across the range, and the downstream evolution of channel gradient along the main, trans-Himalayan rivers (Finlayson *et al.*, 2002). Focused erosion of the south flank of the mountain range is associated with steep, incised river sections, the most dramatic examples of which are found where the Tsangpo and Indus rivers flow from the elevated trans-Himalayan valleys into the south flank of the range. These syntaxes contain young, metamorphic massifs with some of the highest recorded exhumation rates. Zeitler *et al.* (2001) have proposed that tectonic and isostatic rock uplift beneath the syntaxial erosion foci has advected hot and therefore weak material from the mid-crust to the surface, leading to a positive feedback in which rock uplift is locally accelerated, thereby producing and maintaining high topography in the face of rapid erosion.

Although appealing in scope and simplicity, erosion index studies ignore the effect of varying rock mass properties within orogenic systems. The importance of this control on the topographic expression of the interaction of tectonic and climatic processes has come to light in an analysis of a rare and comprehensive dataset on sediment flux from an active mountain belt, assembled in Taiwan (WRA, 1970–2002). There, fluvial suspended sediment discharge has been measured since 1970 and on average every 14 days at around 150 hydrometric stations across the island. These measurements have provided the basis for estimates of fluvial erosion rates (Li, 1976; Fuller *et al.*, 2003). Between 1970 and 1999, the annual average suspended sediment discharge from Taiwan has been around 380 Mt, or 1.9% of the global suspended sediment discharge to the oceans (Milliman and Syvitski, 1992) from 0.024% of Earth's subaerial surface. Up to 30% bedload discharge may be added to this figure, according to bathymetric surveys of high mountain reservoirs. Notably, the sum of suspended and bedload transport is approximately equal to the accretionary mass flux into the orogen (around 480 Mt a^{-1}) because of convergence of the Philippine Sea plate and the Asian continental margin. This implies that the Taiwan orogen is close to flux steady state (Willett and Brandon, 2002).

The adequacy of stream power-based estimates of erosion rate can now be tested against detailed measurements of erosion. The spatial pattern of stream power, ω, in Taiwan emphasizes the area of high elevation and orographic precipitation along the Central Range. It does not predict the first-order pattern of measured erosion, with high erosion at the mountain fronts and an order of magnitude increase from north to south. Moreover, there is little correlation between the average stream power computed for each catchment and its erosion rate (Figure 5.9(A)). This suggests that, when applied at the orogen scale and over timescales much longer than that of a typical flood event, stream power is not the unmoderated, rate-limiting control on the erosion of Taiwan.

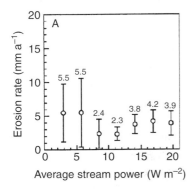 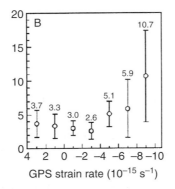 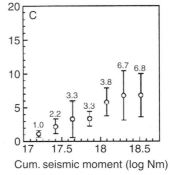

Figure 5.9 *Controls on erosion rates in 130 drainage basins in the Taiwan orogen. Drainage basin erosion rate as a function of (A) basin-wide mean stream power; (B) basin-wide mean dilatation rate, computed as the divergence of a velocity solution inverted to minimize jointly the strain rate and the misfit to observed GPS velocities measured between 1990 and 1995; and (C) cumulative scalar seismic moment, M_0. Error bars indicate 95% confidence interval*

While stream power may determine the erosivity of a river, its efficacy is moderated by the erodibility of the substrate. In Taiwan, compressive rock strengths vary by over three orders of magnitude. It is therefore not reasonable to assume uniform substrate properties across the mountain belt. Given a linear stream power model, $E = k_e \omega$, measurements of erosion rate, E, and stream power, ω, can be used to calculate the value of the constant of proportionality, k_e. Strictly, k_e represents an erodibility, and comparison of its map pattern with the geological map of Taiwan shows that k_e reflects rock mass properties. In particular, the range of values of k_e is similar to that of measured rock strengths and, generally, high values of k_e coincide with low values of compressive rock strength. Weak rocks will never build steep and high relief, but they do permit very high erosion rates at low topographic gradients. This is not a subtlety but a significant characteristic of orogenic systems. It dictates that, in active mountain belts, the link between topography and erosion is overshadowed by substrate properties, that steep mountain relief can only form once weak cover rocks have been stripped off a deforming orogenic pile, and that the erosional mass flux may be determined solely by the tectonic mass flux into the orogen.

This finding provides an orogen-scale, quantitative demonstration of the behaviour of a limit landscape at steady state, in which there is a balance between crustal uplift and fluvial erosion, with efficient adjustment of hillslopes through bedrock landsliding. Constraints on the pattern of tectonic forcing can be obtained, among others, from GPS measurements of surface deformation and the record of historic seismicity. In Taiwan, both measures of tectonic forcing track, in some considerable detail, the present pattern of erosion. Areas of rapid crustal shortening and elevated seismicity correspond well with areas of rapid erosion, and orogen-scale trends in all three variables coincide. A plot of contraction rate and erosion rate (Figure 5.9(B)) shows a linear, four-fold increase in erosion rate over the range of contraction rates. Similarly, a plot of cumulative seismic moment release and erosion rate (Figure 5.9(C)) shows a six-fold linear increase in erosion rate over the moment range. In the Taiwan orogen, erosion matches the pattern and rate of tectonic advection of rock mass.

7 Erosion and orogenesis

Erosion plays a key role in modifying the geologic expression of orogenic processes. It is frequently the principal mechanism for exhuming deeper levels of a convergent orogenic belt, thereby providing the surface expression of metamorphism at depth. In addition, a system with feedback exists in which deformation leads to surface uplift, surface uplift leads to an increase in erosion forced by the impact of orographically enhanced precipitation on steepened relief, and erosion in turn changes the excess topography and modifies the crustal deformation field (Willett, 1999; Willett et al., 2001).

Critical wedge models (Davis et al., 1983; Dahlen, 1984) are a convenient medium for the evaluation of the role of erosion in orogenesis. In these models, plastic (frictional) material in front of a moving 'backstop' deforms to attain the minimum taper angle at which it can slide on its base. The large-scale topographic shape of this wedge reflects a balance between gravitational stresses arising from the surface slope and basal shear stresses resulting from underthrusting of stronger, more rigid material beneath it. Surface slope, and hence excess topography, develop to increase stresses until material throughout the wedge is at the point of plastic failure. Continued accretion induces deformation throughout the wedge leading to an increase in size of the wedge without a change in the cross-sectional shape. Although these critical wedge models were developed for application to accretionary wedges and fold-and-thrust belts, they have been used to model orogenic belts at a larger scale (Jamieson and Beaumont, 1988; Willett et al., 1993). At the scale of an orogen, the deforming domain does not take the simple form of a wedge, but has a sectional shape that reflects the strength and physical properties of the entire deforming crust. However, the principle of critical topographic form still holds even at this largest scale (Willett, 1999).

Critical wedge theory has several important implications. First, the crustal rheology determines the topographic form of a developing orogen. Consider, for example, the case of convergence without erosion, approximated in the arid central Andes. Deformation is initially localized at the plate boundary. With increasing convergence, the region of deformation propagates outward with the formation of back-to-back critical wedges, and crustal thickening leads to warming of the lower crust and reduction of its viscosity. Ultimately, lateral, viscous flow in the lower crust prevents further thickening of the orogenic pile and formation of a high-elevation plateau in the orogen interior ensues.

Second, the balance between the tectonic mass flux and the erosional mass flux determines the growth of the orogen. When the two are equal, the orogen is at flux steady state, as approximated by Taiwan. Then, the orogen is unlikely to change in size and its deformation fronts may be stationary. When the tectonic flux is greater than the erosional flux, the orogen must grow by outward propagation of one or both deformation fronts. In contrast, when the tectonic flux is smaller than the erosional flux, the orogen will contract its zone of active deformation and previous deformation fronts will cease.

Third, erosion causes localization of deformation and rock uplift within an active orogen (cf. Beaumont et al., 2001). Where focused erosion reduces the surface taper of an orogen, the crustal deformation field will adjust so that critical taper is restored through enhanced rock uplift. This may give rise to narrow windows in mountain belts through which rock mass is advected and eroded, as is the case in the Southern Alps of New Zealand (Walcott, 1998; Willett, 1999). The Southern Alps lie east of the Alpine Fault, a transform Fault which marks the boundary between the Australian and Pacific plates (cf. Williams, this volume). They form a topographic barrier across the moisture-laden winds coming off the Indian Ocean and the Tasman Sea, and orographically forced precipitation rates on the west flank of the mountain belt are among the highest in the world ($>10 \text{ m a}^{-1}$). Precipitation rates are lower by up to an order of magnitude across the main divide, and a strong contrast in landscape and erosion rates is the result. Erosion rates of up to 10 mm a^{-1} have been recorded in the steep west flank of the Southern Alps (Griffiths, 1979), and exhumation of rocks from a maximum depth of 20 km has occurred at similar rates for several million years (Tippett and Kamp, 1993). Oligoclase- and garnet-bearing schists are exposed in a narrow zone immediately east of the Alpine Fault. Further to the east, the metamorphic grade of rocks exposed at the surface decreases rapidly over a distance of a few tens of kilometres to low-grade metasediments. This outcrop pattern, and the accompanying landforms, indicate that rock uplift and erosion-driven exhumation occur in a narrow zone in the west flank of the orogen, and essentially on a single, long-lived structure, the Alpine Fault. This is the result of the persistent and opposite directionality of the tectonic and atmospheric fluxes in the region.

Finally, changes in climatic and tectonic boundary conditions give rise to a predictable deformational and topographic response. For example, a decrease in precipitation would give rise to a decrease in erosion, and an increase in the steady-state width and height of an orogen. An increase in tectonic flux for a steady climate would have a similar topographic effect.

The largest, systematic perturbation to impact the Earth's surface in the recent geological past has been the global shift toward a cooler, and supposedly more erosive, climate (Molnar, 2001) in the late Tertiary and Quaternary. Molnar and England (1990) have proposed that this shift has enhanced topographic relief in mountain ranges. They argued that the isostatic response to increased erosion of valley floors would raise mountain peaks higher, promoting further cooling. This positive feedback mechanism introduces the question: which came first, global cooling or uplift? The existence of this feedback depends critically on whether or not the transition to a more erosive climate enhances relief in mountain ranges (Small, 1999). Whipple et al. (1999) evaluated this question in the context of limit landscapes. They argued that in many mountain belts critically steep hillslopes had been attained before the late Cenozoic climate deterioration. Thus, accelerated valley lowering would draw down the mountain ridges and leave local relief unchanged. Glaciation of fluvially sculpted valleys could have resulted in an increase of local relief by up to several hundred metres, through buttressing and formation of hanging valleys. Away from glaciers, Whipple *et al.* (1999) proposed that increased fluvial erosivity would give rise to an increased drainage density, producing in turn a decrease in hillslope relief. They concluded that the late Cenozoic climate change has not resulted in elevation of mountain peaks, but ignored the fact that fluvial erosivity was enhanced as a result of increased storminess of the Quaternary climate superimposed on an overall decrease in precipitation. Increased aridity may have resulted in a decrease in drainage density and a commensurate lengthening of hillslopes. A detailed, modelling-driven analysis of landscape response to changes in average conditions and climate variability is required to resolve this debate.

8 Outlook

Mountain geomorphology and geodynamics is a field that does not consist of one mountain that can be scaled with systematic, expeditionary effort. Rather, it is made of many peaks, each presenting a unique set of problems and requiring specific skills. Among the outstanding challenges we recognize the following. First, there is an ever-growing capability of capturing and rendering surface topography. Regional DEMs are now common currency and have driven rapid progress in terrain analysis. Most analyses have stuck closely with the area–slope paradigm. Much stands to be gained from the exploration of other topographic attributes and alternative methods (cf. Stark and Stark, 2001). Moreover, in the trade-off between coverage and resolution, we have so far sided with coverage. Very high-resolution, local DEMs can be made with equipment readily available in engineering and the medical sciences. Such DEMs can support comprehensive, real-time process studies, and will be especially useful in constraining river channel dynamics and erosion by rivers and debris flows.

Second, many orogens have a long history of submarine construction and erosion before they emerge. The seeds of the subareal evolution of such orogens are sown below sea level, but the emergence and persistence of submarine relief have remained undocumented. In addition, many emergent mountain belts have a submarine continuation, and their dynamics can only be understood in full by considering the connectivity of onshore and offshore domains. This is especially important in the context of sediment supply to basins associated with active orogens.

Third, there is very little collective knowledge of the dissolved fluxes from active orogens (for a review see Caine, this volume). Although of subsidiary importance in terms of the erosional mass budget of orogens, weathering and dissolved transport are major controls on the composition of oceans and the atmosphere. Moreover, an intimate, but poorly resolved coupling exists between chemical and physical processes in erosional environments. To complement the existing suspended sediment load data, systematic and comprehensive measurements of dissolved loads in mountain rivers should be given priority. Similarly, it is desirable to have constraints on the production and routing of particulate and dissolved organic matter in mountain belts.

Fourth, it is now clear that substrate properties are a first-order control on the rate and style of erosion, and the building of topography, but it is less clear which properties matter when. Traditionally, we have looked at compressive rock strength, but tensile strength may determine rates of abrasion (Sklar and Dietrich, 2001). On a larger scale, discontinuities dominate the properties of rock mass, and control its bulk shear strength. Expertise of substrate properties can be found in engineering geology and should be brought into studies of geomorphic processes and landscape evolution.

Finally, we point out the need for optimization of existing formulations of fluvial channel wear, glacial valley excavation, erosion by debris flows and bedrock landsliding. In addition, the interaction between these processes and the role of thresholds in process systems remains poorly constrained. Improved understanding of these aspects should continue to be fed into landscape evolution and geodynamic models. Ultimately, these models will serve to run the experiments that nature has not currently lined up, and to explore the future consequence of present trends and changes of the erosion of mountain belts.

References

Abbott, L.D., Silver, E.A., Anderson, R.S., Smith, R., Ingle, J.C., Kling, S.A., Haig, D., Small, E., Galewsky, J. and Sliter, W., 1997. Measurement of tectonic surface uplift in a young collisional mountain belt. *Nature*, **385**: 501–507.

Ahnert, F., 1970. Functional relationship between denudation, relief, and uplift in large, mid-latitude drainage basins. *American Journal of Science*, **268**: 243–63.

Beaumont, C., Fulsack, P. and Hamilton, J., 1991. Erosional control of active compressional orogens. In McClay, K.R. (ed.), *Thrust tectonics*. New York: Chapman and Hall, 1–18.

Beaumont, C., Jamieson, R.A., Nguyen, M.H. and Lee, B., 2001. Himalayan tectonics explained by extrusion of a low-viscosity crustal channel coupled to focused surface denudation. *Nature*, **414**: 738–42.

Benda, L. and Dunne, T., 1997. Stochastic forcing of sediment supply to channel networks from landsliding and debris flow. *Water Resources Research*, **33**: 2849–63.

Bernet, M., Zattin, M., Garver, J.I., Brandon, M.T. and Vance, J.A., 2001. Steady-state exhumation of the European Alps. *Geology*, **29**: 35–38.

Brandon, M.T. and Calderwood, A.R., 1990. High-pressure metamorphism and uplift of the Olympic subduction complex. *Geology*, **18**: 1252–55.

Brozovic, N., Burbank, D.W. and Meigs, A.J., 1997. Climatic limits on landscape development in the northwestern Himalaya. *Science*, **276**: 571–74.

Burbank, D.W., Leland, J., Fielding, E., Anderson, R.S., Brozovic, N., Reid, M.R. and Duncan, C., 1996. Bedrock incision, rock uplift and threshold hillslopes in the northwestern Himalayas. *Nature*, **379**: 505–10.

Clift, P. and Gaedicke, C., 2002. Accelerated mass flux to the Arabian Sea during the middle to late Miocene. *Geology*, **30**: 207–10.

Dahlen, F.A., 1984. Noncohesive critical Coulomb wedges: an exact solution. *Journal of Geophysical Research*, **89**: 10125–33.

Davis, D., Suppe, J. and Dahlen, F.A., 1983. Mechanics of fold-and-thrust belts and accretionary wedges. *Journal of Geophysical Research*, **88**: 1153–72.

Densmore, A.L., Anderson, R.S., McAdoo, B.G. and Ellis, M.A., 1997. Hillslope evolution by bedrock landslides. *Science*, **275**: 369–72.

Dietrich, W.E., Wilson, C.J., Montgomery, D.R., McKean, J. and Bauer, R., 1992. Erosion thresholds and land surface morphology. *Geology*, **20**: 675–79.

Fielding, E.J., Isacks, B.L., Barazangi, M. and Duncan, C., 1994. How flat is Tibet? *Geology*, **22**: 163–67.

Finlayson, D.P., Montgomery, R.R. and Hallett, B., 2002. Spatial coincidence of rapid inferred erosion with young metamorphic massifs in the Himalayas. *Geology*, **30**: 219–22.

Fuller, C.W., Willett, S.D., Hovius, N. and Slingerland, R.L., 2003. Erosion rates from Taiwan mountain basins: new determinations from suspended sediment records and a stochastic model of their temporal variation. *Journal of Geology*, **111**: 71–87.

Gaillardet, J., Dupré, B., Louvat, P. and Allègre, C.J., 1999. Global silicate weathering and CO_2 consumption rates deduced from the chemistry of large rivers. *Chemical Geology*, **159**: 3–30.

Gilbert, G.K., 1877. *Geology of the Henry Mountains*. US Geological and Geographical Survey of the Rocky Mountain Region. Washington DC: Government Printing Office.

Gomez, B., Page, M., Bak, P. and Trustrum, N., 2002. Self-organized criticality in layered, lacustrine sediments formed by landsliding. *Geology*, **30**: 519–22.

Griffiths, G.A., 1979. High sediment yields from major rivers of the western Southern Alps, New Zealand. *Nature*, **282**: 61–63.

Hall, R. and Nichols, G., 2002. *Cenozoic sedimentation and tectonics in Borneo: climatic influences on orogenesis*. Geological Society Special Publication 191. London: Geological Society of London, 5–22.

Hallett, B., Hunter, L. and Bogen, J., 1996. Rates of erosion and sediment evacuation by glaciers: a review of field data and their implications. *Global and Planetary Change*, **12**: 213–35.

Hancock, G.S., Anderson, R.S. and Whipple, K.X., 1998. Beyond power: bedrock river incision process and form. In Tinkler, K.J. and Wohl, E.E. (eds), *Rivers over rock: fluvial processes in bedrock channels*. American Geophysical Union, Geophysical Monograph 107, 35–60.

Harbor, D.J., 1998. Dynamic equilibrium between an active uplift and the Sevier river, Utah. *Journal of Geology*, **106**: 181–94.

Harp, E.L. and Jibson, R.W., 1996. Landslides triggered by the 1994 Northridge, California, earthquake. *Bulletin of the Seismological Society of America*, **86**: S319–S332.

Hartshorn, K., Hovius, N., Dade, W.B. and Slingerland, R.L., 2002. Climate-driven bedrock incision in an active mountain belt. *Science*, **297**: 2036–38.

Heimsath, A.M., Dietrich, W.E., Nishizumi, K. and Finkel, R.C., 1997. The soil production function and landscape equilibrium. *Nature*, **388**: 358–61.

Hovius, N., Stark, C.P. and Allen, P.A., 1997. Sediment flux from a mountain belt derived by landslide mapping. *Geology*, **25**: 231–34.

Hovius, N., Stark, C.P., Chu, H.T. and Lin, J.C., 2000. Supply and removal of sediment in a landslide-dominated mountain belt: Central Range, Taiwan. *Journal of Geology*, **108**: 73–89.

Howard, A.D., 1980. Thresholds in river regimes. In Coates, D.R. and Vitek, J.D. (eds), *Thresholds in geomorphology*. Winchester: Allen and Unwin, 227–58.

Howard, A.D., 1994. A detachment-limited model of drainage basin evolution. *Water Resources Research*, **30**: 2261–85.

Howard, A.D., 1998. Long profile development of bedrock channels: interaction of weathering, mass wasting, bed erosion, and sediment transport. In Tinkler, K.J. and Wohl, E.E. (eds), *Rivers over rock: fluvial processes in bedrock channels*. American Geophysical Union, Geophysical Monograph 107, 297–319.

Howard, A.D. and Kerby, G., 1983. Channel changes in badlands. *Geological Society of America Bulletin*, **94**: 739–52.

Howard, A.D., Seidl, M.A. and Dietrich, W.E., 1994. Modeling fluvial erosion on regional to continental scales. *Journal of Geophysical Research*, **99**: 13971–86.

Ijjasz-Vasquez, E.J. and Bras, R.L., 1995. Scaling regimes of local slope versus contributing area in digital elevation models. *Geomorphology*, **12**: 299–311.

Jamieson, R.A. and Beaumont, C., 1988. Orogeny and metamorphism: a model for deformation and pressure–temperature–time paths with applications to the central and southern Appalachians. *Tectonics*, **7**: 417–45.

Kamp, P.J.J. and Tippett, J.M., 1993. Dynamics of Pacific plate crust in the South Island (New Zealand) zone of oblique continent–continent convergence. *Journal of Geophysical Research*, **98**: 16107–18.

Kirby, E. and Whipple, K., 2001. Quantifying differential rock-uplift rates via stream profile analysis. *Geology*, **29**: 415–18.

Koons, P.O., 1989. The topographic evolution of collisional mountain belts: a numerical look at the Southern Alps, New Zealand. *American Journal of Science*, **289**: 1041–69.

Kutzbach, J.E., Guetter, P.J., Ruddiman, W.F. and Prell, W.L., 1989. The sensitivity of climate to Late Cenozoic uplift in southern Asia and the American west: numerical experiments. *Journal of Geophysical Research*, **94**: 18393–407.

Lague, D. and Davy, P., 2003. Constraints on the long-term colluvial erosion law by analyzing slope–area relationships at various tectonic uplift rates in the Siwaliks Hills (Nepal). *Journal of Geophysical Research*, **108** (B2): 2129.

Lague, D., Davy, P. and Crave, A., 2000. Estimating uplift rate and erodibility from the area–slope relationship: examples from Brittany (France) and numerical modelling. *Physics and Chemistry of the Earth*, **25**: 543–48.

Lavé, J. and Avouac, J.P., 2001. Fluvial incision and tectonic uplift across the Himalayas of central Nepal. *Journal of Geophysical Research*, **106**: 26561–91.

Li, Y.H., 1976. Denudation of Taiwan island since the Pliocene epoch. *Geology*, **4**: 105–107.

Lyons, W.B., Nezat, C.A., Carey, A.E. and Hicks, D.M., 2002. Organic carbon fluxes to the ocean from high-standing islands. *Geology*, **30**: 443–46.

Mackin, J.H., 1963. Rational and empirical methods of investigation in geology. In Albritton, C.C. (ed.), *The fabric of geology*. Stanford, CA: Freeman-Cooper.

Métivier, F., Gaudemer, Y., Tapponnier, P. and Klein, M., 1999. Mass accumulation rates in Asia during the Cenozoic. *Geophysical Journal International*, **137**: 280–318.

Milliman, J.D. and Syvitski, J.P.M., 1992. Geomorphic/tectonic control of sediment discharge to the ocean: the importance of small mountainous rivers. *Journal of Geology*, **100**: 525–44.

Millot, R., Gaillardet, J., Dupré, B. and Allègre, C.J., 2002. The global control of silicate weathering rates and the coupling with physical erosion: new insights from rivers of the Canadian Shield. *Earth and Planetary Science Letters*, **196**: 83–98.

Molnar, P., 2001. Climate change, flooding in arid environments, and erosion rates. *Geology*, **29**: 1071–74.

Molnar, P. and England, P., 1990. Late Cenozoic uplift of mountain ranges and global climate change. *Nature*, **346**: 29–34.

Montgomery, D.R., 2001. Slope distributions, threshold hillslopes, and steady-state topography. *American Journal of Science*, **301**: 432–54.

Montgomery, D.R. and Brandon, M.T., 2002. Topographic controls on erosion rates in tectonically active mountain ranges. *Earth and Planetary Science Letters*, **201**: 481–89.

Montgomery, D.R. and Foufoula-Georgiou, E., 1993. Channel network representation using digital elevation models. *Water Resources Research*, **29**: 1178–91.

Montgomery, D.R. and Gran, K.B., 2001. Downstream variations in the width of bedrock channels. *Water Resources Research*, **37**: 1841–46.

Montgomery, D.R., Abe, T.B., Buffington, J.M., Peterson, N.P., Schmidt, K.M. and Stock, J.D., 1996. Distribution of bedrock and alluvial channels in forested mountain drainage basins. *Nature*, **381**: 587–89.

Montgomery, D.R., Balco, G. and Willett, S.D., 2001. Climate, tectonics, and the morphology of the Andes. *Geology*, **29**: 579–82.

Pacala, S.W., Hurtt, G.C., Baker, D., Peylin, P., Houghton, R.A., Birdsey, R.A., Heath, L., Sundquist, E.T., Stallard, R.F., Ciasis, P., Moorcroft, P., Caspersen, J.P., Shevliakova, E., Moore, B., Kolmaier, G., Holland, E., Gloor, M., Harmon, M.E., Fan, S.M., Sarmiento, J.L., Goodale, C.L., Schimel, D. and Field, C.B., 2001. Consistent land- and atmosphere-based U.S. carbon sink estimates. *Science*, **292**: 2316–20.

Pelletier, J.D., Malamud, B.D., Blodgett, T.A. and Turcotte, D.L., 1997. Scale-invariance of soil moisture variability and its implications for the frequency–size distribution of landslides. *Engineering Geology*, **48**: 254–68.

Prell, W.L. and Kutzbach, J.E., 1992. Sensitivity of the Indian monsoon to forcing parameters and implications for its evolution. *Nature*, **360**: 647–52.

Raymo, M.E. and Ruddiman, W.F., 1992. Tectonic forcing of late Cenozoic climate. *Nature*, **359**: 117–22.

Rinaldo, A., Dietrich, W.E., Rigon, R., Vogel, G.K. and Rodriguez-Iturbe, I., 1995. Geomorphological signatures of varying climate. *Nature*, **374**: 632–35.

Schmidt, K.M. and Montgomery, D.R., 1995. Limits to relief. *Science*, **270**: 617–20.

Seidl, M.A. and Dietrich, W.E., 1992. The problem of channel erosion into bedrock. *Catena, Supplement*, **23**: 101–24.

Seidl, M.A., Dietrich, W.E. and Kirchner, J.W., 1994. Longitudinal profile development into bedrock: an analysis of Hawaiian channels. *Journal of Geology*, **102**: 457–74.

Sklar, L. and Dietrich, W.E., 1998. River longitudinal profiles and bedrock incision models: stream power and the influence of sediment supply. In Tinkler, K.J. and Wohl, E.E. (eds), *Rivers over rock: fluvial processes in bedrock channels*. American Geophysical Union, Geophysical Monograph, **107**, 237–60.

Sklar, L. and Dietrich, W.E., 2001. Sediment and rock strength controls on river incision into bedrock. *Geology*, **29**: 1087–90.

Small, E., 1999. Does global cooling reduce relief? *Nature*, **401**: 31–33.

Snyder, N.P., 2001. *Bedrock channel response to tectonic, climatic, and eustatic forcing*. Unpublished PhD Thesis. Cambridge MA: Massachusetts Institute of Technology.

Snyder, N.P., Whipple, K.X., Tucker, G.E. and Merritts, D.J., 2000. Landscape response to tectonic forcing: digital elevation model analysis of stream profiles in the Mendocino triple junction region, northern California. *Geological Society of America Bulletin*, **112**: 1250–63.

Stark, C.P. and Hovius, N., 2001. The characterization of landslide size distributions. *Geophysical Research Letters*, **28**: 1091–94.

Stark, C.P. and Stark, G.J., 2001. A channalization model of landscape evolution. *American Journal of Science*, **301**: 486–512.

Stock, J.D. and Montgomery, D.R., 1999. Geologic constraints on bedrock river incision using the stream power law. *Journal of Geophysical Research*, **104**: 4983–93.

Tinkler, K.J. and Wohl, E.E. (eds), 1998. *Rivers over rock: fluvial processes in bedrock channels*. American Geophysical Union, Geophysical Monograph 107.

Tippett, J.M. and Kamp, P.J.J., 1993. Fission track analysis of the Late Cenozoic vertical kinematics of continental Pacific crust, South Island, New Zealand. *Journal of Geophysical Research*, **98**: 16119–48.

Trustrum, N.A., Gomez, B., Page, M.J., Reid, L.M. and Hicks, D.M., 1999. Sediment production, storage and output: the relative role of large magnitude events in steepland catchments. *Zeitschrift für Geomorphologie, Supplementband*, **115**: 71–86.

Tucker, G.E. and Bras, R.L., 1998. Hillslope processes, drainage density, and landscape morphology. *Water Resources Research*, **34**: 2751–64.

Tucker, G.E. and Bras, R.L., 2000. A stochastic approach to modeling the role of rainfall variability in drainage basin evolution. *Water Resources Research*, **36**: 1953–64.

Tucker, G.E. and Slingerland, R.L., 1996. Predicting sediment flux from fold and thrust belts. *Basin Research*, **8**: 329–49.

Tucker, G.E. and Whipple. K.X., 2003. Topographic outcomes predicted by stream erosion models: sensitivity analysis and intermodel comparison. *Journal of Geophysical Research*, in press.

Vance, D., Bickle, M.J., Ivy-Ochs, S. and Kubik, P.W., 2003. Erosion and exhumation in the Himalaya from cosmogenic isotope inventories of river sediments. *Earth and Planetary Science Letters*, **206**: 273–88.

Walcott, R.I., 1998. Modes of oblique compression; late Cenozoic evolution of New Zealand. *Royal Astronomical Society Geophysical Journal*, **52**: 137–64.

Water Resources Agency, 1970–2002. *Hydrological year book of Taiwan Republic of China.* Taipei: Ministry of Economic Affairs, Taiwan.

West, J.A., Bickle, M.J., Collins, R. and Brassington, J., 2002. Small-catchment perspective on Himalayan weathering fluxes. *Geology*, **30**: 355–58.

Whipple, K.X., 2001. Fluvial landscape response time: how plausible is steady-state denudation? *American Journal of Science*, **301**: 313–25.

Whipple, K.X. and Tucker, G.E., 1999. Dynamics of the stream-power river incision model: implications for height limits of mountain ranges, landscape response timescales, and research needs. *Journal of Geophysical Research*, **104**: 17647–61.

Whipple, K.X., Hancock, G.S. and Anderson, R.S., 2000a. River incision into bedrock: mechanics, and relative efficacy of plucking, abrasion, and cavitation. *Geological Society of America Bulletin*, **112**: 490–503.

Whipple, K.X., Kirby, E. and Brocklehurst, S.H., 1999. Geomorphic limits to climate-induced increases in topographic relief. *Nature*, **401**: 39–43.

Whipple, K.X., Snyder, N.P. and Dollenmayer, K., 2000b. Rates and processes of bedrock incision by the Upper Ukak River since the 1912 Novarupta ash flow in the valley of Ten Thousand Smokes, Alaska. *Geology*, **28**: 835–38.

Willett, S.D., 1999. Orogeny and orography: the effects of erosion on the structure of mountain belts. *Journal of Geophysical Research*, **104**: 28957–81.

Willett, S.D. and Brandon, M.T., 2002. On steady states in mountain belts. *Geology*, **30**: 175–78.

Willett, S.D., Beaumont, C. and Fulsack, P., 1993. Mechanical model for the tectonics of doubly vergent compressional orogens. *Geology*, **21**: 371–74.

Willett, S.D., Slingerland, R.L. and Hovius, N., 2001. Uplift, shortening, and steady state topography in active mountain belts. *American Journal of Science*, **301**: 455–85.

Wolman, M.G. and Miller, J.P., 1960. Magnitude and frequency of forces in geomorphic processes. *Journal of Geology*, **68**: 54–74.

Zeitler, P.K., Meltzer, A.S., Koons, P.O., Craw, D., Hallett, B., Chaimberlain, C.P., Kidd, W.S.F., Park, S.K., Seeber, L., Bishop, M. and Shroder, J., 2001. Erosion, Himalayan geodynamics, and the geomorphology of metamorphism. *GSA Today*, **11**: 4–9.

Zhang, P., Molnar, P. and Downs, W.R., 2001. Increased sedimentation rates and grain sizes 2–4 Myr ago due to the influence of climate change on erosion rates. *Nature*, **410**: 891–97.

6
Mechanical and chemical denudation in mountain systems

Nel Caine

1 Introduction

The term *denudation* was introduced into the Earth sciences by Charles Lyell in the early nineteenth century to refer to the process whereby removal of surface material leads to the exposure of underlying strata and rock units (Fairbridge, 1968). Now, *denudation* refers to all the weathering and erosional processes that contribute to lowering of the land surface. Since most of the mountain landscape is the product of weathering and erosion, with only relatively small areas occupied by depositional forms, denudation rates and the varied processes which produce them are important to the understanding of mountain geomorphology. Perhaps even more important is the way in which rates of denudation have varied in the recent geologic past, which has clearly contributed to the variety of the mountain landscape. Here, I summarize studies of contemporary denudation rates in mountain landscapes, noting that they have a significance which goes far beyond the mountains themselves, for they prove to be the source of a disproportionate mass of the sediment transported to the Earth's oceans (Milliman and Syvitski, 1992).

2 Procedures

Rates of sediment and solute flux through stream and river systems, representing the mass removed from the continental surfaces, have been used to estimate rates of denudation for more than a century (Geikie, 1868; Dole and Stabler, 1909). In all of this time, the basic procedures for estimating denudation rates have changed little, though there has been considerable improvement in the records on which they are based. In essence, they require records or estimates of water discharge through the river and the concentration of sediments and solutes in the water. The product of water volume and concentration then yields the mass of rock material removed through the river channel.

Usually, records of river flow are maintained as quasi-continuous time series, depending on the time interval used in digitizing the record. In contrast, sampling for solute and sediment concentrations is much less frequent, usually on a daily, weekly or longer interval. This means that short-lived flow events with high sediment concentrations are often not represented in the record and so an empirical rating relationship between water discharge and sediment concentration may be used to estimate sediment yields (Syvitski *et al.*, 2000). Because of the power-law form of the rating, the resulting estimates of yields may be underestimated, though that can be corrected (Ferguson, 1986). More seriously, the sediment concentration–discharge rating is usually weak and so not useful where high sediment concentrations are not associated with high discharges (Caine, 1992a; Syvitski *et al.*, 2000). However, the recent development of turbidity and conductance meters which provide quasi-continuous time series, equivalent to those of discharge, promises improved accuracy and precision (though still with the need for empirical calibration). As yet, such records for mountain basins remain few in number and of short duration.

The product of water discharge and sediment (or solute) concentration yields an estimate of the rate of mass removal from the drainage basin (mass per unit time). When corrected by the average density of rock material (often assumed to be 2.5 Mg m^{-3}) this is converted to a volume of rock material removed which is then divided by the drainage basin area to give a specific yield (mass per unit time per unit area or length per unit time). The result is an average rate of lowering of the land surface (in units of mm ka^{-1}, sometimes referred to as Bubnoff units; Fischer, 1969). Since clastic sediment and solute concentrations are derived separately in sample analyses, the results are usually differentiated as *mechanical denudation* and *chemical denudation*. The former should include both suspended and bedload transport but is often estimated from the suspended sediment yield alone, or with an added fixed proportion (10–15%) to account for bedload. The latter includes material removed as solutes within river discharges and needs to be corrected for those constituents which are derived from the atmosphere: salts cycled from the world's oceans or carbonates and bicarbonates produced in weathering reactions. For this reason, chemical denudation is often reported as a cation or dissolved silica yield, i.e., those constituents derived from bedrock and soil weathering.

Estimates of water discharge in natural channels may be accurate to ±10–15% and solute/sediment concentrations probably have equivalent accuracy at the time of measurement or sampling. Errors in interpolation and statistical estimation from rating curves of both volume discharge and sediment (or solute) concentrations compound these and suggest that annual sediment yields may be estimated to ±40% (Griffiths, 1981; Walling and Webb, 1981, 1986; Robertson and Roerish, 1999; Fuller et al., 2003). When they are derived for short periods of record, for small drainage basins or from sparse sampling, errors may be appreciably higher than this. However, it is worth noting that the recorded variability in denudation rates based on sediment yields from drainage basins often amounts to two or more orders of magnitude, i.e., is much greater than the associated errors. Further, because of the difference in the relationship between concentration and discharge, errors on chemical denudation rates may be lower than those on mechanical denudation.

A further concern arises from the use of denudation rates in extrapolation to longer time intervals. Most records of river flow and sediment and solute yields are relatively short (few longer than 50 years) and so rarely include extreme events. This is a particular concern in mountain environments where large catastrophic mass failures, such as those on Huascaran and Mt Cook, and seismic and volcanic events are important landforming processes. Allied to this is the concern that erosion is influenced by land use, especially agriculture (Pimental and Skidmore, 1999) and hydrologic manipulation, especially reservoirs (Graf, 1999), which generally increase and decrease sediment yield, respectively.

It is important to recognize that the discharge of sediment and solutes from river basins does not represent all the geomorphic work that occurs within them. This was acknowledged by early workers on basin dynamics who defined sediment budgets which included sediment production and its transfer and storage within the basin in their research (Jäckli, 1957; Rapp, 1960a; Leopold et al., 1966; Trimble, 1975). Nevertheless, basin yields continue to be widely used, in part because they allow relatively consistent comparisons that can be derived from data available from hydrologic monitoring. The lack of a steady state in sediment fluxes within the basin is often represented by an empirical 'sediment delivery ratio': the proportion of sediment mobilized within the basin that is measured at its outlet. Depending on the topography of the basin and the potential storage within it, this may be <5%. It may even be reflected over long relaxation times, perhaps as long as the Holocene in river systems affected by Pleistocene glaciation (Church and Slaymaker, 1989; Ballantyne, 2002). Within high mountain basins, the accumulation of coarse clastic debris during the Holocene has been recognized for >40 years as representing erosion which is not reflected in sediment yields (Rapp, 1960b). In fact, the debris in alpine talus slopes and rock glaciers may remain stored through an entire interglacial (Wahrhaftig, 1987).

Denudation rates have also been estimated by a variety of other, not always independent, observations. The most obvious of these, and the one with the longest history, involves the inverse of the denudation process by considering sedimentation records in lakes and reservoirs (Souch and Slaymaker, 1986; Dearing and Foster, 1993; Owens and Slaymaker, 1993) and in the coastal and continental shelf environments (Leeder, 1997; Carter et al., 2002). These approaches appear to involve error estimates that are equivalent to those associated with denudation rates derived from stream sediment yields, depending on the accuracy of surveys, the trapping efficiency of the lake and the density of the accumulating sediments (Verstraeten and Poesen, 2002). These studies often show lower average rates of surface lowering over geologic time than those estimated for the present (Gardner et al., 1987) though that pattern may not always be true (Kirchner et al., 2001).

Estimates of denudation rates based on the dissection of surfaces of known age, such as glacial till sheets (Clayton, 1997), limestone pavements (Sweeting, 1965; André, 1996a) and tombstones (Cooke et al., 1995) all provide estimates that are difficult to extrapolate to longer time periods, although the extrapolation to larger areas has greatly improved with the use of digital terrain models and absolute dating techniques.

However, developments in the dating of bedrock surfaces and surficial material in the last decade represent the greatest improvement in defining rates of denudation during the late Cenozoic era. K-Ar dating of lavas and the magnitude of dissection on volcanic cones had been used earlier to define erosion rates (Ruxton and McDougall, 1967). More recently, exposure dating of bedrock through the accumulation of cosmogenic nuclides produced *in situ* (Harbor, 1999) is a development that allows the estimation of bedrock lowering rates over long time intervals and the definition of denudation rates on a drainage basin scale (Bierman and Steig, 1996; Kirchner et al., 2001). On even longer timescales, fission-track analysis of apatite has been successfully used to define the long-term tectonic and denudational history of continental margins and mountain chains (Pazzaglia and Kelley, 1998; Bernet et al., 2001).

3 Continental denudation rates

Even when estimated for large drainage basins, rates of denudation are highly variable (Milliman and Syvitski, 1992; Summerfield and Hulton, 1994; Hovius, 1998). Mechanical denudation in the world's largest river systems varies by two orders of magnitude, from 4 to >500 mm ka^{-1} (Summerfield and Hulton, 1994; Hovius, 1998). Chemical denudation rates on a global scale are both lower and less variable than this (Figure 6.1) but still range across an order of magnitude, falling within the range 1–30 mm ka^{-1}. This variability is greatly increased when denudation rates are calculated for small drainage basins, especially those in regions of high relief. Environmental controls on rates of denudation have often been sought in climate, as reflected in effective precipitation, runoff or vegetation (Wilson, 1973; Jansson, 1988). Other studies emphasize the importance of topography and tectonics as controls of denudation rates (Schumm, 1963; Ahnert, 1970; Pinet and Souriau, 1988; Milliman and Syvitski, 1992), especially in tectonically active mountain systems where the interaction of uplift and climate remains a subject of speculation (Molnar and England, 1990; Raymo and Ruddiman, 1992; Whipple et al., 1999; Hovius et al., this volume).

On the basis of data from drainage basins >25000 km^2, Hovius (1998) defines five empirical controls of denudation rates by linear regression: maximum elevation, mean annual temperature, temperature range, annual precipitation (which all have a direct influence on rates) and basin area (an

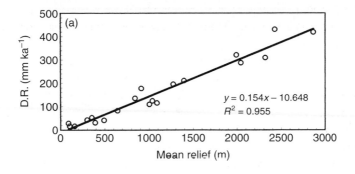

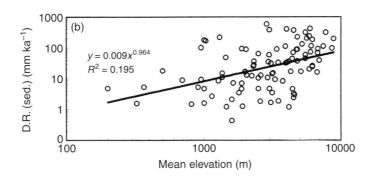

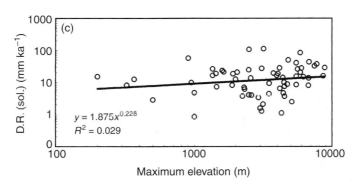

Figure 6.1 *Mountain relief and elevation and denudation rates. (a) Basin relief and denudation rates. Data from Ahnert (1970). (b) Mechanical denudation and maximum elevation within the basin. Data from Milliman and Syvitski (1992). (c) Chemical denudation and maximum elevation within the basin. Data from Milliman and Syvitski (1992)*

inverse effect). The effect of maximum elevation or basin relief on clastic sediment yields in these basins indicates the importance of mountain systems as sources of sediment even though this is not reflected in specific solute yields (Figure 6.1). The inverse relationship between basin area and denudation rates (Figure 6.2) is evident in most global reviews and is usually interpreted as a reflection of increased sediment storage in large basins. Climatic factors of temperature and precipitation are usually collinear with elevation and relief and so their effects are not readily separated. Also clearly influential in global data on sediment yields is the contrast between drainage basins in different tectonic contexts. Hovius (1998) shows that large drainage basins in zones of plate convergence and crustal contraction, i.e., those draining the major mountain chains, show specific denudation rates an

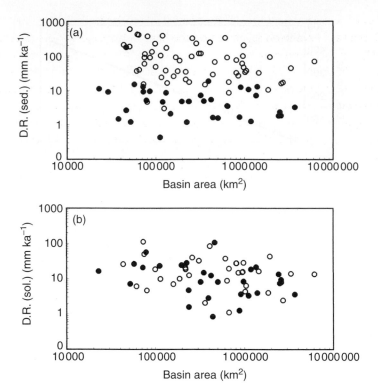

Figure 6.2 *(a) Mechanical and (b) chemical denudation rates versus basin area in contrasting tectonic settings (after Hovius, 1998). Open symbols: tectonic contractional zones; solid symbols: old orogens and cratonic settings Data from Hovius (1998).*

order of magnitude higher than those in more stable tectonic settings (Figure 6.2). This contrast was defined by Ahnert (1970, table 2) who compiled empirical evidence to contrast erosion in regions having uplift rates of more than 100 mm ka^{-1} with those having effectively no recent uplift. Pinet and Souriau (1988) show that the slope of the regression of mechanical denudation rate on mean basin elevation increases six-fold in active tectonic settings.

4 Denudation in mountain systems

4.1 Mechanical denudation

Average surface lowering rates of >1000 mm ka^{-1}, with maxima >10000 mm ka^{-1}, are recorded in the high mountain (>3000 m elevation) and mountain environments of Asia and Oceania (Milliman and Syvitski, 1992). In the western Himalaya, Ferguson (1984) calculates rates of 1800 mm ka^{-1} for the Hunza River, corresponding to high sedimentation in the Tarbela Reservoir on the Indus River (equivalent to a denudation rate of 450 mm ka^{-1}). Within the Indus basin, Burbank and Beck (1991) record rates of land surface lowering at 1000–15000 mm ka^{-1} that have been maintained over 1.5 Ma. For the entire Himalayan system, recent estimates of 2100 mm ka^{-1} for the Ganges drainage and 2900 mm ka^{-1} for the Brahmaputra (Galy and France-Lanord, 2001) are appreciably greater than those based solely on suspended sediment yield (Milliman and Syvitski, 1992; Hovius, 1998). Even higher rates have been estimated for small basins within the Himalaya where denudation rates based on suspended sediment yields range up to 5000 mm ka^{-1} (Ives and Messerli, 1989; West et al., 2002).

Such rapid rates of sediment yield have often been linked to landslide activity and rates as high as 10000–20000 mm ka^{-1} have been suggested on the basis of landslide surveys (Starkel, 1972; Caine and Mool, 1982), though these may be overestimated by a factor of two or three (Ramsey, 1985, quoted in Ives and Messerli, 1989). The importance of agriculture in accelerating erosion has often been suggested but it is noteworthy that equally rapid rates of denudation appear to have been maintained over geologic time (Burbank and Beck, 1991) and so must be driven by a natural system which involves a steep, highly dissected topography, monsoonal precipitation and active tectonics (Ives and Messerli, 1989). They are also associated with high rates of river incision (2–12000 mm ka^{-1}), approximately in equilibrium with uplift rates (Burbank et al., 1996) and with steep bedrock slopes on which catastrophic landslides capable of damming the drainage are a major hazard (Hewitt, 2001, this volume).

Elsewhere in southern and eastern Asia and Oceania, Milliman and Syvitski (1992) record equivalently high rates of denudation in mountain drainage basins. In the Southern Alps of New Zealand, average rates of 2800 ± 2600 mm ka^{-1} in basins draining to the west contrast with lower rates in basins draining to the east (up to 220 mm ka^{-1}) (Griffiths, 1981; see also Figure 4.9 in Williams, this volume). Equivalent rates appear to have been maintained over millions of years of exhumation history in the southern part of the South Island (House et al., 2002). The highest rates are measured in the Cleddau, Haast and Hokitika drainages, where rates of 5000–6600 mm ka^{-1} correspond closely to the estimates of surface lowering due to landsliding in their headwaters (Hovius et al., 1997). Again, this correspondence has been interpreted as representing a steady-state landscape in which steep slopes are modelled by landsliding, from which debris is rapidly evacuated by rivers (Adams, 1980). Similarly high denudation rates have been estimated for rivers draining to Hawke Bay in the North Island which average 2900 mm ka^{-1}, with a maximum of 8000 mm ka^{-1} in the Waiapu River (Griffiths, 1982).

Small basins draining the mountains of Indonesia, New Guinea and the Philippines also show some of the highest denudation rates in the world, with a mean of 2430 ± 2700 mm ka^{-1} (Milliman and Syvitski, 1992). In New Guinea, earlier estimates based on the dissection of K-Ar dated lava surfaces in the Hydrographers Range show a marked response to local relief with lowering rates varying up to 500 mm ka^{-1}, and an extrapolated maximum of 700 mm ka^{-1} over 650000 years (Ruxton and McDougal, 1967). In the Torricelli Mountains, Simonett (1967) estimates surface lowering of between 1000 and 1400 mm ka^{-1} on the basis of landslide surveys from aerial photography. Once more, these rates of mechanical denudation are associated with relatively high levels of landslide activity, seismically triggered on steep slopes which feed directly to the river channels (Simonett, 1967).

Records from the Central Range, Taiwan (Milliman and Syvitski, 1992) also define some of the highest denudation rates in the world (average 2960 with SD 50000 mm ka^{-1}) and these results have been corroborated by recent work. Hovius et al. (2000) estimate a rate of 7700 mm ka^{-1} for the montane part of the Hualien catchment and note that even this is an underestimate since it does not account for bedload transport. Petley and Reid (1999) suggest rates of 4900 mm ka^{-1} and Fuller et al. (2003) report values of 2200–8300 mm ka^{-1} for 11 drainage basins in the eastern Central Range. Once more, the sediment flux represented by such a rate is explained by the importance of landsliding in a system which allows slide debris to reach the river channel and be rapidly exported. These rates of surface lowering are matched by a high rate of tectonic uplift (3000–6000 mm ka^{-1}) and an equivalently high rate of bedrock channel incision: c.3000–4200 mm ka^{-1} (Hartshorn et al., 2002).

On a continental scale in South America, Montgomery et al. (2001) suggest linked patterns of climate, tectonics and erosion along the entire length of the Andes, with rates of excess erosion (derived from the volume of rock above sea level) that vary up to 3000 mm ka^{-1}. Compared with this, estimates based on suspended sediment yields from river systems are small. In the Colombian Andes, Restrepo and Kjerfve (2000) record rates between 500 and 700 mm ka^{-1} in relatively large basins draining steep slopes to the Pacific. Also in the northern Andes, basin-wide erosion rates in the Matacora and Tuy Rivers, Venezuela, amount to 850 and 690 mm ka^{-1}, respectively. These rates are corroborated by other records from the Andes which show a mean rate of mechanical denudation of 400 ± 300 mm ka^{-1} (Milliman and Syvitski, 1992).

In the Caribbean region, denudation rates in the mountains of Puerto Rico are equivalent to those of the northern Andes. In basins of 200 km^2, Milliman and Syvitski (1992) report rates of 650 and 690 mm ka^{-1}. Larsen and Torres-Sanchez (1992) estimate a denudation rate of 160 mm ka^{-1} from landslide surveys following Hurricane Hugo in September 1989 and this is corroborated by the estimate of 140 mm ka^{-1} (Larsen and Parkes, 1997) which corresponds to 80% of the sediment flux from small headwater basins in which sediment storage is slight (230 mm ka^{-1}). This adds support to the conclusion that landsliding is an important connection between the hillslopes and stream channels. In contrast, other processes of mass wasting supply lower volumes of sediment from steep forested hillslopes in the Luquillo Mountains: equivalent to 10–100 mm ka^{-1} for slopewash, and even then the highest rates derive from landslide scars (Larsen et al., 1999). These rates correspond reasonably well to those derived from ^{10}Be studies, such as 43 mm ka^{-1} based on river sediments and 25 and 50 mm ka^{-1} for ridge crest sites (Brown et al., 1995).

For unglaciated drainage basins on the Pacific coast of North America, Milliman and Syvitski (1992) report high rates of denudation: up to 800 mm ka^{-1} with an average of 330 ± 230 mm ka^{-1}. However, natural rates in this environment are slight by comparison with those in catchments disturbed by logging activities. Watersheds 1, 3 and 10 in the Andrews Experimental Forest on the western slope of the Cascade Range show a denudation rate of 24 mm ka^{-1}, of which 75% is associated with debris flows that recur on an interval of centuries (Swanson and Fredriksen, 1982; Caine and Swanson, 1989).

Further south on the Pacific coast, Kelsey (1980) reports rates of up to 120 mm ka^{-1} in the Van Duzen River basin, depending on the estimated residence times of landslide debris, and suggests that they may be as much as two or three times higher in the headwaters of the basin. These rates are associated with episodic landslide failures during intense rainfall events which feed debris to the drainage system from steep hillslopes. Also in northern California, Lehre (1982) derived rates of 10–570 mm ka^{-1}, with an average of 280 mm ka^{-1}, on the basis of a sediment budget for the 1.74 km^2 Lone Tree Creek basin in 1971–74. The three years of record covered different weather conditions with the highest sediment yields in the year with greatest water discharge (20-year recurrence interval) when sediment that had accumulated in the channel was evacuated. Heimsath et al. (1997) estimate long-term landscape lowering rates from cosmogenic nuclides in stream sediments of similar magnitude (60–100 mm ka^{-1}) and show that these correspond to rates of soil production on exposed bedrock over the same kind of timescale. Riebe et al. (2001a) report equivalent rates of long-term erosion (20–60 mm ka^{-1}) in seven basins on granitic rock in the Sierra Nevada.

From studies in drainage basins in British Columbia affected by late Pleistocene glaciation, Slaymaker (1987) was the first to show that denudation rates in mountain catchments do not always show an inverse relationship to basin area and elevation (e.g., Milliman and Syvitski, 1992). He shows that small basins (<100 km^2) have the lowest denudation rates; intermediate-size basins (10^3–10^5 km^2) give yields that are an order of magnitude higher, and the largest basins yield sediment at intermediate rates (Figure 6.3). This pattern is also described by Church and Slaymaker (1989) who interpret it in terms of the effects of past glaciation on contemporary sediment storage and remobilization in a paraglacial context (Church and Ryder, 1972; Ballantyne, 2002). The later study of Owens and

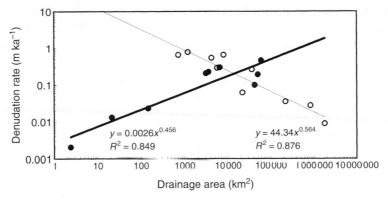

Figure 6.3 *Contrasting patterns of the denudation rate–drainage area relationship in western North America. Open symbols: drainage basins in the Pacific Northwest and Interior Ranges; solid symbols: Pacific drainages in British Columbia*
Data from Slaymaker (1987) and Milliman and Syvitski (1992).

Slaymaker (1993) confirms these results and suggests denudation rates of less than 0.1 mm ka^{-1} in basins <0.2 km^2 area, increasing to 14 mm ka^{-1} at 21.6 km^2 and 200 mm ka^{-1} in the Squamish River (3600 km^2). In the Alaska Range, rates of Holocene bedrock lowering on alpine slopes at about 1000 mm ka^{-1} are indicated by the debris mass accumulated in rock glaciers (Wahrhaftig and Cox, 1959; Barsch, 1996). However, it is important to note that this represents sediment stored within the alpine environment, rather than exported from it.

In the Canadian Rockies, McPherson's (1975) study of 36 basins suggests denudation rates that are appreciably lower than those found in the coastal ranges, described above, with a maximum of 40 mm ka^{-1}. This is corroborated by reviews of denudation in the Canadian Cordillera (Slaymaker and McPherson, 1977; Luckman, 1981) which suggest rates in the range of 40–75 mm ka^{-1}. Further south, Kirchner *et al.* (2001) have compared contemporary estimates of denudation rates in drainage basins of the Northern Rockies (Clayton and Megahan, 1997) with long-term (up to 10^8 years) estimates based on ^{10}Be in stream sediments and apatite fission-track analysis. They show that the geologic estimates are an order of magnitude higher than modern denudation rates (up to 10 mm ka^{-1}) and suggest that this indicates the importance of large episodic events with a frequency too long to be represented in modern studies, even those based on decades of record. In forested environments, infrequent stand-clearing fires such as those in Yellowstone National Park in 1988 are capable of producing an equivalent increase in sediment discharge in subsequent years (Ewing, 1996).

In the alpine environments of the southern Rockies, contemporary denudation rates appear to be generally low. In the Colorado Front Range, sediment flux from the alpine zone suggests denudation rates of less than 0.2 mm ka^{-1} and accounting for lake sedimentation only increases this by a factor of two or three (Caine, 1976, 2001). Such low denudation rates appear to have been maintained over geologic time on the broad upland surfaces of the southern Rockies. During the Pleistocene, Dethier (2001) estimates river incision rates in the Rocky Mountains to range between 20 and 300 mm ka^{-1} in the 640000 years since deposition of the Lava Creek B tephra. These rates are matched by valley incision rates estimated from cosmogenically produced ^{10}Be and ^{26}Al, and contrast with lower rates of bedrock erosion on interfluve crests (8$\pm$4 mm ka^{-1} which increases by a factor of two under a regolith: Small *et al.*, 1997, 1999; Anderson, 2002). On even longer timescales, Pazzaglia and Kelley (1998) suggest average exhumation rates of 120–170 mm ka^{-1} in the Front Range during the Cenozoic. At the present time, geomorphic activity is most active on valley walls where the storage of coarse sediment, largely as talus and rock glaciers, is important (Caine and Swanson, 1989; Caine, 2001). Over the Holocene, coarse debris accumulation within the Green Lakes system is approximately equivalent to a lowering rate of 100 mm ka^{-1}, with rates that are probably an order of magnitude greater on the steepest bedrock cliffs (Caine, 1986). At lower elevations in the Front Range, erosion rates on hillslopes and sediment release to the drainage system increases greatly, as a reflection of vegetation and land-use history (Bovis, 1978).

In the European Alps, Jäckli (1957) provides a classic statement of a sediment budget for the upper Rhine catchment (4307 km^2), including estimates of internal transfers effected by landslides and glacial transport. His data suggest a denudation rate of 930 mm ka^{-1}. Later work, using a variety of procedures, supports this general rate of denudation. For example, Muller and Forstner (1968) report a rate of 220 mm ka^{-1} for the Rhine at Lustenau, above Lake Constance, on the basis of suspended sediment concentrations and river discharge. In high unglacierized catchments in the Alps, recent work continues to emphasize the importance of sediment storage within the valley systems. Thus, Barsch (1996) estimates rates of landscape modification at 2500–3200 mm ka^{-1} on the basis of rock glacier and talus volumes in alpine valleys. In similar fashion, retreat rates on valley walls of the Dolomites have been estimated at between 260 and 5560 mm ka^{-1}, with a best estimate of 1100 mm ka^{-1}, on the basis of debris accumulation (Schrott and Adams, 2002). In the Bavarian Alps, Schrott et al. (2002) record even more rapid rates of valley wall retreat in recent centuries (18000–27000 mm ka^{-1}) but, again, little of the material represented by this activity is exported to lower elevations. On a geologic timescale, these local rates compare with range-wide estimates of exhumation at a quasi-steady state at 400–700 mm ka^{-1} over the past 17 Ma (Bernet et al., 2001).

In northern Europe, the work of Rapp (1960a), like that of Jäckli (1957) in the Alps, represents an early evaluation of sediment budgets in a mountain system. It defines low rates of sediment export from the mountain basin of Kärkevagge in northern Scandinavia and recognizes that clastic sediment flux through streams is not always representative of geomorphic activity within the catchment. In this case, rates of hillslope lowering, largely through internal exchanges within the basin, amount to no more than 50 mm ka^{-1}. The uplands of Britain show equivalently low denudation rates, with averages of 34 ± 19 mm ka^{-1} for the rivers summarized in Milliman and Syvitski (1992).

Denudation rates in basins that are presently glacierized are summarized by Hallet et al. (1996).

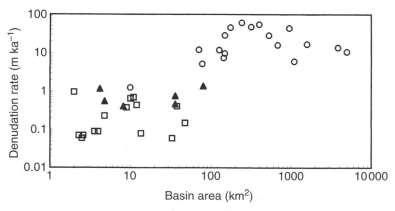

Figure 6.4 *Denudation rates in glacierized basins. Solid triangles: Alps; open squares: Norway; open circles: southeastern Alaska*
Data from Hallet, B., Hunter, L. and Bogen, J., Rates of erosion and sediment evacuation by glaciers: a review of field data and their implications. Global and Planetary Change, *12: 213–225. © 1996, with permission from Elsevier.*

These data show a high variability but define a pattern in which denudation rates increase with total catchment area, and so with the volume of glacier ice in the basin (Figure 6.4). This contrasts the inverse pattern found in many nonglacierized systems (Figure 2; Milliman and Syvitski, 1992). Despite the problems of accounting for bedload transport in proglacial channels (Gomez, 1987; Pearce et al., 2003), the highest rates of denudation in the world are derived from the glacierized mountain catchments of southeastern Alaska and British Columbia, where rapidly sliding, temperate glaciers occupy a large part of the basins. In these catchments, denudation rates of up to 100 m ka^{-1} have been recorded (Hallet et al., 1996). Similarly high rates have been estimated for Himalayan systems, e.g., 5000–7000 mm ka^{-1} for the Rakiot Glacier (Hallet et al., 1996). In contrast, small glaciers with relatively low activity give denudation rates that are one or two orders of magnitude lower. On crystalline rocks in Norway they average only 100–500 mm ka^{-1} (Bogen, 1989) and on varied bedrock in the Alps they are typically within the range 1000–2000 mm ka^{-1} (Bezinge, 1987; Gurnell, 1987; Small, 1987). The small glaciers of the Rocky Mountains match this pattern with relatively low rates of erosion and sediment transport at the present time. Reheis (1975) estimates present rates of denudation on the Arapaho Glacier at 95–165 mm ka^{-1}, largely from supra-glacial sources, which is much lower than the estimate for the Little Ice Age (>5000 mm ka^{-1}), when the volume of ice in the glacier was greater than it is now (Reheis, 1975). The basin of the Ivory Glacier in the Southern Alps, New Zealand, yields a denudation rate of 5600±300 mm ka^{-1}, high on a global scale but not significantly different from rates recorded in nearby unglacierized basins on similar bedrock with equally high precipitation amounts (Hicks et al., 1990).

4.2 Chemical denudation

Globally, chemical denudation rates in mountain environments are usually lower and less variable than those for mechanical denudation (Figures 6.1 and 6.2), though comprehensive studies of both components in the same catchments remain few (Walling and Webb, 1986; Pinet and Souriau, 1988). The clear exception to this occurs in carbonate terrains where solute removal may exceed 100 mm ka^{-1} (Corbel, 1959; Bögli, 1980). Despite low ambient temperatures, rates are frequently higher than the global average, even in glacier systems: a cationic flux of 380 mEq m^{-2} a^{-1} (Souchez and Lemmens, 1987) and a net solute denudation rate of 8 mm ka^{-1} (Hallet et al., 1996).

In the Himalaya, West et al. (2002) estimate net cation fluxes from four small basins, two in the Middle Hills and two in the High Himalaya. Their results suggest denudation rates between 5 and 11 mm ka^{-1} which are produced predominantly by silicate weathering in the Middle Hills and carbonate reactions in the High Himalaya. These authors also point to the importance of continued weathering reactions during downstream transport through the Ganges basin and to the role of mechanical weathering in accounting for high rates of solute denudation. By comparison with the magnitude of clastic sediment fluxes from the range, these denudation rates are low but they are still high on a global scale.

In western North America, Riebe et al. (2001b) have estimated long-term Si-weathering rates in small granitic basins in the Sierra Nevada. On the basis of the enrichment in the regolith of insoluble elements such as zirconium and estimates of mechanical denudation rates from cosmogenic isotopes, they suggest chemical denudation rates of 1–40 mm ka^{-1}, with little correspondence between these rates and present mean temperatures and precipitation (Riebe et al., 2001b). Dethier (1986) provides estimates of contemporary chemical denudation from 41 catchments up to 30000 km^2 in the mountains of the Pacific Northwest. The results suggest cationic fluxes corresponding to denudation rates of 5–20 mm ka^{-1} and dissolved Si denudation rates of 2–13 mm ka^{-1}, which vary directly with runoff up to 4000 mm a^{-1}. In British Columbia, Slaymaker (1987) summarizes total solute yields which correspond to denudation rates of about 80 mm ka^{-1}. In the Yukon River, rates are about 10 mm ka^{-1}. The contrast between the coastal and continental basins corresponds to that defined earlier by McPherson (1975) and Slaymaker and McPherson (1977), who derived solute denudation rates of <60 mm ka^{-1} in the Rocky Mountains of Alberta. In contrast, higher rates have been derived for the karst systems of the Rocky Mountains, where Ford et al. (1981) estimate rates of valley floor lowering at 130 mm ka^{-1}, and perhaps as much as 2000 mm ka^{-1}, on the basis of U-series dating of speleothems.

In the continental mountains of the Central and Southern Rockies, rates of chemical denudation on granitic and metamorphic terrains are also generally low. In Colorado, cation yields from Green Lakes Valley are equivalent to denudation rates of 1 mm ka^{-1} and Si yields are of the same magnitude (Caine and Thurman, 1990; Caine, 2001), resulting in a total chemical denudation rate of 2 mm ka^{-1}. In the same mountain range, preliminary work by Baron and Bricker (1987) suggested a net cation flux from Loch Vale that approximates 3.5 mm ka^{-1} and a Si flux of c.4.5 mm ka^{-1}. Later estimates for the same basin reduce these to a cationic denudation rate of 0.3 mm ka^{-1} and a Si rate of 0.4 mm ka^{-1} (Mast et al., 1990). Early work in two small alpine basins of the San Juan Mountains also suggested low rates of chemical denudation (1 mm ka^{-1} and 0.4 mm ka^{-1} in basins on volcanic and quartzitic bedrock, respectively) but these are likely underestimated since they are based on records for only part of the year (Caine, 1976). Elsewhere in the Southern Rockies, Stottlemyer and Troendle (1987) report geochemical balances for the Deadhorse Creek and Lexen Creek basins, Fraser Experimental Forest, which suggest a cationic denudation rate of 2 mm ka^{-1} and a net solute yield equivalent to 4 mm ka^{-1}. In all of these cases, a direct relationship between solute yields and water fluxes, both between catchments and between years is noted and has been proposed to account for local variations in weathering rates associated with the variable depths of winter snow accumulation (Thorn, 1975; Caine, 1979, 1992b).

In Europe, general rates of solute denudation remain low at c.2 mm ka^{-1} (Walling and Webb, 1986). Small headwater basins show higher rates: Paces (1986) reports values of 9 mm ka^{-1}, 14 mm ka^{-1} and 32 mm ka^{-1} in Bohemia, and they are higher again in mountainous carbonate terrains where denudation rates of up to 300 mm ka^{-1} are estimated from river loads (Sweeting, 1972). André (1996a) suggests that rates of lowering on glaciated limestone surfaces in the Alps have averaged up to 15 mm ka^{-1} over the Holocene. In the uplands of Britain, equivalent rates of chemical denudation have frequently been reported: for example, Walling and Webb (1986) suggest rates of up to 10 mm ka^{-1}, which are confirmed by studies in small catchments in Scotland where Creasey et al. (1986) report rates of 5–7 mm ka^{-1} and Bain et al. (1990) rates of 4–12 mm ka^{-1}. Again, reported rates of limestone surface lowering are appreciably higher than the denudation rates for basins on other lithologies (Sweeting, 1965).

Net denudation by solute transport in the arctic-alpine Kärkevagge basin, northern Scandinavia, approximates 8 mm ka^{-1} in recent work (Darmody et al., 2000; Thorn et al., 2001), which is comparable with Rapp's (1960a) preliminary estimate (10 mm ka^{-1}). In the same area, André (1996b) has estimated rates of Holocene surface lowering on exposed glaciated bedrock which are lower than these basin-wide estimates: 5 mm ka^{-1} on dolomite to 0.2 mm ka^{-1} on quartzite.

Glacierized basins also show rates of chemical denudation that exceed the global average and that are correlated to rates of mechanical denudation (Hallet et al., 1996; Anderson et al., 1997). Large Himalayan glacier systems exhibit cationic denudation rates as high as 30 mm ka^{-1} (Chauhan and Hasnain, 1993; Collins, 1999; Hodson et al., 2000). High rates of cationic flux have also been reported from the active glacier systems of western North America: 13 mm ka^{-1} for the Worthington Glacier (Hodson et al., 2000); 8 mm ka^{-1} for the Berendon Glacier (Eyles et al., 1982); and 7 mm ka^{-1} for the

South Cascade Glacier (Reynolds and Johnson, 1972). In the Alps, the removal of cations from glacier basins tends to be lower but still exceeds the global average (Hallet et al., 1996). Sharp et al. (1995) suggest a denudation rate of 5 mm ka^{-1} for the Haut Glacier d'Arolla during 1989–90; Souchez and Lemmens (1987) a rate of 4 mm ka^{-1} for the Tsidjiore Nouve Glacier; and 4 mm ka^{-1} for the Gornergletscher (Collins, 1983). Equivalent data for the small glaciers of the Rocky Mountains, such as the Arikaree Glacier in Colorado Front Range (Caine, 2001), suggest a cationic flux equivalent to 1 mm ka^{-1}, clearly lower than rates for larger, more active glacier systems. All of these data suggest the significance of chemical denudation in contemporary glacial systems and also point to its correspondence with rates of mechanical denudation and water fluxes (Anderson et al., 1997).

5 Discussion: patterns of mountain denudation

These empirical data show that contemporary denudation rates in mountain systems are high by comparison with global average rates, as suggested by Milliman and Syvitski (1992), and this appears true of both mechanical and chemical denudation. There are, of course, exceptions to this: the Transantarctic Mountains, for example, have probably experienced very low rates (<1 mm ka^{-1}) throughout the late Cenozoic (Summerfield et al., 1999). Rapid rates of erosion are also associated with a wide variability (often amounting to more than an order of magnitude) on local and regional scales as well as between mountain systems. From this variability, a number of conclusions that point to general controls of denudation may be drawn.

A common conclusion from work of the last few decades is that geochemical denudation in mountain environments is more significant than had been recognized previously. This was first suggested by the preliminary surveys of Rapp (1960a) and has since been defined in a wide range of mountain environments. More recently, this has been augmented by the recognition that solute denudation and mechanical denudation rates are linked through the effect of rock comminution which provides reactive sites and surfaces for chemical weathering. This is clearly defined in glacierized environments (Anderson et al., 1997) but is also evident in rapidly eroding catchments that are not presently glacierized (West et al., 2002).

Globally, the highest rates of denudation (dominated by clastic sediment fluxes) are associated with temperate glacier systems in mountains such as those of southeastern Alaska and British Columbia. In these glacial systems, sliding is an important component of glacial movement and the volume of water flux through the glaciers is high (Hallet et al., 1996; Anderson et al., 1997). This pattern is probably also true of Himalayan glacier systems though there are fewer long-term datasets to support such a conclusion. Small mountain-glaciers are less effective in producing higher than ambient sediment yields (Hicks et al., 1990), perhaps because of the relative significance of sediment storage (Harbor and Warburton, 1993). On the other hand, high denudation rates in the glacier zone have been suggested as an effective long-term limit to relief and elevation in mountain ranges (Whipple et al., 1999; Brocklehurst and Whipple, 2002).

High rates of denudation are also associated with rapid tectonic uplift in the Karakoram, Taiwan, New Guinea and New Zealand (Hovius, 1998). These are collisional zones in which rapid rates of plate convergence and crustal shortening are associated with an approximate balance between rates of uplift and rates of denudation. They are also mountain systems in which the confounding influence of tectonics and topography on glaciation, precipitation, runoff and vegetation may have global environmental effects (Molnar and England, 1990; Raymo and Ruddiman, 1992). Many of these mountain landscapes appear to be adjusted to the rapid evacuation of erosional products: they have slopes which are close to instability and which abut directly to stream and river channels (Schmidt and Montgomery, 1995). Landslide and debris flow activity is, thus, relatively frequent and steep river gradients allow the efficient removal of debris supplied to them. Such a landscape, in which slopes and channels are well integrated, has often been described as steady state or time-independent, i.e., one in which form may be maintained over geologic time by a balance of rock input through tectonics and its removal by erosion (Thornes and Brunsden, 1977; Adams, 1980; Hovius et al., 1997).

Other mountain systems, especially those of mid- and low-latitudes that were not glaciated during the late Pleistocene, exhibit a similar balance of regolith production and sediment evacuation by stream and river systems (Heimsath et al., 1997; Schlunegger et al., 2002). However, in these systems contemporary rates of denudation are much lower, though they may have been maintained over long intervals (Kirchner et al., 2001). Landsliding associated with intense rainstorms or other catastrophic events with long return intervals, supplies material to the fluvial system that may be removed during the intervals between instability events giving a steady-state landscape.

Even with high relief and precipitation, mountain landscapes affected by glacial action during the late Pleistocene often show contemporary denudation rates that are orders of magnitude lower than the global maxima and that reflect only a small part of the geomorphic work performed in them (Caine, 2001). In many of these systems, the effects of past glaciation have left a high potential for sediment storage on valley walls and floors in the glaciated area that has dominated post-glacial activity (Schrott and Adams, 2002; Schrott et al., 2002). Consequently, measured denudation rates based on river loads are often low, with relatively high geochemical fluxes. This suggests a slowly decaying or time-dependent mountain landscape (Thornes and Brunsden, 1977) in which the landforms of Pleistocene glaciation survive. At lower elevations, denudation rates may increase in response to the release of sediment from valley-floor storage that dates from the Pleistocene (Church and Slaymaker, 1989). In the Rocky Mountains, the disjunction between fluvial systems and the surrounding slopes and inter-fluves is reflected in an order of magnitude difference between rates of valley and ridge lowering (Small et al., 1997; Anderson, 2002).

Apart from the association of the highest denudation rates with contemporary glacierization, climatic controls on erosion rates are not well demonstrated by empirical evidence (Wilson, 1973; Riebe et al., 2001a, b). The failure to define a simple climate–denudation correlation may reflect multiple causes: (1) denudation rates, themselves, have a high stochastic variability (Fuller et al., 2003); (2) large-scale effects of tectonics and relief are often confounded with precipitation and temperature (Summerfield and Hulton, 1994; Hovius, 1998) which makes the discrimination of an independent effect difficult and is further confounded with the influence of vegetation on erosion; (3) the inadequacy of simple climatic characteristics, such as mean annual precipitation and temperature, as descriptors of a 'geomorphic climate'.

Anthropogenic influences on denudation are also not clearly demonstrated in the empirical records reviewed here (Ives and Messerli, 1989). This is supported by the correspondence between contemporary rates of denudation, measured by fluvial loads, and long-term rates derived from geologic and cosmogenic evidence (Burbank and Beck, 1991; Small et al., 1997; Kirchner et al., 2001). It suggests that, at least on a regional scale, land-use changes are not capable of inducing orders of magnitude change in erosion rates and that the results reported here may be taken as a fair reflection of natural conditions. The situation may be different on a local scale.

Finally, it is important to recognize the significance of extreme geomorphic events in mountain landscapes. This was recognized explicitly in the first studies of sediment budgets in mountain drainages. Jäckli (1957) shows the relative insignificance of contemporary denudation in the Rhine catchment by comparison with the magnitude of the historic Flims slide. Rapp (1960a) records the equivalent significance of two large rock avalanches in Kärkevagge, one of which impounded Rissajaure in the early post-glacial period. These and equivalent catastrophic events, such as the major rock avalanches from

Huascaran and the volcanic collapse of Mt St Helens, remain difficult to accommodate in the sediment budgets of the drainage basins they impact and are not represented in most of the work reviewed here (but see Thouret, this volume).

6 Conclusion: prospects

The study of denudation rates has been part of geomorphology for more than 150 years, initially as a test of uniformitarian concepts and later as a means of evaluating geomorphic processes and landscape dynamics. Here, I suggest four possible developments in denudation studies of mountain environments. (1) Continued instrumental development should improve the accuracy and precision of estimated denudation rates. This includes improvements in river and water quality monitoring, in surveying, remote sensing and data handling techniques, and in the dating of surfaces and sediments. These developments should reduce error and uncertainty in estimation, which will remain as a major concern. (2) Further research should include better identification of the sources of water, sediment and solutes within drainage basins and, thus, lead to better sediment and solute budgets for a variety of drainage basins. This would allow some opening of the black box, which represents the drainage basin in classic studies that treat only the material exports from it. It may allow the use of measures other than average bedrock lowering in evaluating geomorphic activity (Caine, 1976; Warburton, 1993). (3) There will continue to be the need to reconcile denudation rates estimated from contemporary processes with those based on geologic or equivalent long-term records, or at least explain discrepancies among them (Kirchner et al., 2001). (4) The influence of global environmental changes on mountain denudation rates will remain, particularly as mountain glaciers and permafrost (and the erosion associated with them) respond to changed climates (cf. Slaymaker and Owens, this volume).

References

Adams, J., 1980. Contemporary uplift and erosion of the Southern Alps, New Zealand. *Geological Society of America Bulletin*, **91**: 1–114.

Ahnert, F., 1970. Functional relationships between denudation, relief and uplift in large mid-latitude drainage basins. *American Journal of Science*, **268**: 243–63.

Anderson, R.S., 2002. Modeling the tor-dotted crests, bedrock edges, and parabolic profiles of high alpine surfaces of the Wind River Range, Wyoming. *Geomorphology*, **46**: 35–58.

Anderson, S.P., Drever, J.I. and Humphrey, N.F., 1997. Chemical weathering in glacial environments. *Geology*, **25**: 399–402.

André, M.-F., 1996a. Vitessses de dissolution aréolaire postglaciaire dans les karsts polaires et haut-alpins: de l'Arctique scandinave aux Alpes de Nouvelle-Guinée. *Revues d'Analyse Spatiale Quantitative et Appliquée*, **38–39**: 99–107.

André, M.-F., 1996b. Rock weathering rates in arctic and subarctic environments (Abisko Mountains, Swedish Lappland). *Zeitschrift für Geomorphologie*, **40**: 499–517.

Bain, D.C., Mellor, A., Wilson, M.J. and Duthie, D.M.L., 1990. Weathering in Scottish and Norwegian catchments. In Mason, B.J. (ed.), *The surface waters acidification programme*. Cambridge: Cambridge University Press, 223–36.

Ballantyne, C.K., 2002. A general model of paraglacial landscape response. *The Holocene*, **12**: 371–76.

Baron, J. and Bricker, O.P., 1987. Hydrologic and chemical flux in Loch Vale watershed, Rocky Mountain National Park. In Averett, R.C. and McKnight, D.M. (eds), *Chemical quality of water and the hydrologic cycle*. Michigan, WI: Lewis Publishers, 141–55.

Barsch, D., 1996. *Rockglaciers*. Berlin: Springer-Verlag.

Bernet, M., Zattin, M., Garver, J.I., Brandon, M.T. and Vance, J.A., 2001. Steady-state exhumation of the European Alps. *Geology*, **29**: 35–38.

Bezinge, A., 1987. Glacial meltwater streams, hydrology and sediment transport: the case of the Grande Dixence hydroelectricity scheme. In Gurnell, A.M. and Clark, M.J. (eds), *Glacio-fluvial sediment transfer: an alpine perspective*. Chichester: Wiley, 473–98.

Bierman, P. and Steig, E.J., 1996. Estimating rates of denudation using cosmogenic isotope abundances in sediment. *Earth Surface Processes and Landforms*, **21**: 125–39.

Bogen, J., 1989. Glacial sediment production and development of hydro-electric power in glacierized areas. *Annals of Glaciology*, **13**: 6–11.

Bögli, A., 1980. *Karst hydrology and physical speleology*. Berlin: Springer-Verlag.

Bovis, M.J., 1978. Soil loss in the Colorado Front Range: sampling design and areal variation. *Zeitschrift für Geomorphologie, Supplementband*, **29**: 10–21.

Brocklehurst, S.H. and Whipple, K.X., 2002. Glacial erosion and relief production in the eastern Sierra Nevada, California. *Geomorphology*, **42**: 1–24.

Brown, E.T., Stallard, R.F., Larsen, M.C., Raisbeck, G.M. and Yiou, F., 1995. Denudation rates determined from the accumulation of in-situ-produced [10]Be in the Luquillo Experimental Forest, Puerto Rico. *Earth and Planetary Science Letters*, **129**: 193–202.

Burbank, D.W. and Beck, R.A., 1991. Rapid, long-term rates of denudation. *Geology*, **19**: 1169–72.

Burbank, D.W., Leland, J., Fielding, E., Anderson, R.S., Brozovic, N., Reid, M.R. and Duncan, C., 1996. Bedrock incision, rock uplift, and threshold hillslopes in the northwestern Himalaya. *Nature*, **379**: 505–10.

Caine, N., 1976. A uniform measure of subaerial erosion. *Geological Society of America Bulletin*, **87**: 137–40.

Caine, N., 1979. Rock weathering rates at the soil surface in an alpine environment. *Catena*, **6**: 131–44.

Caine, N., 1986. Sediment movement and storage on alpine hillslopes in the Colorado Rocky Mountains. In Abrahams, A.D. (ed.), *Hillslope processes*. Winchester, MA: Allen and Unwin, 115–37.

Caine, N., 1992a. Sediment transfer on the floor of the Martinelli Snowpatch, Colorado Front Range, USA. *Geografiska Annaler*, **74A**: 133–44.

Caine, N., 1992b. Spatial patterns of geochemical denudation in a Colorado alpine environment. In Dixon, J.C. and Abrahams, A.D. (eds), *Periglacial geomorphology*. Chichester: Wiley, 63–88.

Caine, N., 2001. Geomorphic systems of Green Lakes Valley. In Bowman, W.D. and Seastedt, T.R. (eds), *Structure and function of an alpine ecosystem*. New York: Oxford University Press, 45–74.

Caine, N. and Mool, P.K., 1982. Landslides in the Kolpu Khola drainage, Middle Mountains, Nepal. *Mountain Research and Development*, **2**: 157–73.

Caine, N. and Swanson, F.J., 1989. Geomorphic coupling of hillslope and channel systems in two small mountain basins. *Zeitschrift für Geomorphologie*, **33**: 189–203.

Caine, N. and Thurman, E.M., 1990. Temporal and spatial variations in the solute content of an alpine stream, Colorado Front Range. *Geomorphology*, **4**: 55–72.

Carter, L., Manighetti, B., Elliot, M., Trustrum, N. and Gomez, B., 2002. Source, sea level and circulation effects on the sediment flux to the deep ocean over the past 15 ka off eastern New Zealand. *Global and Planetary Change*, **33**: 339–55.

Chauhan, D.S. and Hasnain, S.I., 1993. Chemical characteristics, solute and suspended sediment loads in the meltwaters draining Satopanth and Bhagirath Kharak Glaciers, Ganga basin, India. In Young, G.J. (ed.), *Snow and glacier hydrology*. AHS Publication 218. Wallingford: IAHS Press, 403–10.

Church, M. and Ryder, J.M., 1972. Paraglacial sedimentation: a consideration of fluvial processes conditioned by glaciation. *Geological Society of America Bulletin*, **83**: 3059–72.

Church, M. and Slaymaker, O., 1989. Disequilibrium of Holocene sediment yield in glaciated British Columbia. *Nature*, **337**: 452–54.

Clayton, J.L. and Megahan, W.F., 1997. Natural erosion rates and their prediction in the Idaho batholith. *Journal of the American Water Resources Association*, **33**: 689–703.

Clayton, K.M., 1997. The rate of denudation of some British lowland landscapes. *Earth Surface Processes and Landforms*, **22**: 721–31.

Collins, D.N., 1983. Solute yield from a glacierized high mountain basin. In Webb, B.W. (ed.), *Dissolved loads of rivers and surface water quantity/quality relationships*. IAHS Publication 141. Wallingford: IAHS Press, 41–50.

Collins, D.N., 1999. Solute flux in meltwaters draining from a glacierized basin in the Karakoram Mountains. *Hydrological Processes*, **13**: 3001–15.

Cooke, R.U., Inkpen, R.J. and Wiggs, G.F.S., 1995. Using gravestones to assess changing rates of weathering in the United Kingdom. *Earth Surface Processes and Landforms*, **20**: 531–46.

Corbel, J., 1959. Vitesse de l'érosion. *Zeitschrift für Geomorphologie*, **3**: 1–28.

Creasey, J., Edwards, A.C., Reid, J.M., MacLeod, D.A. and Cresser, M.S., 1986. The use of catchment studies for assessing chemical weathering rates in two contrasting upland areas in northeast Scotland. In Colman, S.M. and Dethier, D.P. (eds), *Rates of chemical weathering of rocks and minerals*. London: Academic Press, 467–502.

Darmody, R.G., Thorn, C.E., Harder, R.L., Schlyter, J.P.L. and Dixon, J.C., 2000. Weathering implications of water chemistry in an arctic-alpine environment, northern Sweden. *Geomorphology*, **34**: 89–100.

Dearing, J.A. and Foster, I.D.L., 1993. Lake sediments and geomorphological processes: some thoughts. In McManus, J. and Duck, R.W. (eds), *Geomorphology and sedimentology of lakes and reservoirs*. Chichester: Wiley, 5–14.

Dethier, D.P., 1986. Weathering rates and the chemical flux from catchments in the Pacific Northwest, U.S.A. In Colman, S.M. and Dethier, D.P. (eds), *Rates of chemical weathering of rocks and minerals*. London: Academic Press, 503–30.

Dethier, D.P., 2001. Pleistocene incision rates in the western United States calibrated using Lava Creek B tephra. *Geology*, **29**: 783–86.

Dole, R.B. and Stabler, H., 1909. Denudation. *U.S. Geological Survey Water Supply Paper*, **234**: 78–93.

Ewing, R., 1996. Postfire suspended sediment from Yellowstone National Park, Wyoming. *Water Resources Bulletin*, **32**: 605–27.

Eyles, N., Sasseville, D.R., Slatt, R.M. and Rogerson, R.J., 1982. Geochemical denudation rates and solute transport mechanisms in a maritime temperate glacier basin. *Canadian Journal of Earth Sciences*, **19**: 1570–81.

Fairbridge, R.W., 1968. Denudation. In Fairbridge, R.W. (ed.), *Encyclopedia of geomorphology*. New York: Reinhold Book Corporation, 261–71.

Ferguson, R.I., 1984. Sediment load of the Hunza River. In Miller, K.J. (ed.), *International Karakoram Project*. Cambridge: Cambridge University Press, 581–98.

Ferguson, R., 1986. River loads underestimated by rating curves. *Water Resources Research*, **22**: 74–76.

Fischer, A.G., 1969. Geologic time–distance rates: the Bubnoff unit. *Geological Society of America Bulletin*, **80**: 549–51.

Ford, D.C., Schwarcz, H.P., Drake, J.J., Gascoyne, M., Harmon, R.S. and Latham, A.G., 1981. Estimates of the age of the existing relief within the southern Rocky Mountains of Canada. *Arctic and Alpine Research*, **13**: 1–10.

Fuller, C.W., Willett, S.D., Hovius, N. and Slingerland, R., 2003. Erosion rates for Taiwan mountain basins: new determinations from suspended sediment records and a stochastic model of their temporal variation. *Journal of Geology*, **111**: 71–87.

Galy, A. and France-Lanord, C., 2001. Higher erosion rates in the Himalaya: geochemical constraints on riverine fluxes. *Geology*, **29**: 23–26.

Gardner, T.W., Jorgensen, D.W., Shuman, C. and Lemieux, C.R., 1987. Geomorphic and tectonic process rates: effects of measured time interval. *Geology*, **15**: 259–61.

Geikie, A., 1868. On denudation now in progress. *Geological Magazine*, **5**: 249–54.

Gomez, B., 1987. Bedload. In Gurnell, A.M. and Clark, M.J. (eds), *Glacio-fluvial sediment transfer: an alpine perspective*. Chichester: Wiley, 355–76.

Graf, W.L., 1999. Dam nation: a geographic census of American dams and their large-scale hydrologic impacts. *Water Resources Research*, **35**: 1305–11.

Griffiths, G.A., 1981. Some suspended sediment yields from South Island catchments, New Zealand. *Water Resources Bulletin*, **17**: 662–71.

Griffiths, G.A., 1982. Spatial and temporal variability in suspended sediment yields of North Island Basins, New Zealand. *Water Resources Bulletin*, **18**: 575–84.

Gurnell, A.M., 1987. Suspended sediment. In Gurnell, A.M. and Clark, M.J. (eds), *Glacio-fluvial sediment transfer: an alpine perspective*. Chichester: Wiley, 305–54.

Hallet, B., Hunter, L. and Bogen, J., 1996. Rates of erosion and sediment evacuation by glaciers: a review of field data and their implications. *Global and Planetary Change*, **12**: 213–35.

Harbor, J. (ed.), 1999. Cosmogenic isotopes in geomorphology. *Geomorphology*, **27**: 1–172.

Harbor, J. and Warburton, J., 1993. Relative rates of glacial and nonglacial erosion in alpine environments. *Arctic and Alpine Research*, **25**: 1–7.

Hartshorn, K., Hovius, N., Dade, W.B. and Slingerland, R.L., 2002. Climate-driven bedrock incision in an active mountain belt. *Science*, **297**: 2036–38.

Heimsath, A.M., Dietrich, W.E., Nishiizumi, K. and Finkel, R.C., 1997. The soil production function and landscape equilibrium. *Nature*, **388**: 358–61.

Hewitt, K., 2001. Catastrophic rockslides and the geomorphology of the Hunza and Gilgit River valleys, Karakoram Himalaya. *Erdkunde*, **55**: 72–93.

Hicks, D.M., McSaveney, M.J. and Chinn, T.J.H., 1990. Sedimentation in proglacial Ivory lake, Southern Alps, New Zealand. *Arctic and Alpine Research*, **22**: 26–42.

Hodson, A., Tranter, M. and Vatne, G., 2000. Contemporary rates of chemical denudation and atmospheric CO_2 sequestration in glacier basins: an arctic perspective. *Earth Surface Processes and Landforms*, **25**: 1447–71.

House, M.A., Gurnis, M., Kamp, P.J.J. and Sutherland, R., 2002. Uplift in the Fiordland region, New Zealand: implications for incipient subduction. *Science*, **297**: 2038–41.

Hovius, N., 1998. Controls on sediment supply by large rivers. In Shanley, K.W. and McCabe, P.J. (eds), *Relative role of eustasy, climate and tectonism in continental rocks*. Society of Economic Palaeontologists and Mineralogists, Special Publication No. 59, 3–16.

Hovius, N., Stark, C.P. and Allen, P.A., 1997. Sediment flux from a mountain belt derived by landslide mapping. *Geology*, **25**: 231–34.

Hovius, N., Stark, C.P., Cho, H.T. and Lin, J.C., 2000. Supply and removal of sediment in a landslide-dominated mountain belt: Central range, Taiwan. *Journal of Geology*, **108**: 73–89.

Ives, J.D. and Messerli, B., 1989. *The Himalyan dilemma: reconciling development and conservation*. London: Routledge.

Jäckli, H., 1957. Gegenwartsgeologie des bunderischen Rheingebeites: ein Beitrager zur exogenen Dynamik alpiner Gebirgslandschaften. *Beitrage zur Geologie der Schweiz, Geotechnische Serie*, **36**, 135 pp.

Jansson, M.B., 1988. A global survey of sediment yield. *Geografiska Annaler*, **70A**: 81–98.

Kelsey, H.M., 1980. A sediment budget and analysis of geomorphic process in the Van Duzen River basin, north coastal California, 1941–1975. *Geological Society of America Bulletin*, **91**: 190–95.

Kirchner, J.W., Finkel, R.C., Riebe, C.S., Granger, D.E., Clayton, J.E. and Megahan, W.F., 2001. Mountain erosion over 10 yr, 10 k.y., and 10 m.y. time scales. *Geology*, **29**: 591–94.

Larsen, M.C. and Parkes, J.E., 1997. How wide is a road? The association of roads and mass wasting in a forested montane environment. *Earth Surface Processes and Landforms*, **22**: 835–48.

Larsen, M.C. and Torres-Sanchez, A.J., 1992. Landslides triggered by Hurricane Hugo in eastern Puerto Rico, September 1989. *Caribbean Journal of Science*, **28**: 113–25.

Larsen, M.C., Torres-Sanchez, A.J. and Concepcion, I.M., 1999. Slopewash, surface runoff and fine-litter transport in forest and landslide scars in humid-tropical steeplands, Luquillo Experimental Forest, Puerto Rico. *Earth Surface Processes and Landforms*, **24**: 481–502.

Leeder, M.R., 1997. Sedimentary basins: tectonic recorders of sediment discharge from drainage catchments. *Earth Surface Processes and Landforms*, **22**: 229–37.

Lehre, A., 1982. Sediment budget of a small Coast Range drainage basin in north-central California. In Swanson, F.J., Janda, R.J., Dunne, T. and Swanston, D.N. (eds), *Sediment budgets and routing in forested drainage basins*. U.S. Department of Agriculture, Pacific Northwest Forest and Range Experiment Station General Technical Report PNW-141, 67–77.

Leopold, L.B., Emmett, W.W. and Myrick, R.M., 1966. Channel and hillslope processes in a semiarid area, New Mexico. *US Geological Survey, Professional Papers*, 352-G: 193–253.

Luckman, B.H., 1981. The geomorphology of the Alberta Rocky Mountains: a review and commentary. *Zeitschrift für Geomorphologie, Supplementband*, **37**: 91–119.

Mast, M.A., Drever, J.I. and Baron, J., 1990. Chemical weathering in the Loch Vale watershed, Rocky Mountain National Park, Colorado. *Water Resources Research*, **26**: 2971–78.

McPherson, H.J., 1975. Sediment yields from intermediate-sized stream basins in southern Alberta. *Journal of Hydrology*, **25**: 243–57.

Milliman, J.D. and Syvitski, J.P.M., 1992. Geomorphic/tectonic control of sediment discharge to the ocean: the importance of small mountainous rivers. *Journal of Geology*, **100**: 525–44.

Molnar, P. and England, P., 1990. Late Cenozoic uplift of mountain ranges and global climate change: chicken or egg? *Nature*, **246**: 29–34.

Montgomery, D.R., Balco, G. and Willett, S.D., 2001. Climate, tectonics and the morphology of the Andes. *Geology*, **29**: 579–82.

Muller, G. and Forstner, U., 1968. General relationship between suspended sediment concentration and water discharge in the Alpenrhein and some other rivers. *Nature*, **217**: 244–45.

Owens, P. and Slaymaker, O., 1993. Lacustrine sediment budgets in the Coast Mountains of British Columbia, Canada. In McManus, J. and Duck, R.W. (eds), *Geomorphology and sedimentology of lakes and reservoirs*. Chichester: Wiley, 105–23.

Paces, T., 1986. Rate of weathering and erosion derived from mass balance in small drainage basins. In Colman, S.M. and Dethier, D.P. (eds), *Rates of chemical weathering of rocks and minerals*. London: Academic Press, 531–50.

Pazzaglia, F.J. and Kelley, S.A., 1998. Large-scale geomorphology and fission track thermochronology in topographic and exhumation reconstructions of the Southern Rocky Mountains. *Rocky Mountain Geology*, **33**: 229–57.

Pearce, J.T., Pazzaglia, F.J., Evenson, E.B., Lawson, D.E., Alley, R.B. and Germanoski, D., 2003. Bedload component of glacially discharged sediment: insights from Matanuska Glacier, Alaska. *Geology*, **31**: 7–10.

Petley, D.N. and Reid, S., 1999. Uplift and landscape stability at Taroko, eastern Taiwan. In Smith, B.J., Whalley, W.B. and Warke, P.A. (eds), *Uplift, erosion and stability: perspectives on long-term landscape development*. Geological Society Special Publication, 162, London: Geological Society of London, pp. 169–81.

Pimental, D. and Skidmore, E.L., 1999. Rates of soil erosion. *Science*, **286**: 1477.

Pinet, P. and Souriau, M., 1988. Continental erosion and large-scale relief. *Tectonics*, **7**: 563–82.

Rapp, A., 1960a. Recent developments of mountain slopes in Kärkevagge and surroundings, northern Scandinavia. *Geografiska Annaler*, **42A**: 71–200.

Rapp, A., 1960b. Talus slopes and mountain walls at Tempelfjorden, Spitsbergen: a geomorphological study of the denudation of slopes in an Arctic locality. *Norsk Polarinstitutt, Skrifter*, **119**, 96 pp.

Raymo, M.E. and Ruddiman, W.E., 1992. Tectonic forcing of late Cenozoic climate. *Nature*, **359**: 117–22.

Reheis, M.J., 1975. Source, transportation and deposition of debris on Arapaho Glacier, Front Range, Colorado, U.S.A. *Journal of Glaciology*, **14**: 407–20.

Restrepo, J.D. and Kjerfve, B., 2000. Water discharge and sediment load from the western slopes of the Colombian Andes with focus on Rio San Juan. *Journal of Geology*, **108**: 17–33.

Reynolds, R.C., Jr and Johnson, N.M., 1972. Chemical weathering in the temperate glacial environment of the northern Cascade Mountains. *Geochimica et Cosmochimica Acta*, **36**: 537–54.

Riebe, C.S., Kirchner, J.W., Granger, D.E. and Finkel, R.C., 2001a. Minimal climatic control on erosion rates in the Sierra Nevada, California. *Geology*, **29**: 447–50.

Riebe, C.S., Kirchner, J.W., Granger, D.E. and Finkel, R.C., 2001b. Strong tectonic and weak climatic control of long-term chemical weathering rates. *Geology*, **29**: 511–14.

Robertson, D.M. and Roerish, E.D., 1999. Influence of various water quality sampling strategies on load estimates for small streams. *Water Resources Research*, **35**: 3747–59.

Ruxton, B.P. and McDougall, I., 1967. Denudation rates in northeast Papua from potassium dating of lavas. *American Journal of Science*, **265**: 5450–61.

Schlunegger, F., Detzner, K. and Olsson, D., 2002. The evolution towards steady state erosion in a soil-mantled drainage basin: semi-quantitative data from a transient landscape in the Swiss Alps. *Geomorphology*, **43**: 55–76.

Schmidt, K.M. and Montgomery, D.R., 1995. Limits to relief. *Science*, **270**: 617–20.

Schrott, L. and Adams, T., 2002. Quantifying sediment storage and Holocene denudation in an alpine basin, Dolomites, Italy. *Zeitschrift für Geomorphologie, Supplementband*, **128**: 129–45.

Schrott, L., Niederheide, A., Hankammer, M., Hufschmidt, G. and Dikau, R., 2002. Sediment storage in a mountain catchment: geomorphic coupling and temporal variability (Reintal, Bavarian Alps, Germany). *Zeitschrift für Geomorphologie, Supplementband*, **127**: 175–96.

Schumm, S.A., 1963. The disparity between present rates of denudation and orogeny. *US Geological Survey, Professional Paper*, 454-H. Washington, DC: Government Printing Office.

Sharp, M., Tranter, M., Brown, G.H. and Skidmore, M., 1995. Rates of chemical denudation and CO_2 drawdown in a glacier-covered alpine catchment. *Geology*, **23**: 61–64.

Simonett, D.S., 1967. Landslide distribution and earthquakes in the Bewani and Torricelli Mountains, New Guinea, statistical analysis. In Jennings, J.N. and Mabbutt, J.A. (eds), *Landform studies from Australia and New Guinea*. Canberra: A.N.U. Press, 64–84.

Slaymaker, O., 1987. Sediment and solute yields in British Columbia and Yukon: their geomorphic significance reexamined. In Gardiner, V. (ed.), *International geomorphology*. Chichester: Wiley, 925–45.

Slaymaker, O. and McPherson, H.J., 1977. An overview of geomorphic processes in the Canadian Cordillera. *Zeitschrift für Geomorphologie*, **21**: 169–86.

Small, E.E., Anderson, R.S. and Finkel, R.C., 1997. Erosion rates of summit flats using cosmogenic radionuclides. *Earth and Planetary Science Letters*, **150**: 413–25.

Small, E.E., Anderson, R.S., Hancock, G.S. and Finkel, R.C., 1999. Estimates of regolith production from ^{10}Be and ^{26}Al: evidence for steady state alpine hillslopes. *Geomorphology*, **27**: 131–50.

Small, R.J., 1987. Englacial and supraglacial sediment: transport and deposition. In Gurnell, A.M. and Clark, M.J. (eds), *Glacio-fluvial sediment transfer: an alpine perspective*. Chichester: Wiley, 111–45.

Souch, C. and Slaymaker, O., 1986. Temporal variability of sediment yield using accumulations in small ponds. *Physical Geography*, **7**: 140–53.

Souchez, R.A. and Lemmens, M.M., 1987. Solutes. In Gurnell, A.M. and Clark, M.J. (eds), *Glacio-fluvial sediment transfer: an alpine perspective*. Chichester: Wiley, 285–303.

Starkel, L., 1972. The role of catastrophic rainfall in the shaping of the relief of the lower Himalaya (Darjeeling Hills). *Geographica Polonica*, **23**: 151–73.

Stottlemyer, R. and Troendle, C., 1987. *Trends in streamwater chemistry and input–output balances, Fraser Experimental Forest, Colorado*. U.S. Department of Agriculture, Rocky Mountain Forest and Range Experimental Station Research Paper RM-275.

Summerfield, M.A. and Hulton, N.J., 1994. Natural controls of fluvial denudation rates in major world drainage basins. *Journal of Geophysical Research*, **99**: 13871–83.

Summerfield, M.A., Stuart, F.M., Cockburn, H.A.P., Sugden, D.E., Denton, G.H., Dunai, T. and Marchant, D.R., 1999. Long-term rates of denudation in the Dry Valleys, Transantarctic Mountains, southern Victoria Land, Antarctica based on in-situ-produced cosmogenic ^{21}Ne. *Geomorphology*, **27**: 113–29.

Swanson, F.J. and Fredriksen, R.L., 1982. Sediment routing and budgets: implications for judging impacts of forestry practices. In Swanson, F.J., Janda, R.J., Dunne, T. and Swanston, D.N. (eds), *Sediment budgets and routing in forested drainage basins*. U.S. Department of Agriculture, Pacific Northwest Forest and Range Experiment Station General Technical Report PNW-141, 129–37.

Sweeting, M.M., 1965. The weathering of limestones, with particular reference to the Carboniferous limestones of northern England. In Dury, G. (ed.), *Essays in geomorphology*. London: Heinemann, 177–210.

Sweeting, M.M., 1972. *Karst landforms*. London: Macmillan.

Syvitski, J.P., Morehead, M.D., Bahr, D.B. and Mulder, T., 2000. Estimating fluvial sediment transport: the rating parameters. *Water Resources Research*, **36**: 2747–60.

Thorn, C.E., 1975. Influence of late-lying snow on rock-weathering rinds. *Arctic and Alpine Research*, **7**: 373–78.

Thorn, C.E., Darmody, R.G., Dixon, J.C. and Schlyter, P., 2001. The chemical weathering regime of Kärkevagge, arctic-alpine Sweden. *Geomorphology*, **41**: 37–52.

Thornes, J.B. and Brunsden, D., 1977. *Geomorphology and time*. New York: Halsted Press.

Trimble, S.W., 1975. Denudation studies: can we assume stream steady state? *Science*, **188**: 1207–208.

Verstraeten, G. and Poesen, J., 2002. Using sediment deposits in small ponds to quantify sediment yield from small catchments: possibilities and limitations. *Earth Surface Processes and Landforms*, **27**: 1425–39.

Wahrhaftig, C., 1987. Foreword. In Giardino, J.R., Shroder, J.F. and Vitek, J.D. (eds), *Rock glaciers*. Winchester, MA: Allen and Unwin, 7–42.

Wahrhaftig, C. and Cox, A., 1959. Rock glaciers in the Alaska Range. *Geological Society of America Bulletin*, **70**: 383–436.

Walling, D.E. and Webb, B.W., 1981. The reliability of suspended sediment load data. IAHS Publication 133, Wallingford: IAHS Press, 177–94.

Walling, D. and Webb, B., 1986. Solutes in river systems. In Trudgill, S.T. (ed.), *Solute processes*. Chichester: Wiley, 251–327.

Warburton, J., 1993. Energetics of alpine proglacial geomorphic processes. *Transactions of the Institute of British Geographers*, **18**: 197–206.

West, A.J., Bickle, M.J., Collins, R. and Brasington, J., 2002. Small-catchment perspective on Himalayan weathering fluxes. *Geology*, **30**: 355–58.

Whipple, K.X., Kirby, E. and Brocklehurst, S.H., 1999. Geomorphic limits to climate-induced increases in topographic relief. *Nature*, **401**: 39–43.

Wilson, L., 1973. Variations in mean annual sediment yield as a function of mean annual precipitation. *American Journal of Science*, **273**: 335–49.

7

Hillslope hydrology and mass movements in the Japanese Alps

Yuichi Onda

1 Introduction

The mountain ranges of Japan receive intense and large amounts of precipitation, which, despite dense vegetation, results in high rates of surface and subsurface runoff, and consequently high rates of denudation (Oguchi *et al.*, 2001a). These effects cause the Japanese Alps to be deeply incised and promote mass movement events. Furthermore, Japan is tectonically very active, experiencing pronounced vertical displacement and numerous large earthquakes, which also promote numerous mass movement events, which in turn have a pronounced effect on the shape of the landscape in mountainous areas (Yoshikawa *et al.*, 1981). This chapter describes the various types of mass movement in parts of the Japanese Alps with respect to bedrock lithologies and hillslope hydrological processes.

2 Neotectonics and plate movement in and around Japan

The mountains of Japan are associated with the Circum-Pacific orogenic belt (Figure 7.1; cf. Owen, this volume). The central mountain ranges of Japan (Japanese Alps) lie at the junction of northeastern Japan, southwestern Japan, and the Izu-Ogasawara Island Arc. The Philippine plate subducts at the Ryukyu Trench and Nankai Trough, while the Pacific plate subducts at the Japan Trench and Izu-Ogasawara Trench.

The Japanese Alps are the highest mountain ranges in Japan. They consist of three major ranges: the Northern Alps, the Central Alps and the Southern Alps. The Central Alps are mostly underlain by late Cretaceous granitic rocks, the Southern Alps by Mesozoic sedimentary rocks, and the Northern Alps by granitic and Mesozoic sedimentary rocks. This chapter focuses on mass movement events in the Central and Southern Alps.

Oguchi *et al.* (2001b) briefly summarize the morphogenesis of the Japanese Alps. The Central Alps are characterized by low-angle thrust faults with high slip rates (Figure 7.2). The range began to uplift around 0.5 Ma, which indicates that the average uplift rate is >4 mm a^{-1} (Ikeda, 1990). The horizontal slip rate of the main fault below the range is even greater, resulting in horizontal shortening of >4 km between the range and the Ina Valley (Figures 7.2 and 7.3), which has displaced late Quaternary alluvial fan surfaces in the Ina Valley (Ikeda, 1990). The Southern Alps have been uplifting since the early Quaternary.

3 Mass movement processes

Because the Ina area (cf. Figure 7.3) has a temperate climate (annual average temperature of 10°C) and a monsoon climate (precipitation of about 2000 mm a^{-1}), denudation processes are very active. The processes vary among the mountain ranges in Japan, according to various factors such as geology

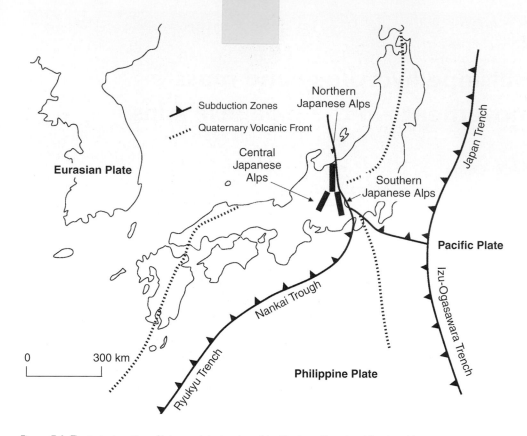

Figure 7.1 *The tectonic setting of Japan and the location of the Northern, Central and Southern Alps*

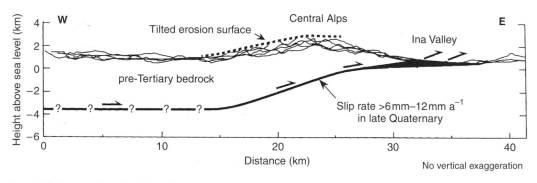

Figure 7.2 *Cross-section of the Central Alps*
Adapted from Ikeda (1990), reproduced with the permission of Yasutaka Ikeda.

and relief. Various types of mass movement are controlled by geology and subsurface water movement, and these are described below.

3.1 Weathering processes in the Japanese Alps

In the Central Alps, the dominant geology is late Cretaceous granitic rocks (hornblende-biotite granite). In the Southern Alps, the bedrock is composed of Mesozoic (mainly Cretaceous) sedimen-

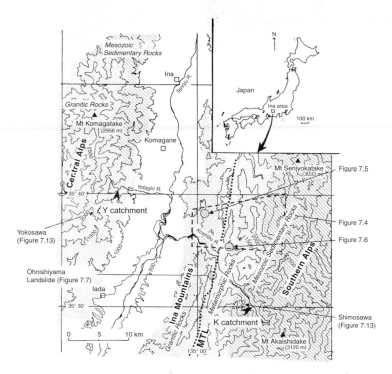

Figure 7.3 *Location of the photographs and experimental catchments in the Ina valley*

tary rocks, with Sambagawa metamorphic rocks (mostly composed of schist) to the east of the Median Tectonic Line (MTL). The MTL is an active tectonic line and a major structural and lithologic boundary (Figure 7.3). The Sambagawa metamorphism is a high-pressure intermediate-type regional metamorphism (Toriumi, 1990).

In the Central Alps, the main weathering product of the granite is a sandy material called 'gruss', produced by the weathering of minerals such as mica, hornblende, quartz and feldspar. A marked boundary between the gruss layer (regolith) and bedrock is found in granitic rocks at Obara (Iida and Okunishi, 1983). The boundary is a potential sliding plane for shallow landslides (Okimura, 1983).

In the steep mountainous region in the Ina area, most regolith is a mixture of sand and gravel formed by both chemical and mechanical weathering. Mechanical weathering includes unloading processes, because of the high rates of uplifting and denudation.

The weathering products of granite and sedimentary rocks were described by Yamada (1955). In most cases, granitic rock has larger fissure spacings and smaller opening distances, so that intensive weathering is limited to the narrow area of the fissure zone. In contrast, in sedimentary rocks, because of the presence of bedding planes and numerous joints, weathering tends to be deeper and wider. Different weathering processes operate between different bedrock lithologies. For example, shale has lower resistance to chemical weathering than sandstone, and this may enhance the formation of fissures in sedimentary bedrock.

The weathering products of the Sambagawa metamorphic rocks are mostly clay minerals, which are produced by chemical weathering. The weathering processes occur predominantly along the foliation zones.

3.2 Mass movement types and geology

The forests on the mountains of the Ina area (Figure 7.3) are dominated by secondary-growth Japanese larch (*Larix kaempferi*) plantations and Japanese beech (*Fagus crenata*). Figure 7.4 is a photograph of the Ina Mountains (location shown in Figure 7.3). In the lower part of the photograph, the underlying geology is granite. The mountain landscape is much dissected and has a very high drainage density. In the upper part of the photograph, however, the mountain landscape has a smaller number of valleys without marked dissection. Here, the geology is dominated by schist. The hillslopes appear smooth and gentle. The marked landform differences are caused by different modes of mass movement that are controlled by different modes of weathering, which in turn reflect the different underlying geology.

Figure 7.5 shows a photograph of shallow landslides common in the granite area. The depths of the landslides are <1 m, which is typical of landslides in areas underlain by granite (Iida and Okunishi, 1983). Debris flows triggered by the shallow landslides occur in many places, causing injury and damage to people and property.

3.3 Translational landslides underlain by schist

Figure 7.6 is a photograph of the Iriya landslide, which is underlain by schist, in the Sambagawa formation. A series of translational landslides can be seen. The surfaces of the sliding blocks are used for paddy fields and villages, suggesting that the movement of the sliding blocks is fairly slow. In cross-section, the translational landslides are estimated to have 30-m deep circular slip faces (Tenryu River Upstream Sabo Office, 1998), and the gradient of the sliding blocks is approximately 28°. Current mitigation work is aimed at stopping the movement of those areas shaded on Figure 7.6 by pumping out groundwater via wells and numerous underground drainage networks.

3.4 Shallow landslides in granitic rocks and deep-seated landslides in sedimentary rocks

Table 7.1 shows the landslide density in two catchments underlain by contrasting lithologies after a storm event in 1961 (Tenryu River Upstream Sabo Office, 1964). The Yotagiri catchment, underlain by granite, has a much higher density of shallow landslides (32.7 km^{-2}) than the Koshibu catchment,

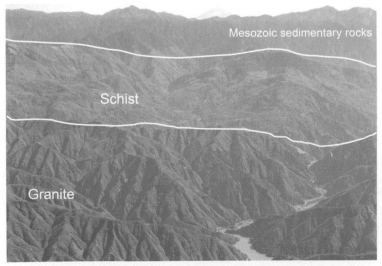

Figure 7.4 *Oblique aerial photograph of the Ina Mountains, taken in October 1990, indicating underlying granite, schist, and Mesozoic sedimentary rocks. The approximate location of the area covered is shown in Figure 7.3*
After Tenryu River Upstream Sabo Office (1998), reproduced with permission.

Figure 7.5 *Photograph taken in July 1961 of shallow landslides triggered by the 1961 storm in the area underlain by granite*
Source: *after Tenryu River Upstream Sabo Office (1998), reproduced with permission.*

River Sio-kawa

Figure 7.6 *Iriya landslides and their sliding blocks. The shaded area (Blocks B and C) is the subject of work to stabilize the other blocks*
After Tenryu River Upstream Sabo Office (1998), reproduced with permission.

Table 7.1 – Landslides in 1961 in the Yotagiri and Koshibu catchments

Catchment	Number of landslides	Landslide density (km^{-2})	Average landslide mass (m^3)
Yotagiri (granite)	837	32.7	1490
Koshibu (shale)	487	13.1	1710

Figure 7.7 *Ohnishiyama landslide (white arrow)*
After Tenryu River Upstream Sabo Office (1998), reproduced with permission.

which is underlain by Mesozoic sedimentary rocks (13.1 km^{-2}). Several deep-seated landslide scars or linear depressions related to mass rock creep (Chigira, 1992) can be found in the Koshibu area. Although the landslide density in the Koshibu catchment is smaller than that in the Yotagiri catchment (Table 7.1), the average landslide mass in the Koshibu catchment is larger than that in the Yotagiri catchment (Tenryu River Upstream Sabo Office, 1964).

Most of the landslides occurred on 29 June 1961 during the peak of heavy rainfall in the granite area, although a deep-seated landslide (Ohnishiyama landslide, Figure 7.7) occurred two days after the rainfall peak in an area underlain by metamorphic rocks, killing 42 people (Irasawa, 1986; Onda *et al.*, 1999). Such deep-seated landslides are common in the area underlain by shale, such as the Chausyama landslide, which occurred in 1903. Lag times of a few days between rainfall and deep-seated landslide initiation are typical (Sidle *et al.*, 1985).

4 Mountain development and hillslope hydrology

4.1 Hillslope hydrology and bedrock in the Central and Southern Alps

Subsurface water is one of the most important triggers of mass movements. Hydrometric research is an effective tool for analysing the causes of mass movements, focusing on the subsurface water flow path through the underlying geology. The information in this section is based on research in the Central and Southern Alps (Onda et al., 2001; in press).

Hydrological monitoring was carried out in the Koshibu (K) and Yotagiri (Y) catchments in the Ina region (see Figure 7.3). The Y1 (granite) subcatchment is located approximately 10 km southwest of Komagane city, at an elevation of 1200–1620 m asl (Figure 7.8). The K1 (shale) subcatchment is located 20 km east of Komagane city, at an elevation of 1090–1480 m asl. Soils have developed from the underlying bedrock but, because of high erosion rates, gravel-sized particles are widely distributed in both catchments. Bedrock in the Y catchment consists of relatively unweathered granite with frequent fissures. The Y1 subcatchment (6.3 ha) is drained by a second-order stream. Several drainage areas contribute to the main channel. The adjacent, smaller, Y2 subcatchment (0.88 ha) consists of a zero-order subcatchment and a bedrock seep. Both subcatchments have large catchment relief; the relief ratios are 0.67 in Y1 and 0.97 in Y2. Several bedrock springs were found flowing directly from bedrock fissures. The Yotagiri and Koshibu catchments have steep slopes (>40°) and are covered with a soil mantle 0.6–1.5 m thick. Laboratory measurements of saturated hydraulic conductivity of soils are of the order of 10^{-5} m s^{-1} for these catchments.

4.2 Bedrock spring distribution and runoff

Figure 7.8 shows the distribution and approximate discharge of the springs in the Yotagiri and Koshibu catchments for both relatively dry (October) and wet (July) periods. Photographs of two bedrock

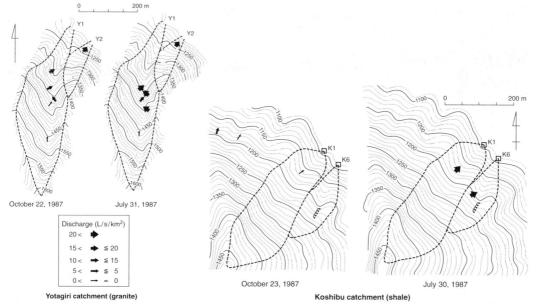

Figure 7.8 Bedrock spring distributions and discharge during the dry (October) and the wet (July) periods in the Yotagiri and Koshibu catchments

Source: Onda, Y., Tsujimura, M. and Tabuchi, H., The role of subsurface water flow paths on hillslope hydrological processes, landslides and landform development in steep mountains of Japan. Hydrological Processes, in press © John Wiley & Sons Limited. Reproduced with permission.

**Y2 spring
(granite)**

**K6-U spring
(shale)**

Figure 7.9 *Typical bedrock springs in the Y2 and K6 subcatchments*

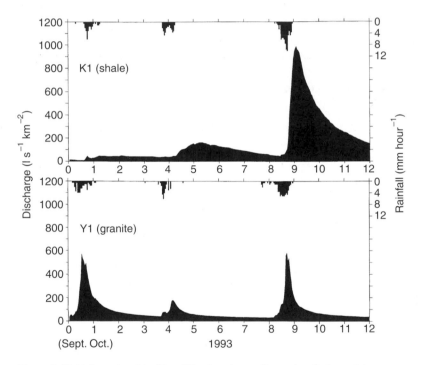

Figure 7.10 *Hydrographs of the K1 and Y1 subcatchments for a series of rain events in September and October 1993*

springs are shown in Figure 7.9. Springs in the granitic Yotagiri catchment were located at relatively higher elevations than those in the Koshibu catchment, and a small amount of water flowed from them even during dry periods. In contrast, little runoff was observed from springs in the dry period in the Koshibu catchment.

Stream hydrographs for Y1 and K1 show distinctly different runoff responses to the same storm events. Figure 7.10 shows hydrographs from both subcatchments during the beginning of October

1993, when a series of rainstorms occurred, each followed by several dry days (API$_{30}$ (Mosley, 1979) was 15.5 mm in K1 and 15.6 mm in Y1). The runoff in K1 did not respond sensitively to the initial rain event (37 mm), exhibiting only a small initial peak and a delayed increase in discharge. After the next rain event (32 mm), a delayed runoff peak (more than eight times larger than the first), characterized by a slow runoff peak with convex recession limb, was observed. After the third rain event (73.5 mm), a much larger runoff peak was observed (more than 20 times larger than the first event), having a small initial peak discharge followed by a maximum discharge with a delay time of approximately 12 hours from the rainfall peak.

In the Y1 subcatchment, in contrast, a high and sharp runoff response coinciding with each rainfall peak was detected, even after a dry period. The runoff peak at the second event (47 mm rainfall) was lower than that at the first event (72 mm). The third event (73.5 mm) produced a similar peak runoff to the first. The similarity of runoff responses between events in Y1 indicates that the peak runoff rates are proportional to the rainfall amount, regardless of antecedent rainfall.

4.3 Mass movement mechanics

To evaluate the susceptibility of landsliding in both shale (K6) and granitic rock (Y1) hillslopes, direct shear tests were performed on disturbed soil materials in the laboratory. The diameter of the shear box was 24 cm and the depth was 10 cm. The soil samples were taken just above the soil–bedrock boundary. Soil samples were passed through a 20 mm sieve, and direct shear tests were conducted at natural water content, with normal stresses of 4, 10, 15 and 20 kN m^{-2}. The tests were conducted under undrained conditions. The displacement rate was set at 1.9 mm s^{-1}. The shear tests continued until the peak shear stress appeared; if peak stress was not apparent, the test was conducted until a vertical displacement of 22 mm occurred.

The infinite slope stability model can be used to calculate a safety factor (F), in order to determine the relative landslide hazard for shallow soil mantles (e.g., Selby, 1983):

$$F = \frac{c' + (\gamma - m\gamma_w)z \cos^2 \beta \tan \phi'}{\gamma z \cos \beta \sin \beta}, \quad (1)$$

where ϕ' is the friction angle, c' is the cohesion of the soil at natural water content, γ is the unit weight of the soil at natural water content, γ_w is the unit weight of water, m is the relative groundwater level (defined as the height of the watertable above the slide plane as a fraction of the regolith thickness above the shear plane), z is the depth of the shear plane, and β is the slope angle. The measured values were $c' = 1.6$ kN m^{-2} and $\beta' = 41.3°$ in the Koshibu catchment, and $c' = 1.5$ kN m^{-2} and $\beta' = 42.6°$ in the Yotagiri catchment. The soil thickness, z, is the average soil thickness above the potential sliding surface: 0.69 and 0.49 m in the Koshibu and Yotagiri catchments, respectively.

A plot of relative water level against the safety factor is shown in Figure 7.11. At any water level, the safety factor was higher in the Yotagiri (granite) catchment than in the Koshibu (shale) catchment. However, landslide density is larger in areas underlain by granite. When $F = 1$ (i.e., dictating failure conditions), the groundwater level was calculated as 0.29 m in Koshibu and 0.34 m in Yotagiri.

The maximum groundwater levels in the monitored hillslopes of the Koshibu and Yotagiri catchments were 0.08 m and 0.28 m, respectively. It is interesting to compare these with the above groundwater levels needed to cause slope failure. The values suggest that, although the shear strength of the soil is greater in the Yotagiri catchment, the hillslopes are more unstable because of the higher groundwater level.

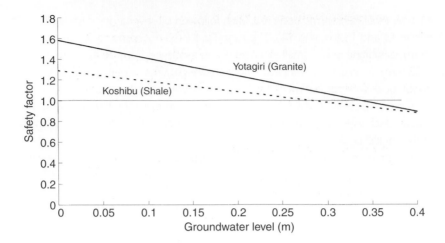

Figure 7.11 *The relationship between groundwater level and safety factor based on infinite slope stability analysis*
Source: *Onda, Y., Tsujimura, M. and Tabuchi, H., The role of subsurface water flow paths on hillslope hydrological processes, landslides and landform development in steep mountains of Japan. Hydrological Processes, In press © John Wiley & Sons Limited. Reproduced with permission.*

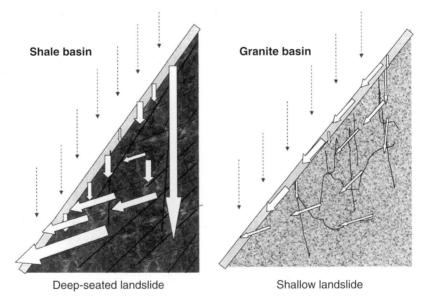

Figure 7.12 *Schematic diagram showing the flow of subsurface water and groundwater in shale and granite catchments. The upper layer represents the soil.*

4.4 Links and feedback mechanisms between hillslope hydrology and landforming processes

From the hillslope hydrometric measurements and the stability analysis, a schematic diagram of groundwater flow paths is shown in Figure 7.12. In the area underlain by granite, subsurface water flows along the soil–bedrock interface, and the proportion of water flow through bedrock is small. In the area underlain by shale, most soil water infiltrates into the bedrock, and bedrock springs are the major source of groundwater outflow. These different subsurface flow paths should trigger the differ-

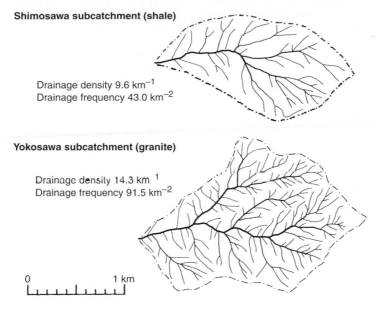

Shimosawa subcatchment (shale)

Drainage density 9.6 km^{-1}
Drainage frequency 43.0 km^{-2}

Yokosawa subcatchment (granite)

Drainage density 14.3 km^{-1}
Drainage frequency 91.5 km^{-2}

0   1 km

Figure 7.13 *Maps of the drainage in the Shimosawa (shale) and Yokosawa (granite) subcatchments*

ent modes of mass movement described earlier: i.e., in the granitic area shallow landslides occur because of the perched groundwater at the soil–bedrock interface; but in the shale area, deep-seated landslides occur because of the rise of groundwater level in the fractured bedrock.

Figure 7.13 shows the drainage densities and frequencies in subcatchments of the Koshibu (Shimosawa) and Yotagiri (Yokosawa) catchments. First-order streams were identified on medium-scale (1:25 000) topographic maps from topographic convergences (Takayama, 1972). The patterns of the drainage networks show a marked contrast between the two bedrock types. The granitic Yokosawa subcatchment has many short valleys, whereas the shale Shimosawa subcatchment has fewer but longer valleys. The drainage density in the Yokosawa subcatchment is higher than that in the Shimosawa subcatchment. The drainage frequency, defined as the total number of channels per unit area, also is higher in the Yokosawa than in the Shimosawa subcatchment. The differences are greater for drainage frequency than for drainage density. Similar characteristics due to differences in underlying geology have been reported by other researchers (e.g., Tanaka, 1957; Oguchi, 1988; Onda, 1994).

5 Conclusion

In the Ina area of central Japan, various kinds of mass movement occur. In this area, because of the high uplift and denudation rates, the regolith tends to be very thin. Consequently, the type of mass movement is controlled by the type of bedrock. The flow of subsurface water exerts an important control on the type and frequency of mass movements, resulting in landforms with different drainage densities in areas of different geology. The Ina area of central Japan is, therefore, a good example of the relation between geomorphic processes and landform development in a mountainous environment.

References

Chigira, M., 1992. Long-term gravitational deformation of rocks by mass rock creep. *Engineering Geology*, **32**: 157–84.

Iida, T. and Okunishi, K., 1983. Development of hillslopes due to landslides. *Zeitschrift für Geomorphologie, Supplementband*, **46**: 67–77.

Ikeda, Y., 1990. Erosion and uplift: observational basis for modeling mountain building processes. *Journal of the Seismological Society of Japan*, **43**: 137–52.

Irasawa, M., 1986. Ohnishiyama landslide. *Journal of the Japan Society of Erosion Control Engineering*, **39**: 30–32.

Matsushima, T., 1991. *Flood and debris flows in Ina Valley.* Iida, Japan: Ina Valley Natural Science Research Group.

Mosley, M.P., 1979. Streamflow generation in a forested watershed, New Zealand. *Water Resources Research*, **15**: 795–806.

Oguchi, T., 1988. Differences in landform development during the late glacial and the post-glacial ages among drainages around the Matsumoto basin, central Japan. *Geographical Review of Japan*, **61A**: 872–93.

Oguchi, T., Aoki, T. and Katsube, K., 2001b. Geomorphology of the Central Japan Alps. In Onda, Y. (ed.), *Field Trip B3 Guidebook, 5th International Conference of Geomorphology*. Tokyo: Japan Geomorphological Union, 3–5.

Oguchi, T., Saito, K., Kadomura, H. and Grossman, M., 2001a. Fluvial geomorphology and paleohydrology in Japan. *Geomorphology*, **39**: 3–19.

Okimura, T., 1983. A slope stability method for predicting rapid mass movements on granite. *Journal of Natural Disaster Science*, **5**: 13–30.

Onda, Y., 1994. Contrasting hydrological characteristics, slope processes and topography underlain by Paleozoic sedimentary rocks and Granite. *Transactions of the Japanese Geomorphological Union*, **15A**: 49–65.

Onda, Y., Komatsu, Y., Tsujimura, M. and Fujiwara, J., 1999. Possibility for predicting landslide occurrence by analyzing the runoff peak response time. *Journal of the Japan Society of Erosion Control Engineering*, **51**: 48–52.

Onda, Y., Komatsu, Y., Tsujimura, M. and Fujiwara, J., 2001. The role of subsurface runoff through bedrock on storm flow generation. *Hydrological Processes*, **15**: 1693–706.

Onda, Y., Tsujimura, M. and Tabuchi, H., in press. The role of subsurface water flow paths on hillslope hydrological processes, landslides and landform development in steep mountains of Japan. *Hydrological Processes*, **17**.

Selby, M.J., 1983. *Hillslope materials and processes.* Oxford: Oxford University Press.

Sidle, R.C., Pearce, A.J. and O'Loughlin, C.L., 1985. *Hillslope stability and land use.* Water Resources Monograph 11. Washington, DC: American Geophysical Union.

Takayama, S., 1972. Map scale effect on the stream order analysis. *Geographical Review of Japan*, **45**: 112–19.

Tanaka, S., 1957. The drainage density and rocks (granitic and Paleozoic) in the Setouchi Sea coast region, western Japan. *Geographical Review of Japan*, **30**: 564–78.

Tenryu River Upstream Sabo Office, 1964. *Landslides by 1961 heavy rainstorms.* Disaster Research Report, Ministry of Construction, Japan.

Tenryu River Upstream Sabo Office, 1998. *Mitigation works of Iriya landslide.* Technical Report, Ministry of Construction, Japan.

Toriumi, M., 1990. The transition from brittle to ductile deformation in the Sambagawa metamorphic belt, Japan. *Journal of Metamorphic Geology*, **8**: 457–66.

Yamada, M., 1955. *Forest physiographical studies on the micro-topographical analysis.* Tokyo: RINYA-KYOSAI-KAI.

Yoshikawa, T., Kaizuka, S. and Ota, Y., 1981. *The landform of Japan.* Tokyo: University of Tokyo Press.

8
Glacial lake outburst floods in mountain environments

Helgi Björnsson

1 Introduction

Floods can happen where glaciers dam lakes in mountainous areas. Occurring in both the temperate and subpolar regions of the Earth, such floods are called *débâcles* in the European Alps, *aluviones* in South America and *jökulhlaups* in Iceland. A dam of ice or sediment blocks the water to form a lake, while drainage is initiated by an opening of the hydraulic seal, which can be broken either suddenly or gradually. The impounded water is released directly into river channels, and has a typical discharge that is orders of magnitude higher than when running as a direct result of intense ablation. In regions with active, ice-covered volcanoes, meltwater is repeatedly released in floods from lakes that collect at subglacial hydrothermal areas. Occasionally, these floods take place without any significant prior storage of water, since ice melts instantaneously in volcanic eruptions. During the largest glacial floods in history, discharges reached 10^6 m^3 s^{-1}. For comparison, the meltwater that was temporarily stored in lakes at the edges of downwasting Pleistocene ice sheets was released in bursts that were only one order of magnitude larger than this, even though the enormous volumes involved may have altered the circulation of deep water in the North Atlantic Ocean of the late Pleistocene era.

Glacial outbursts can have pronounced geomorphological impacts, since they scour river courses and inundate floodplains. Outbursts result in enormous erosion, for they carry huge loads of sediment and imprint the landscape, past and present, with deep canyons, channelled scabland, ridges standing parallel to the direction of flow, sediment deposited on outwash plains, coarse boulders strewn along riverbanks, kettleholes where massive ice blocks have become stranded and have melted, and breached terminal moraines. Some modern outbursts have produced flood waves in coastal waters (tsunamis). In the North Atlantic, outburst sediments dumped onto the continental shelf and slope have been transported great distances by turbid currents. Outburst floods wreak havoc along their paths, threatening people and livestock, destroying vegetated lowlands, devastating farms, disrupting infrastructure such as roads, bridges and power lines, and threatening hydroelectric plants on glacially fed rivers.

Knowledge of jökulhlaup behaviour is essential for recognizing potential or imminent hazards, predicting and warning of occurrences, enacting preventive measures, assessing consequences and responding for the purpose of civil defence. The goal of this chapter is to outline and describe the following: (1) the location and geometry of glacial flood sources, including the properties of dams that impound the water; (2) the accumulation of water leading to outbursts and the conditions in which they begin; (3) the mechanisms and discharge characteristics of outbursts; and (4) case histories of floods, illustrating potential hazards.

2 Anatomy of glacial lakes and their outburst floods

2.1 Sources of glacial outbursts

Glacial lakes are found in various topographic settings. Meltwater may be impounded within the glacier (englacially and subglacially), on the surface of the glacier (supraglacially) or in lakes formed by dams at the glacier's margin (proglacially). The general characteristics, geometry and filling of glacial lakes can be described by the basic physics of hydrology. The movement of a thin film of water along the glacier base follows the gradient in the fluid potential

$$\phi_b = \rho_w \, g \, z_b + p_w, \qquad (1)$$

which is the sum of terms expressing the gravitational potential and the water pressure, p_w, assuming the effects of kinetic energy to be negligible. The symbol ρ_w represents the density of water and equals $1000 \ \text{kg m}^{-3}$, whereas g equals $9.81 \ \text{m s}^{-2}$ and represents the acceleration due to gravity, while z_b is the elevation of the glacier substrate above sea level (Figure 8.1). When describing the regional flow of basal water, it has proved useful to present a static approximation of water pressure, equal to the overburden pressure from the ice, p_i: $p_w \approx p_i = \rho_i \, g \, H$, where ρ_i equals $916 \ \text{kg m}^{-3}$ and represents the density of ice, H equals $z_s - z_b$ and is the thickness of the glacier and z_s is the elevation of the ice surface in relation to sea level (Shreve, 1972). Thus the gradient driving the water is

$$\nabla \phi_b = (\rho_w - \rho_i) \, g \, \nabla z_b + g \, \rho_i \, \nabla z_s. \qquad (2)$$

This formula predicts that the surface slope of the glacier is some ten times more influential than the slope of its bed in determining the flow of basal water.

Glacial lakes may be situated at points where water flows towards a minimum in the total fluid potential. Trapped in such a lake, water is surrounded by a hydraulic seal. As the overlying glacier floats in a state of static equilibrium, and the vertical force balance is reflected by the shape of the lake. The roof of a subglacial lake slopes approximately ten times more steeply than the surface of the ice, and in the opposite direction (Figure 8.2). While resembling a grounded ice shelf, the ice on the perimeter of the lake essentially comprises an ice dam. The state in which there is no gradient driving water inside the lake, $\nabla \phi_b = 0$, can be used to define the location and geometry of a subglacial reservoir. Hence

$$\nabla z_b = -\left[\rho_i / (\rho_w - \rho_i) \right] \nabla z_s, \qquad (3)$$

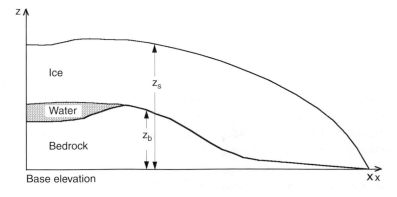

Figure 8.1 *Parameters for describing the principles of water flow in glaciers*

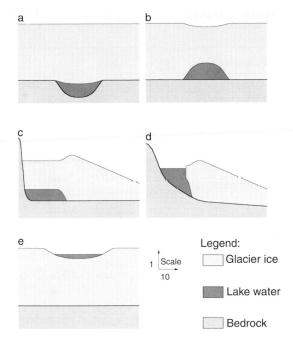

Figure 8.2 *Schematic diagram of the main types of glacial lake*

describes the relationship of the slope in the ice/water outline of the lake to the slope at the upper surface of the glacier (note that there might also be lakes with equal inflow and outflow which would not meet the condition of a zero gradient in relation to fluid potential). Given the glacier's surface and the bedrock geometry, it is possible to assess whether a water reservoir is likely to exist on the glacier bed.

The general relationship in eq. (3) applies to all types of glacial lakes (Figure 8.2, Table 8.1). Water may gather in a bedrock hollow beneath a relatively flat or dome-shaped glacial surface (Figure 8.2(a)) – in a type of reservoir which has been delineated beneath today's Antarctic Ice Sheet (Oswald and Robin, 1973; Ridley et al., 1993; Siegert et al., 2001) and was presumably widespread during the Quaternary glaciations. Another type of lake can form a cupola above the glacier bed, accompanied by a depression in the glacier's surface (Figure 8.2(b),(c)). Subglacial water reservoirs of this sort are common in Iceland, owing to volcanic and geothermal activity which causes the ice to melt and thereby creates depressions in the glacier surface (Björnsson, 1974, 1988). Meltwater flows along the bed and accumulates beneath the depression. Subaerial glacial lakes (marginal or proglacial lakes) are confined by ice on one side and by topographic barriers on the other, for instance by the edges of ravines, riverbeds or main valleys (Figure 8.2(d)). The glacier's surface slopes toward the lake, which stores meltwater from the glacier along with runoff from the surrounding terrain. The lakes formed during the summer at the margins of subpolar glaciers may be sealed by an ice dam frozen to the bed (e.g., Maag, 1969). In other instances, a proglacial lake may lie in a depression behind moraine at the front of the glacier, which is typically retreating rapidly, while sediment is deposited too slowly to fill up the over-deepened bed. This happened in the case of the receding late Pleistocene ice sheets and applies to many shrinking twentieth century glaciers. Supraglacial lakes are isolated in depressions on the glacier surface (Figure 8.2(e)), hydrologically separated from the basal drainage system. A final addition to our list might be water stored in cavities located here and there inside the glacier (Kamb et al., 1985; Walder, 1986; Kamb, 1987).

2.2 Drainage of glacial lakes

In general, glacial lakes of every type can drain either continuously or in episodic bursts. Marginal and proglacial lakes sometimes drain constantly via subaerial spillways over the rock or sediment barrier that contains them. Such barriers might consist of moraine, volcanic material or landslides. Nevertheless, lakes held back by these materials may break forth abruptly if the dams fail. Floods released suddenly when sediment or ice impoundments are breached rise almost linearly over a period of minutes or hours. They reach a high, sharp discharge peak and then fall along a steep recession limb (Figure 8.3(b)). Sediment dams can fail because of overtopping or accelerated fluvial erosion, which results in retrogressive incision enlarging the point of outflow. Moreover, a break in

Table 8.1 – Sources of glacial floods and characteristics of drainage[a]. p_w stands for water pressure and p_i for ice overburden pressure

Source/type of lake	Flood initiation and drainage routes				
	Subglacial drainage	Englacial drainage	Subaerial drainage (overtopping, downcutting, dam failure)		
Marginal lake[1] Subglacial eruption, often in the absence of significant storage[2] Subglacial lakes in hydrothermal areas[3]	Through ice tunnels at $p_w{<}p_i$ or in a sheet flow at $p_w{=}p_i$, propagating a flood wave at $p_w{>}p_i$	Upwards from the base, creating supraglacial outlets by hydrofracturing the ice and retrofeeding moulins at $p_w{>}p_i$	Drainage over a subaerial breach	Fluvial erosion of sediment dams; piping	Mechanical failure caused by tectonic activity; rockfalls; landslides
Supraglacial lakes in cauldrons and sink holes[4] Englacial storage[5]		Downward to the base through englacial tunnels			
Linked subglacial cavities[6]	Distributed drainage system switches to a tunnel system				

[a] Superscript numbers refer to publications as follows:

[1] Bretz (1925, 1969); Thorarinsson (1939); Liestøl (1956); Stone (1963); Maag (1969); Post and Mayo (1971); Clague and Mathews (1973); Björnsson (1976); Lliboutry et al. (1977); Haeberli (1983); Sturm and Benson (1985); Yamada (1998).

[2] Thorarinsson (1957, 1958); Sturm et al. (1986); Björnsson (1988); Thouret (1990); Pierson et al. (1990); Trabant and Meyer (1992); Guðmundsson et al. (1997).

[3] Björnsson (1974, 1988, 2002; Guðmundsson et al. (1995).

[4] Björnsson (1976); Russell (1993).

[5] Haeberli (1983); Driedger and Fountain (1989); Walder and Driedger (1995).

[6] Kamb et al. (1985); Kamb (1987); Björnsson (1998).

sediment dams can be caused by seepage and piping (progressive groundwater movement) within the sediment barrier, resulting in liquefied flows and embankment slips that weaken or disintegrate the dam. Tectonic activity and landslides are further threats to sediment dams, while breaching may also be triggered by flood water entering the lake or by heavy rain leading to a rapid rise in the lake level. Finally, dam failure can be generated by waves from landslides, rock falls, ice avalanches, calving icebergs or glacier surges into the lake. Providing the topographic barriers hold, on the other hand,

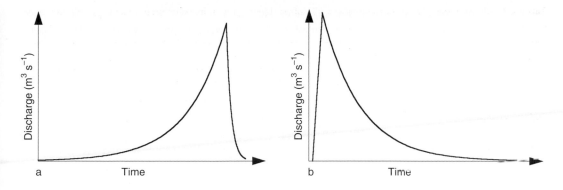

Figure 8.3 Hypothetical graphs for two types of jökulhlaup. (a) Typical shape of the hydrograph when a single basal ice tunnel enlarges because of melting; (b) abruptly peaking curve not explained by the classical theory of outbursts

proglacial lakes serve to dampen jökulhlaups from other areas. Any outburst that does occur may be a one-time event if it ruins the dam. Where the barrier is a wall of ice, it may break down mechanically, similarly to a sediment dam. Marginal lakes at subpolar glaciers, where the ice barricade is frozen to the bed, are typically breached as lake water spills over the top of the dam into a supraglacial channel that melts into a bigger breach – commonly at the juxtaposition of the glacier and a rock wall (e.g., Schytt, 1956; Maag, 1969). The rate of ice melt, the level of the lake in relation to the outlet and the hypsometry of the water reservoir dictate the progress of lake drainage. As the lake surface falls, the breach may broaden and deepen, even undercutting the ice wall so as to cause calving.

When subglacial lakes are situated in bedrock hollows under flat or dome-shaped glacial surfaces, they are not expected to expand and drain in sudden outbursts, except perhaps during rapid deglaciation which drastically alters glacier geometry and drainage (Goodwin, 1988). In contrast, when ice-dammed lakes, regardless of whether they are positioned subglacially or marginally, receive water inflow and are gradually made to expand, basal water pressure will increase and the overlying ice will be raised. Eventually, the hydraulic seal of the ice dam will be ruptured, so that water will begin to drain from the lake via a basal pathway. Seepage beginning beneath the ice blockage causes enlargement of the drainage system, initiating a flood under the surrounding ice. In a typical case, though for reasons not yet fully understood, leakage starts before the lake has reached the level which would cause flotation of the ice dam (Björnsson, 1975, 1988). After discharge has begun, pressure from the ice constricts the passageway, and water flow at an early stage in the jökulhlaup correlates primarily with enlargement of the ice tunnel owing to heat from friction against the flowing water and to thermal energy stored in the lake (Nye, 1976; Spring and Hutter, 1981, 1982; Clarke, 1982; Björnsson, 1992). Increasing as an approximate exponent of time over a matter of hours or days, the discharge falls quickly after peaking (Figure 8.3(a)). The recession stage of the hydrograph sets in when tunnel deformation begins to exceed enlargement by melting. The overlying ice may collapse abruptly into the tunnel and seal the lake again, sometimes before it is empty, with new water beginning to accumulate until another jökulhlaup occurs. The timing of bursts depends on what lake level will provide the subglacial water pressure necessary for breaking the hydraulic seal affected by the overburden pressure at the ice dam; therefore, if long-term records of lake levels are available, the elevation at which a jökulhlaup will begin can be predicted with some precision. Since the frequency of outbursts depends on the rate at which a lake is filled, marginal lakes often experience them at the end of the ablation season, when meltwater storage climaxes. Fluctuations in the thickness of the blocking ice, resulting from climatic variations or surges, may modify the outburst cycle or even stop bursts completely.

Occasionally, jökulhlaups are triggered by flotation of the ice dam. Rather than initial drainage from the lake being localized in one narrow conduit, the water is suddenly released as a sheet flow, surging downhill and propagating a subglacial pressure wave, which exceeds the ice overburden and lifts the glacier in order to create space for the water. In this instance, discharge increases faster than can be explained by conduits expanding through melting (Björnsson, 1992, 1997, 2002; Björnsson et al., 2001; Jóhannesson, 2002). The resulting hydrograph (Figure 8.3(b)) presents a rapid rise to its peak, succeeded by a more gentle waning stage. While sheet flows may be followed by drainage through high-capacity conduits, swift floods may instead manage to cause hydrofracturing and force their way englacially from the base of the glacier to its surface (Liestøl, 1977; Roberts et al., 2000; Russell et al., 2000; Roberts, 2002).

Supraglacial lakes, on the other hand, are usually transient features (e.g., Björnsson, 1976), because they form englacial channels and drain down to the subglacial water system. Water that collects in lakes on polar ice masses frequently freezes and unites with the ice during the winter. Where glacial surges occur, floods may come about when a surge terminates and drainage of the water from subglacial cavities switches from a distributed system – conducting water slowly downglacier – to a quickly effective network of tunnels (Kamb et al., 1985; Kamb, 1987). While the outlets for these bodies of water are not fixed, they may be expected to empty into rivers running from the glacier before the surge. As a final case, there have been reports of frequent abrupt releases of water, lasting from minutes to some hours, from diminutive amounts of subglacial or englacial storage, but the nature of these releases is poorly described (Haeberli, 1983; Drieger and Fountain, 1989; Walder and Diedger, 1995).

2.3 Impacts of jökulhlaups

The impact of outbursts corresponds to their discharge rate, the peak and nature of the flow, sediment load, routeway topography and frequency. Depending on the concentration of sediment, jökulhlaups range across a continuum from water floods to hyperconcentrated flows and debris flows. Most jökulhlaups from lakes dammed by ice contain relatively small concentrations of sediment (<40% by weight and 20% by volume) and can be described adequately in terms of standard hydraulics in turbulent water flow (calculating flow by Manning's equation and sediment transport by Einstein's equation). In the most catastrophic floods, intense erosion may occur through cavitation; the enormous velocity of flow reduces absolute pressure, causing air bubbles to form, which collapse and generate shock waves. Conditions for this are favourable in constricted channels at low water depths, producing channelled scabland features, whereas more uniform flow produces a topography of subdued erosion.

Jökulhlaups triggered by the failure of a sediment dam or by volcanic eruptions may temporarily contain high concentrations of sediment. Hyperconcentrated jökulhlaups represent moderately turbulent to laminar flow of non-Newtonian fluids, with sediment concentrations varying between 40% and 70% by weight (20% and 47% by volume). Debris flows are laminar, containing a uniform sediment concentration ranging from 70% to 90% by weight and from 47% to 77% by volume (Costa, 1988). Hyperconcentrated flows produce crudely stratified, less finely sorted deposits than water floods, and the deposits from debris flows are composed of poorly sorted clasts.

River courses and landforms on outwash plains have adjusted to the magnitude of frequent or cyclic jökulhlaups, i.e., those occurring with annual regularity or at intervals of several years. During rare extensive floods, however, erosion, transportation and downstream deposition modify river courses extensively. During the aftermath, flood channels remain destabilized, which may lead to slumping and debris flows.

3 Estimating the discharge of jökulhlaups

3.1 Measured and calculated discharge

Direct measurements of water current velocity and discharge are usually precluded in major jökulhlaups. Conventional stream-gauging stations are rarely set up for recording extreme floods, and even if the streams have been gauged no developed curves are available for comparing stage and streamflow. Large-scale floods overflow the normal riverbanks, distort river courses and damage or destroy instruments. Although a number of jökulhlaups have been gauged some distance downstream from glacier margins, the flood wave may have been significantly attenuated in the intervening interval by passing through such storage bodies as braided channels or lakes.

Indirect methods can be applied to estimate average stream velocity, using hydraulic calculations along with field evidence of channel geometry (including cross-sectional area, heights and dates of flood levels, and channel roughness). This approach employs Manning's equation, which was developed empirically for open, canalized waterways and which relates energy dissipation to the roughness of flood paths. In such an approximation, the surface slope at high water becomes the basis for determining the energy slope. Because of nonuniform flow, computations must take into account the conservation of mass and energy as the flood moves forward (Webb and Jarrett, 2002). The maximum flood stages are reconstructed by inspecting eroded channel margins, finding the highest flotation elevation of the largest boulders that were encased by ice and drifted downstream (ice-rafted erratics), and locating water divides where floodwater spilled over cols. The reconstruction of flood discharge obviously may be complicated when sedimentation at the end of the flood changes the area of the channel cross-section as well as the roughness of the flood course.

3.2 Prediction of discharge by theoretical and empirical methods

For the purpose of risk assessment, both empirical and theoretical methods can be relevant to predicting the probable rate of change of discharge, peak discharge and duration of jökulhlaups.

3.2.1 Subglacial drainage

Theoretical models of jökulhlaups discharging through water tunnels in the glacier base have been derived by Nye (1976), Spring and Hutter (1981) and Clarke (1982). These models are based on the physics of mass continuity, momentum, energy conservation and heat transfer, and they describe how water is driven by a fluid potential gradient through a tunnel of given roughness. From his general model, Nye (1976) derived an analytical solution, which predicted discharge, Q, to rise asymptotically with time as $Q(t) = k(1/t)^4$, if confinement by the overburden was neglected and expansion of the ice tunnel was solely attributed to the instantaneous transfer of frictional heat (i.e., the dissipation of potential energy) from the flowing water to the enclosing ice. Numerical models for simulating the entire hydrograph were derived by Spring and Hutter (1981), who included lake temperature, and Clarke (1982), who added lake geometry. Clarke (1982) successfully simulated jökulhlaups from Lake Hazard (the Yukon, Canada) and Summit Lake (British Columbia, Canada), which are dammed by marginal ice. Björnsson (1992) tested Clarke's (1982) modification of Nye's (1976) general model, applying it to Icelandic jökulhlaups and concluding that the amended model in some respects successfully simulated jökulhlaups from the subglacial lake Grímsvötn in Vatnajökull. In general, the ascending slopes simulated in the graph corresponded to the measured ones. The peak in the computed hydrographs, however, was not as sharp as the actual climaxes, rendering the simulation of the descending limbs unsatisfactory. This presumably happens because, in the model, the tunnel is assumed to be cylindrical. It was only possible to simulate the rapid rise of some outbursts in Iceland (from subglacial as well as marginal lakes) by assuming a lake temperature several degrees above the melting point, which may suggest that thermal energy stored in these bodies of water contributes to tunnel expansion and thereby affects the discharge rate. The values computed for lake temperatures, however, should not be taken seriously, as the theory of heat transfer is questionable. The simulation of such swift jökulhlaups completely failed to illustrate the closure of tunnels and the recession of discharge.

An empirically based regression relation between the peak discharge Q_{max} (in $m^3 s^{-1}$) and the total volume V_t (in $10^6 m^3$) of water passing through subglacial tunnels in jökulhlaups was formulated by Clague and Mathews (1973):

$$Q_{max} = K V_t^b. \qquad (4)$$

For jökulhlaups emerging from ten marginal ice-dammed lakes, which ranged in volume from 10^6 to $10^9 m^3$, Clague and Mathews obtained $K = 75 s^{-1} m^{-0.99}$ and the power coefficient $b = 0.67$. A recent update for 26 marginal ice-dammed lakes throughout the world yielded $K = 46 s^{-1} m^{-1.02}$ and $b = 0.66$ (Walder and Costa, 1996). Referring to eleven jökulhlaups from the Icelandic lake Grímsvötn, Björnsson (1992) found $K = 4.15 \times 10^{-3} s^{-1} m^{-2.52}$ and the power coefficient $b = 1.84$. Clarke (1982) derived theoretical relationships for predicting peak discharge, suggesting that b depended on the lake's contribution of stored thermal energy in proportion to the frictional energy dissipated in the jökulhlaup. The outcome of his calculations was that $b = 0.8$ when the thermal energy originating in the lake was dominant in melting the tunnel and $b = 1.33$ when its share was negligible. Ng and Björnsson (2003) point out that incomplete lake draining, a phenomenon contrasting with common behaviour in marginal lakes dammed by ice, may explain the deviation of Grímsvötn from the statistically derived value.

There is neither a theory nor sufficient empirical data yet available for describing how jökulhlaups drain subglacially at a faster rate than can be explained by the expansion of conduits through melting. This quicker drainage involves the ice overburden being exceeded by water pressure, which lifts the glacier and allows space for the water. Initially propagated as a pressure wave, the water then moves forward in a sheet flow, prior to draining through conduits (Björnsson, 1997, 2002; Björnsson et al., 2001; Jóhannesson, 2002).

3.2.2 Subaerial drainage

A subaerial burst following the sudden breaching of an ice dam is typically over sooner than a flood exiting through a basal ice tunnel. Discharge in the former outburst may increase linearly with time and, for a given lake volume, peak discharges are significantly higher than for drainage through subglacial tunnels. The size of the outlet and the quantity of the impounded water body determine the peak discharge through the breach.

Physical models of drainage over a subaerial breach have been drawn up along similar lines to those for drainage through subglacial tunnels. Walder and Costa (1996) successfully simulated hydrographs for Lake George in Alaska. The widening of the opening is assumed to be controlled by melting, while the lake discharge spilling over the breach is dictated by the hydraulic conditions of critical flow through an open channel. Furthermore, an empirical power-law was derived in the form of eq. (4), with $K = 1100$ s^{-1} $m^{-1.69}$ and $b = 0.44$, i.e., similar figures to those for regression during known outbursts that breached man-made earthen dams, in which instances $K = 1200$ s^{-1} $m^{-1.56}$ and $b = 0.48$ (Costa, 1988). Raymond and Nolan (2000) made the first attempt to identify and describe by a physical model the processes controlling unstable drainage over a spillway with an ice floor; in other words, they formulated criteria for when discharge water would melt the spillway path down faster than the lake level dropped. Their work successfully explained observations of supraglacial drainage from Black Rapids Glacier, Alaska.

4 Case studies

4.1 Floods from sediment-dammed lakes

Floods in which lakes breach sediment dams carry large amounts of debris. Generally unpredictable, they are regarded as the most dangerous kind of outbursts, but have been successfully prevented at many hazardous locations by artificial drains through the sediment dams, which restrict lake depths.

Advances and subsequent retreats of glaciers in the Little Ice Age created many unstable moraine dams, with examples reported from the European Alps, the Himalaya and Peru. An Austrian valley was flooded in 1874 when the terminal moraine of the Madatschferner glacier gave way and the lake escaped from in front of the retreating ice. The French Tête Rousse débâcle of 1892, released 2×10^5 m^3 of water and 8×10^5 m^3 of sediment, killed 175 people (Lliboutry, 1971). In 1926 a flood released from the Himalayan Shyok glacier devastated a village and cultivated land 400 km from the source (Mason, 1929), while the Peruvian Jancarurish outburst of 1950 discharged 2×10^6 m^3 of water and 3 $\times 10^6$ m^3 of sediment (Lliboutry et al., 1977). In British Columbia, debris flows up to 20 m thick have travelled as far as 20 km downvalley after breaching terminal moraines (Evans and Clague, 1994). As still another example, an ice avalanche from the Langmoche glacier in Nepal broke a moraine dam in 1985, resulting in a flood that destroyed bridges, houses and a hydroelectric plant (Vuichard and Zimmermann, 1986, 1987).

4.2 Floods from marginal ice-dammed lakes

Jökulhlaups from ice-dammed lakes at glacial edges have posed a serious threat to roads, power lines and the human population in many countries of Europe, North and South America, Asia and New Zealand. In most countries, however, thinning of the ice dams by the warm climate of the twentieth century has led to smaller and more frequent outbursts from marginal lakes dammed by ice. Walder and Costa (1996) recently compiled a comprehensive review of outburst floods from glacially dammed lakes.

Some of the best-documented modern lakes blocked by ice are found in North America. Maag (1969) identified 125 lakes with ice dams on Axel Heiberg Island, Canada, and studies of several ice-dammed lakes in that country have been of major importance for the scientific understanding of jökulhlaups, for example, investigations of Tulsequah Lake, British Columbia (Marcus, 1960); Summit Lake, British Columbia (Mathews, 1965; Clarke, 1982; Mathews and Clague, 1993); Hazard Lake, Yukon Territory (Clarke, 1982); and Flood Lake, British Columbia (Clarke and Waldron, 1984).

Post and Mayo (1971) identified 750 lakes behind marginal ice dams in Alaska; of these, many have been thoroughly researched, such as Lake George (Stone, 1963; Hulsing, 1981; Lipscomb, 1989), Snow River (Chapman, 1981) and Strandline Lake (Sturm and Benson, 1985). The largest observed terrestrial flood from a breached impoundment took place on 8 October 1986, below the ice-dammed Lake Russell, Alaska. A surge had caused Hubbard Glacier, North America's largest tidewater glacier, to advance across the entrance of Russell Fjord, turning it into Lake Russell. A moraine was pushed up in front of the advancing glacier, raising inland water 25.5 m asl before the lake overflowed and cut the dam at the junction of the ice and the valley wall. The entire ice dam was removed within a few hours, with released water amounting to 5.4×10^9 m^3 and the peak discharge reaching 1.5×10^6 m^3 s^{-1} (Mayo, 1988, 1989). A similar event, though somewhat smaller, occurred in the summer of 2002.

In the Swiss Alps, damage from glacial floods occurs, on average, biennially; 60–70% of the outbursts originate in lakes at glacial margins and 30–40% in pockets of water within glaciers (Haeberli, 1983). Nowadays, about 15 ice-marginal lakes in Iceland drain by means of jökulhlaups (Thorarinsson, 1939; Björnsson, 1997) and frequently cause problems for bridges and other road structures. A pioneering thesis was written by Liestøl (1956) on glacially dammed lakes in Norway. Outburst floods have been scientifically described in Greenland (Schytt, 1956; Lister and Willie, 1957; Dawson, 1983; Russell, 1989), Argentina (Heinsheimer, 1954, 1958; Fernández et al., 1991), Pakistan (Gunn et al., 1930; Hewitt, 1982), and the former USSR (Glazyrin and Sokolov, 1976; Krenke and Kotlyakov, 1985; Konovalov, 1991).

The largest outbursts from ice-marginal lakes known through geological records occurred at the end of the last glaciation (Bretz, 1969; Baker, 1973, 1983; Teller and Clayton, 1983; Waitt, 1984; Clarke et al., 1984; Bond et al., 1992; Shaw, 1996, 2002). Jökulhlaups drained lakes that were dammed by margins of the Laurentide ice sheet in North America (Lake Agassiz and others) and by the Fennoscandinavian ice sheet (situated over the Baltic Sea). The Eurasian ice sheet had blocked off lakes, the outbursts of which emptied into northward-draining rivers in Siberia. Pulses of meltwater amassed in major proglacial lakes were released into the Mississippi and St Lawrence Rivers and the Arctic and Hudson Straits, eventually to enter the North Atlantic Ocean. These floods emitted enormous amounts of freshwater, which may have affected ocean circulation and deep-water production in the North Atlantic (Teller, 1990). Giant jökulhlaups from the Laurentide ice sheet during the last glacial cycle may have played a part in depositing the Heinrich layers in the North Atlantic (Heinrich, 1988; Alley and MacAyeal, 1994; Colman, 2002).

Floods from glacial Lake Missoula at the border of the Laurentide ice sheet repeatedly transported around 2×10^{12} m^3 (2000 km^3) of water in less than four days, with flow velocities of up to 30 m s^{-1} and estimated peak discharges of 2×10^7 m^3 s^{-1} (Baker, 1973; Baker and Bunker, 1985; O'Connor and Baker, 1992), so that they comprise the greatest discharges of freshwater known in geological history. The Lake Missoula floods lasted approximately one week, and the magnitude of their peak discharge indicates either supraglacial release or suddenly failing barriers, because such discharges cannot be explained by the gradual enlargement of basal ice tunnels. The floods from Lake Missoula produced the huge channelled scablands on the Columbia Plateau, Washington State.

4.3 Floods from subglacial lakes

The best-documented floods from subglacial lakes occur in Iceland, where jökulhlaups regularly drain six lakes of this type. They are located in hydrothermal areas, where geothermal activity continuously melts ice on the glacier bed, creating a depression in the glacier surface, under which water accumulates until released in floods. The best-known jökulhlaups drain the lake Grímsvötn in Vatnajökull glacier (Björnsson, 1974, 1988; Thorarinsson, 1974) and occur at 1- to 10-year intervals. Their peak discharge ranges from 600 to $4-5 \times 10^5$ m^3 s^{-1} at the outwash plain, their duration is two days to four weeks and their total volume at each event is $0.5-4.0 \times 10^9$ m^3. The greatest floods have peaked in less than one week and subsided in two days, whereas the smaller ones peak in two to three weeks, after which they usually terminate in approximately one week (Figures 8.4, 8.5 and 8.6). Jökulhlaups from Grímsvötn flow a distance of some 50 km beneath the ice to the glacier terminus by the Skeiðarársandur outwash plain (Figure 8.7). The most violent Grímsvötn jökulhlaups have flooded this

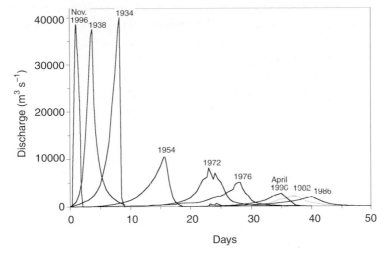

Figure 8.4 *Discharge hydrograph of various jökulhlaups from Grímsvötn, Iceland*

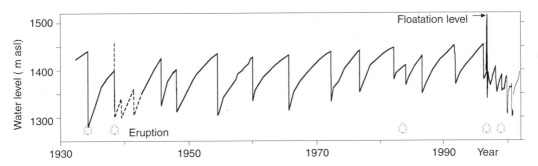

Figure 8.5 *The water level of lake Grímsvötn, 1930–2000. The crater level rises until a jökulhlaup takes place. In 1996 the lake rose to the level required for floatation of the ice dam*

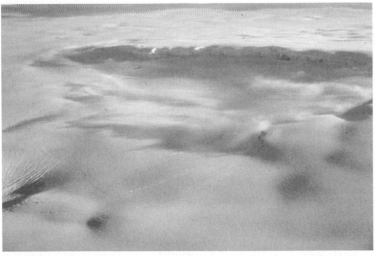

Figure 8.6 *A 10-km-wide depression formed by continuous melting at the subglacial hydrothermal area of Grímsvötn, Vatnajökull, Iceland. View towards the 400-m-high caldera rim, Mt Grímsfjall. Photo: H. Björnsson*

Figure 8.7 *A glacial river coursing through an outwash plain toward the coast (Skeiðarársandur, Iceland); length of the bridge: 900 m.* Photo: H. Björnsson

entire plain, measuring 1000 km² (e.g., Björnsson, 1988). Typically, lake drainage begins at water pressures 6–7 bar lower than those exerted by the overburden at the ice dam. Conduits are enlarged in the course of several days or up to three weeks, so that the floods develop from small initial discharges which have been explained by classic jökulhlaup theories. Occasionally, nevertheless, the lake level rises until the ice dam floats; in this case, discharge increases faster than can be accommodated by the melting of conduits, and the glacier is lifted along the flow path as the water forces open space for itself. An outburst of this type took place in November 1996, when meltwater from the Gjálp eruption collected in Grímsvötn before flowing out in a disastrous flood: this is discussed below in connection with volcanism.

Other well-known jökulhlaups from subglacial Icelandic lakes originate at the Skaftá cauldrons (Figure 8.8), 10–15 km northwest of Grímsvötn, and result in floods of 50–350 × 10⁶ m³. They rise over one to three days to a peak discharge of 200–1500 m³ s⁻¹, before receding slowly for one to two weeks. So far, the hydrographs of these outbursts have not been simulated theoretically. Although their speedy rise suggests a water temperature at the reservoir well above the melting point (10–20°C), this must not be considered definite, because the theory of heat transfer is debatable.

4.4 Floods related to subglacial volcanic eruptions

In active volcanic areas of Iceland and of North and South America, eruptions frequently trigger catastrophic glacial floods (Thorarinsson, 1957, 1958; Sturm *et al.*, 1986; Björnsson, 1988; Trabant and

Figure 8.8 *A 3-km-wide and 150-m-deep cauldron after a jökulhlaup in Vatnajökull, Iceland.*
Photo: H. Björnsson

Meyer, 1992; Pierson, 1995; Guðmundsson *et al.*, 1997; Thouret, this volume). Lava erupts directly into water stored under the ice and the boiling water transfers heat readily. Meltwater created by sub-glacial eruptions may discharge instantly toward the glacier margins, depending on the rate of volcanic melting. The bursts may either travel subglacially in tunnels or spread out in a broad sheet, heaving up the ice. Moreover, the water may overtop the glacier, melt channels on its surface, and crack off giant blocks of ice. Alternatively, the meltwater forming beneath substantial ice masses may accumulate in subglacial lakes, eventually to exit from them in jökulhlaups. A recent example of such a flood occurred when meltwater from the 1996 Gjálp eruption gathered in the subglacial lake Grímsvötn, entering at a rate of up to 5×10^3 m^3 s^{-1}, before draining from the lake in a catastrophic flood, with discharge increasing almost linearly to a peak of 4×10^4 m^3 s^{-1} in 16 hours. Within a period of 40 hours, 3.2×10^9 m^3 of water had left the lake. Even though the outburst had begun as a flood wave, so that the initial discharge was of the order of 5×10^3 to 5×10^4 m^3 s^{-1}, an exponential increase would have taken around two days to reach the estimated peak.

Eruptions of snow-clad volcanoes may generate huge floods when pyroclastic, turbulent fluid melts the icy cover, producing slushy, debris-rich flows which speed down valley slopes. Thouret (this volume) describes how a relatively small eruption from the snow-capped volcano Nevado del Ruiz (5400 m high) in Colombia in 1985 caused disastrous lahars that killed more than 23 000 inhabitants of the town Armero, 72 km away. In 1966–68 an eruption at Mt Redoubt, Alaska, melted or washed away 6×10^7 m^3 of glacial ice (Sturm *et al.*, 1986) and in 1989–90 another eruption achieved double that amount (Trabant and Meyer, 1992).

A contemporary description of the outburst during the 1362 Öræfajökull eruption in Iceland pro-vides an example of abrupt, destructive jökulhlaups from an ice-capped stratovolcano (Thorarinsson 1957, 1958). In less than one day, the flood may have reached its peak of at least 10^5 m^3 s^{-1} (Thorarinsson, 1958). Water originating in the 2000 m high vicinity of the summit streamed sub-glacially down the slopes before bursting out to wash away several farmsteads and to leave dead ice, sediment and hummocks covering the lowlands at a depth of 2–4 m. Outwash plains appeared where sea depths had previously measured 50 m. A contemporary record (report by Rev. Jón Thorláksson, cited by Henderson (1818); Thorarinsson, 1958: p. 31) describes how *several floods of water gushed out, the last of which was the greatest. When these floods were over, the glacier itself slid forwards over the plane ground, just like melted metal poured out of a crucible. The water now rushed down on the earth side without intermission, and destroyed what little of the pasture grounds remained . . . Things now assumed quite a different appearance. The glacier itself burst and many icebergs were run down [sic] quite to the sea, but the thickest remained on the plain at a short distance from the foot of the mountain . . . We could only proceed with the utmost danger, as there was no other way except between the ice-mountain and the glacier that had slid forwards over the plain, where the water was so hot that the horses almost got unmanageable.*

A less devastating eruption took place in AD 1727 (Thorarinsson, 1958).

The swiftest, greatest outbursts of glacial meltwater noted in world history accompany volcanic eruptions in the Katla caldera, which rests under the Icelandic ice cap Mýrdalsjökull. Occurring on average twice per century, the outbursts have durations of 3–5 days, peak discharges estimated at $10^5–10^6\,\mathrm{m}^3\,\mathrm{s}^{-1}$, and total volumes of $1–8 \times 10^9\,\mathrm{m}^3$ (Thorarinsson, 1957, 1975; Maizels, 1989, 1995; Tómasson, 1996; Larsen, 2000). Immense blocks of ice break off the glacial margins, and a mixture of water, ice, volcanic emissions and sediment surges over the outwash plain at velocities of $5–15\,\mathrm{m}\,\mathrm{s}^{-1}$. During a segment of the flow event, the water may consist of a hyperconcentrated fluid–sediment mixture. The amount of volcanic debris produced per event and carried away by the water has been estimated to range from 0.7 to $1.6 \times 10^9\,\mathrm{m}^3$, or be of the order of $2 \times 10^9\,\mathrm{t}$ (Tómasson, 1996; Larsen, 2000). During the last eruption in 1918, the neighbouring coastline advanced 3 km into the sea. Although many loose deposits have been washed away in succeeding years, the affected seashore remains 2.2–2.5 km further south than its position in 1660. Jökulhlaups have played a dominant role in building up the outwash plain, which like other such plains in Iceland is called a *sandur* – a term that has acquired international use. Marine sediments containing debris transported from the eruption site are found in the ocean several hundred kilometres to the south. Jökulhlaups from Katla have threatened local communities, damaged vegetation, ruptured roads on the alluvial flats surrounding the ice cap and even generated flood waves in coastal waters. On the northern side of Mýrdalsjökull, the outbursts have eroded deep canyons in the bedrock.

The largest, most catastrophic jökulhlaups in Iceland, with peak discharges of the order of $10^6\,\mathrm{m}^3\,\mathrm{s}^{-1}$, were caused by prehistoric eruptions in the voluminous, ice-filled calderas of Bárðarbunga and Kverkfjöll in the northern reaches of Vatnajökull (see Tómasson, 1973; Björnsson, 1988; Björnsson and Einarsson, 1991; Knudsen and Russell, 2002; Waitt, 2002). Floods sweeping down Jökulsá á Fjöllum carved a conspicuous scabland and a deep canyon (Jökulsárgljúfur; Figure 8.9); erosion by cavitation is considered to have been a significant factor in their creation.

5 Concluding remarks and future prospects

This chapter has reviewed the current knowledge of glacial floods and their sources in impounded glacial meltwater, and on the dams holding back the water and dam failure. Also, I have described the characteristics of flood discharges, their impact in the proglacial zone, and their significance for society.

Despite some outburst sources lying hidden, field inspection or remote sensing can detect most of them. Potential jökulhlaup hazards can be recognized by examining the properties of reservoir dams and evaluating their stability. While the onset of outbursts cannot be timed exactly, those stemming from ice-dammed lakes may be predicted empirically – though no more precisely than to the nearest month, based on records of lake levels at the beginning of previous outbursts. The discharge rates of change and impacts of jökulhlaups depend on their release mechanisms, i.e., whether a sediment dam suddenly fails, an ice dam is abruptly lifted or overtopped, or the dam seal begins to leak through a subglacial tunnel at gradually increasing rates. The nature of the flood hazard depends on reservoir

Figure 8.9 *A canyon eroded and carved by jökulhlaups (the 300-m-deep Jökulsárgljúfur, north of Vatnajökull, Iceland)*
Photo: Björn Rúriksson, reproduced with permission.

volume and the temperature of the fluid, along with the character and content of the sediment that is entrained in the flow. Empirical and theoretical relations help to forecast the possible magnitude of jökulhlaups. Theoretical models of flood discharge may help to predict the slopes of the ascending limbs of the hydrograph if the drainage occurs through tunnels under an ice dam. However, formulating such models requires information on water temperature and bottom roughness, and these data are rarely available. At present, theoretical models are unreliable for predicting peak discharge, although empirical relationships may provide some estimate, given the total volume of the reservoir at the beginning of the outburst. Once the flood has started, the subsequent discharge rate may be predicted on the grounds of reservoir volume and of flow measured during the initial phase of the flood. No theory is yet available for subglacially draining jökulhlaups, which increase faster than can be explained by conduits widening through melting.

Glacier lakes are ephemeral features, dependent on variable conditions such as the advance or recession of glacial termini. Continuous monitoring of impounded water and inflow rates is needed for assessing risks and delineating danger zones. Outbursts from lakes behind sediment dams at glacial margins can be prevented by installing drainage pipes to control water levels. Some englacial sources of floods may be difficult to detect, and subglacial volcanic eruptions are hard to predict, but maps of bedrock topography and the glacier surface can be used to identify potential hazard zones with regard to jökulhlaups. Field studies of landforms, sediments and other evidence of varying flow processes would improve understanding of the outburst mechanism. At present, both empirical and theoretical calculations suffer from inaccurate estimates of discharges and particularly of peak flows. Further studies are required on the physics of how jökulhlaups begin to enable more precise timing of their onset. Additionally, studies of supraglacial outbursts that melt gorges through ice dams and of subglacial bursts that lift the glacier to propagate a flood wave are still at an early stage.

Studies of the physics of modern glacial floods of every magnitude may fundamentally improve our understanding of palaeofloods in the late Pleistocene and early Holocene. They released enormous amounts of freshwater into the ocean, perhaps altering deep-water circulation and the global climate. In this respect, the glacial floods of today may become a key to the past.

References

Alley, R.B. and MacAyeal, D.R., 1994. Ice-rafted debris associated with binge/purge oscillations of the Laurentide Ice Sheet. *Paleooceanography*, **9**: 503–11.

Baker, V.R., 1973. *Paleohydrology and sedimentology of Lake Missoula flooding in Eastern Washington*. Geological Society of America Special Paper 144.

Baker, V.R., 1983. Large scale paleohydrology. In Gregory, K.J. (ed.), *Background to palaeohydrology*. Chichester: Wiley, 455–78.

Baker, V.R. and Bunker, R.C., 1985. Cataclysmic late Pleistocene flooding from glacial Lake Missoula: a review. *Quaternary Science Reviews*, **4**: 1–41.

Björnsson, H., 1974. Explanation of jökulhlaups from Grímsvötn, Vatnajökull, Iceland. *Jökull*, **24**: 1–26.

Björnsson, H., 1975. Subglacial water reservoirs, jökulhlaups and volcanic eruptions. *Jökull*, **25**: 1–14.

Björnsson, H., 1976. Marginal and supraglacial lakes in Iceland. *Jökull*, **26**: 40–51.

Björnsson, H., 1988. *Hydrology of ice caps in volcanic regions*. Reykjavík: Societas Scientarium Islandica, 45.

Björnsson, H., 1992. Jökulhlaups in Iceland: prediction, characteristics and simulation. *Annals of Glaciology*, **16**: 95–106.

Björnsson, H., 1997. Grímsvatnahlaup fyrr og nú. In Haraldsson, H. (ed.), *Vatnajökull. Gos og hlaup 1996*. Reykjavík, Iceland: Icelandic Public Roads Administration, 61–77.

Björnsson, H., 1998. Hydrological characteristics of the drainage system beneath a surging glacier. *Nature*, **395**: 771–74.

Björnsson, H., 2002. Subglacial lakes and jökulhlaups in Iceland. *Global and Planetary Change*, **35**: 255–71.

Björnsson, H. and Einarsson, P., 1991. Volcanoes beneath Vatnajökull, Iceland. Evidence from radio echo-sounding, earthquakes and jökulhlaups. *Jökull*, **40**: 147–68.

Björnsson, H., Pálsson, F., Guðmundsson, M.T. and Flowers, G.E., 2001. The extraordinary 1996 jökulhlaup from Grímsvötn, Vatnajökull, Iceland. *Eos (Transactions, American Geophysical Union)*, **82**(47): F528.

Bond, G., Heinrich, H., Broecker, W., Labeyrie, L., McManus, J., Andrews, J., Huon, S., Jantschik, R., Clasen, S., Simet, C., Tedesco, K., Klas, M., Bonani, G. and Ivy, S., 1992. Evidence for massive discharges of icebergs into the North Atlantic Ocean during the last glacial period. *Nature*, **360**: 245–49.

Bretz, J.H., 1925. The Spokane flood beyond the Channeled Scablands. *Journal of Geology*, **33**: 97–115.

Bretz, J.H., 1969. The Lake Missoula floods and Channeled Scabland. *Journal of Geology*, **77**: 505–43.

Chapman, D.L., 1981. *Jökulhlaups on Snow River in southeastern Alaska: a compilation of recorded and inferred hydrographs and a forecast procedure*. National Oceanographic and Atmospheric Administration Technical Memorandum NWS AR-31. Anchorage, Alaska: NOAA.

Clague, J.J. and Mathews, W.H., 1973. The magnitude of jökulhlaups. *Journal of Glaciology*, **12**: 501–504.

Clarke, G.K.C., 1982. Glacier outburst flood from "Hazard Lake", Yukon Territory, and the problem of flood magnitude prediction. *Journal of Glaciology*, **28**: 3–21.

Clarke, G.K.C. and Waldron, D.A., 1984. Simulation of the August 1979 sudden discharge of glacier-dammed Flood Lake, British Columbia. *Canadian Journal of Earth Sciences*, **21**: 502–504.

Clarke, G.K.C., Mathews, W.H. and Pack, R.T., 1984. Outburst floods from glacial Lake Missoula. *Quaternary Research*, **22**: 289–99.

Colman, S.M., 2002. A fresh look at glacial floods. *Science*, **296**: 1251–52.

Costa, J.E., 1988. Rheologic, geomorphic, and sedimentologic differentiation of water floods, hyperconcentrated flows, and debris flows. In Baker, V.R., Kochel, R.C. and Patton, P.C. (eds), *Flood geomorphology*. New York: Wiley, 113–22.

Dawson, A.G., 1983. Glacier-dammed lake investigations in the Hullet Lake area, South Greenland. *Meddelelser om Grönland, Geoscience*, **11**, 22 pp.

Driedger, C.L. and Fountain, A.G., 1989. Glacier outburst floods at Mount Rainier, Washington. *Annals of Glaciology*, **13**: 51–55.

Evans, S.G. and Clague, J.J., 1994. Recent climatic change and catastrophic geomorphologic processes in mountain environments. *Geomorphology*, **10**: 107–28.

Fernández, P.C., Fornero, L., Maza, J. and Yañez, H., 1991. Simulation of flood waves from outburst of glacier-dammed lake. *Journal of Hydraulic Engineering*, **117**: 42–53.

Glazyrin, G.E. and Sokolov, L.N., 1976. Forecasting of flood characteristics caused by glacier lake outbursts. *Materialy Gliatsiologicheskikh Issledovanii*, **26**: 78–85.

Goodwin, L.D., 1988. The nature and origin of a jökulhlaup near Casey Station, Antarctica. *Journal of Glaciology*, **34**: 95–101.

Guðmundsson, M.T., Björnsson, H. and Pálsson, F., 1995. Changes in jökulhlaup sizes in Grímsvötn, Vatnajökull, Iceland, 1934–1991, deduced from in situ measurements of subglacial lake volume. *Journal of Glaciology*, **41**: 263–72.

Guðmundsson, M.T., Sigmundsson, F. and Björnsson, H., 1997. Ice–volcano interaction of the 1996 Gjálp subglacial eruption, Vatnajökull, Iceland. *Nature*, **389**: 954–57.

Gunn, J.P., Todd, H.J. and Mason, K., 1930. The Shyok flood, 1929. *Himalayan Journal*, **2**: 35–47.

Haeberli, W., 1983. Frequency and characteristics of glacier floods in the Swiss Alps. *Annals of Glaciology*, **4**: 85–90.

Heinrich, H., 1988. Origin and consequences of cyclic ice rafting in the northeast Atlantic Ocean during the past 130,000 years. *Quaternary Research*, **29**: 142–52.

Heinsheimer, G.J., 1954. Der Durchbruch des Morengletschers, Lago Argentino, Patagonien, 1953. *Zeitschrift für Gletscherkunde und Glaziologie*, **3**: 33–38.

Heinsheimer, G.J., 1958. Zur Hydrologie und Glaziologie des Lago Argentino und Ventisquero Moreno III. *Zeitschrift für Gletscherkunde und Glaziologie*, **4**: 61–72.

Henderson, E., 1818. *Iceland; or the journal of a residence in that island during the years 1814 and 1815, vol. I. Oliphant.* Edinburgh: Waugh and Innes.

Hewitt, K., 1982. Natural dams and outburst floods of the Karakoram Himalaya. In Glen, J.W. (ed.), *Hydrological aspects of alpine and high-mountain areas.* IAHS Publication 138. Wallingford: IAHS Press, 259–69.

Hulsing, H., 1981. *The breakout of Alaska's Lake George.* Popular publications of the U.S. Geological Survey. Reston, VA: USGS and Washington, DC: Government Printing Office.

Jóhannesson, T., 2002. Propagation of a subglacial flood wave during the initiation of jökulhlaup. *Hydrological Sciences Journal*, **47**: 417–34.

Kamb, B., 1987. Glacier surge mechanism based on linked cavity configuration of the basal water conduit system. *Journal of Geophysical Research*, **92**: 9083–100.

Kamb, B., Raymond, C.F., Harrison, W.D., Engelhardt, H., Echelmeyer, K.A., Humphrey, N., Burgman, M.M. and Pfeffer, T., 1985. Glacier surge mechanism: 1982–1983 surge of Variegated Glacier, Alaska. *Science*, **227**: 469–79.

Knudsen, Ó. and Russell, A.R., 2002. Jökulhlaup deposits at Ásbyrgi, northern Iceland: sedimentology and implications of flow type. In Snorrason, A., Finnsdóttir, H.P. and Moss, M.E. (eds), *The extremes of the extremes: extraordinary floods.* IAHS Publication 271. Wallingford: IAHS Press, 107–12.

Konovalov, V.G., 1991. Methods for computing the onset date and daily discharge hydrograph of the outburst from Mertzbacher Lake, Northern Inylchek Glacier, Tien Shan. In Kotlyakov, V.M., Ushakov, A. and Glazovsky, A. (eds), *Glaciers–ocean–atmosphere interactions.* IAHS Publication 208. Wallingford: IAHS Press, 359–66.

Krenke, A.N. and Kotlyakov, V.M., 1985. USSR case study: catastrophic floods. In: Young, G.J. (ed.), *Techniques for prediction of runoff from glacierized areas*. IAHS Publication 149. Wallingford: IAHS Press, 115–24.

Larsen, G., 2000. Holocene eruptions within the Katla volcanic system, Iceland: notes on characteristics and environmental impact. *Jökull*, **50**: 1–28.

Liestøl, O., 1956. Glacier dammed lakes in Norway. *Norsk Geografisk Tidsskrift*, **15**: 122–49.

Liestøl, O., 1977. Setevatnet, a glacier dammed lake in Spitsbergen. *Norsk Polarinstitutt Årbok*, **1975**: 31–35.

Lipscomb, S.W., 1989. *Flow and hydraulic characteristics of the Knik-Matanuska River estuary, Cook Inlet, south-central Alaska*. U.S. Geological Survey Water Resources Investigations Report 89–4064. Denver, CO: USGS, Water Resources Division and Washington, DC: Government Printing Office.

Lister, H. and Wyllie, P.J., 1957. The geomorphology of Dronning Louise Land. *Meddelelser om Grönland*, **158** (1).

Lliboutry, L., 1971. Les Catastrophes glaciaires. *La Recherche*, **2**: 417–25.

Lliboutry, L., Arnao, B.M., Pautre, A. and Schneider, B., 1977. Glaciological problems set by the control of dangerous lakes in the Cordillera Blanca, Peru. I. Historical failures of morainic dams, their causes and prevention. *Journal of Glaciology*, **18**: 239–54.

Maag, H.U., 1969. *Ice dammed lakes and marginal glacial drainage on Axel Heiberg Island, Canadian Arctic Archipelago*. Axel Heiberg Osland Research Reports. Jacobsen-McGill Arctic Research Expedition 1959–1962. McGill University, Montreal.

Maizels, J.K., 1989. Sedimentology and paleohydrology of Holocene flood deposits in front of a jökulhlaup glacier, south Iceland. In Beven, K. and Carling, P. (eds), *Floods. Hydrological, sedimentological and geomorphological implications*. Chichester: Wiley, 239–52.

Maizels, J., 1995. Sediments and landforms of modern proglacial terrestrial environments. In Menzies, J. (ed.), *Modern glacial environments*. Oxford: Butterworth-Heinemann, 365–416.

Marcus, M.G., 1960. Periodic drainage of glacier-dammed Tukequant Lake, British Columbia. *Geographical Review*, **50**: 89–106.

Mason, K., 1929. Indus floods and Shyok glaciers. *Himalayan Journal*, **1**: 10–29.

Mathews, W.H., 1965. Two self-dumping ice-dammed lakes in British Columbia. *Geographical Review*, **55**: 46–52.

Mathews, W.H. and Clague, J.J., 1993. The record of jökulhlaups from Summit Lake, northwestern British Columbia. *Canadian Journal of Earth Sciences*, **30**: 499–508.

Mayo, L.R., 1988. Hubbard Glacier near Yakutat, Alaska – the ice damming and breakout of Russel Fiord/Lake, 1986. In Moody, D.W., *et al.* (compilers), *National water summary 1986; hydrologic events and ground-water quality*. U.S. Geological Survey Water-Supply Paper 2325. Reston, VA: USGS and Washington, DC: Government Printing Office, 42–49.

Mayo, L.R., 1989. Advances of Hubbard Glacier and 1986 outburst of Russel Fiord, Alaska, U. S. A. *Annals of Glaciology*, **13**: 189–94.

Ng, F. and Björnsson, H., 2003. On the Clague-Matthews relation for jökulhlaups. *Journal of Glaciology*, **49**.

Nye, J.F., 1976. Water flow in glaciers: jökulhlaups, tunnels and veins. *Journal of Glaciology*, **17**: 181–207.

O'Connor, J.E. and Baker, V.R., 1992. Magnitudes and implications of peak discharges from glacial Lake Missoula. *Geological Society of America Bulletin*, **104**: 267–79.

Oswald, G.K.A. and Robin, G. de Q., 1973. Lakes beneath the Antarctic Ice Sheet. *Nature*, **245**: 251–54.

Pierson, T.C., 1995. Flow characteristics of large eruption-triggered debris flows at snow-clad volcanoes. Constraints for debris-flow models. *Journal of Volcanology and Geothermal Research*, **66**: 283–94.

Pierson, T.C., Janda, R.J., Thouret, J.-C. and Borrero, C.A., 1990. Perturbation and melting of snow and ice by the 13 November 1985 eruption of Nevado del Ruiz, Colombia, and consequent mobilization, flow and deposition of lahars. *Journal of Volcanology and Geothermal Research*, **41**: 17–66.

Post, A. and Mayo, L.R., 1971. *Glacier dammed lakes and outburst floods in Alaska*. U.S. Geological Survey Hydraulic Investigations Atlas HA-455. Washington, DC: Government Printing Office.

Raymond, C.F. and Nolan, M., 2000. Drainage of a glacial lake through an ice spillway. In Nakawo, M. Raymond, C.F. and Fountain, A. (eds), *Debris-covered glaciers*. IAHS Publication 264. Wallingford: IAHS Press, 199–207.

Ridley, J.K., Cudlip, W. and Laxon, S.W., 1993. Identification of subglacial lakes using ERS-1 radar altimeter. *Journal of Glaciology*, **39**: 625–34.

Roberts, M.J., 2002. Controls on supraglacial outlet development during glacial outburst floods. Unpublished Ph.D. Thesis, Staffordshire University, UK.

Roberts, M.J., Russell, A.J., Tweed, F.S. and Knudsen, Ó., 2000. Ice fracturing during jökulhlaups: implications for englacial floodwater routing and outlet development. *Earth Surface Processes and Landforms*, **25**: 1–18.

Russell, A.J., 1989. A comparison of two recent jökulhlaups from an ice-dammed lake, Söndre Strömfjord, West Greenland. *Journal of Glaciology*, **35**: 157–62.

Russell, A.J., 1993. Supraglacial lake drainage near Söndre Strömfjord, Greenland. *Journal of Glaciology*, **39**: 431–33.

Russell, A.J., Tweed, F.S. and Knudsen, Ó., 2000. Flash flood at Sólheimajökull heralds the reawakening of an Icelandic subglacial volcano. *Geology Today*, **16**: 102–106.

Schytt, V., 1956. Lateral drainage channels along the northern side of the Moltke glacier, North-West Greenland. *Geografiska Annaler*, **38A**: 64–77.

Shaw, J., 1996. A meltwater model for Laurentide subglacial landscapes. In McCann, S.B. and Ford, D.C. (eds), *Geomorphology sans frontières*. Chichester: Wiley, 181–236.

Shaw, J., 2002. The meltwater hypothesis for subglacial landforms. *Quaternary International*, **90**: 5–22.

Shreve, R., 1972. Movement of water in glaciers. *Journal of Glaciology*, **62**: 205–14.

Siegert, M.J., Ellis-Evans, J.C., Tranter, M., Mayer, C., Petit, J.-R., Salamatin, A. and Priscu, J.C., 2001. Physical, chemical and biological processes in Lake Vostok and other Antarctic subglacial lakes. *Nature*, **414**: 603–609.

Spring, U. and Hutter, K., 1981. Numerical studies of jökulhlaups. *Cold Regions Science and Technology*, **4**: 221–44.

Spring, U. and Hutter, K., 1982. Conduit flow of a fluid through its solid phase and its application to interglacial channel flow. *International Journal of Engineering Science*, **20**: 327–63.

Stone, K.H., 1963. Alaskan ice-dammed lakes. *Annals of the Association of American Geographers*, **53**: 332–49.

Sturm, M. and Benson, C.S., 1985. A history of jökulhlaups from Strandline Lake, Alaska, U.S.A. *Journal of Glaciology*, **31**: 272–80.

Sturm, M., Benson, C.S. and MacKeith, P., 1986. Effects of the 1966–68 eruptions of Mount Redoubt on the flow of Drift Glacier, Alaska, U.S.A. *Journal of Glaciology*, **32**: 355–62.

Teller, J.T., 1990. Volume and routing of Late-Glacial runoff from the southern Laurentide ice sheet. *Quaternary Research*, **34**: 12–23.

Teller, J.T. and Clayton, L. (eds), 1983. *Glacial Lake Agassiz*. Canada: Geological Association of Canada Special Paper, 26.

Thorarinsson, S., 1939. The ice dammed lakes of Iceland with particular reference to their values as indicators of glacier oscillations. *Geografiska Annaler*, **21A**: 216–42.

Thorarinsson, S., 1957. The jökulhlaup from the Katla area in 1955 compared with other jökulhlaups in Iceland. *Jökull*, **7**: 21–25.

Thorarinsson, S., 1958. The Öræfajökull eruption of 1362. *Acta Naturalia Islandica*, 2.

Thorarinsson, S., 1974. *Vötnin strið. Saga Skeiðarárhlaups og Grímsvatnagosa*. Menningersjóður, Reykjavík, Iceland.

Thorarinsson, S., 1975. Kalta og Annáll Kötlugosa. *Árbók Ferðafélags Íslands*, **30**: 125–49.

Thouret, J.C. 1990. Effects of the November 13, 1985 eruption on the snow pack and ice cap of Nevado del Ruiz volcano, Colombia. *Journal of Volcanology and Geothermal Research*, **41**: 177–201.

Tómasson, H., 1973. Hamfarahlaup í Jökulsá á Fjöllum. *Náttúrufræðingurinn*, **43**: 12–34.

Tómasson, H., 1996. The jökulhlaup from Katla in 1918. *Annals of Glaciology*, **22**: 249–54.

Trabant, D.C. and Meyer, D.F., 1992. Flood generation and destruction of Drift Glacier by 1989–90 eruption of Redoubt Volcano, Alaska. *Annals of Glaciology*, **16**: 33–38.

Vuichard, D. and Zimmerman, M., 1986. The Langmoche flash-flood, Khumbu Himal, Nepal. *Mountain Research and Development*, **6**: 90–94.

Vuichard, D. and Zimmerman, M., 1987. The 1985 catastrophic drainage of a moraine-dammed lake. Khumbu Himal, Nepal: cause and consequences. *Mountain Research and Development*, **7**: 91–110.

Waitt, R.B., Jr, 1984. Prehistoric jökulhlaups from Pleistocene glacial Lake Missoula – new evidence from varved sediment in Northern Idaho and Washington. *Quaternary Research*, **22**: 46–58.

Waitt, R.B., Jr, 2002. Great Holocene floods along Jökulsá á Fjöllum, north Iceland. In Martini, I.P., Baker, V.R. and Garzón, G. (eds), *Floods and megaflood deposits. Recent and ancient examples*. International Association of Sedimentalogists Special Publication, 32. Malden MA: Blackwell 37–51.

Walder, J.S., 1986. Hydraulics of subglacial cavities. *Journal of Glaciology*, **32**: 439–45.

Walder, J.S. and Costa, J.E., 1996. Outburst floods from glacier-dammed lakes: the effect of mode of lake drainage on flood magnitude. *Earth Surface Processes and Landforms*, **21**: 701–23.

Walder J.S. and Driedger, C.L., 1995. Frequent outburst floods from South Tahoma Glacier, Mount Rainier, U.S.A.: relation to debris flows, meteorological origin and implications for subglacial hydrology. *Journal of Glaciology*, **41**: 1–10.

Webb, R.H. and Jarrett, R.D., 2002. One-dimensional estimation techniques for discharges of paleofloods and historical floods. In House, P.K., Webb, R.H., Baker, V.R. and Levish, D.R. (eds), *Ancient floods, modern hazards. Principles and applications of paleoflood hydrology*. Washington, DC: American Geophysical Union, 111–25.

Yamada, T., 1998. *Glacier Lake and its outburst flood in the Nepal Himalaya*. Monograph No. 1. Tokyo: Data Center for Glacier Research, Japanese Society of Snow and Ice.

PART 4 Applied Mountain Geomorphology

Debris-flow protection channel, Howe Sound, British Columbia, Canada
Photo: P.N. Owens

9
Geomorphic hazards in mountain environments

Kenneth Hewitt

1 Introduction

This chapter examines the geomorphic conditions and events in mountains that endanger humans. It considers the incidence and potential destructiveness of geomorphic processes and ways in which human activity can increase or decrease these dangers. Landform interpretation, in relation to land use in mountain settings, is also shown to be an important key to the map of risk. The mountain context introduces some unique problems at the highest altitudes, because of the rarified atmosphere, intense solar radiation, extreme low temperatures and high winds. However, these conditions mainly become hazardous above and away from zones of permanent settlement and activities other than mountaineering. They can become serious problems in mountain aircraft crashes, high-altitude mining and military activity, but these will not be considered here.

Some geomorphic hazards such as snow avalanches, catastrophic rock slides and natural dam-burst floods are found largely or only in mountain environments. Others, such as earthquakes, debris flows and volcanism, flash floods and threats associated with valley glaciers, are mainly concentrated in the mountains. In global terms earthquakes and floods appear to account for more than half of the more destructive mountain land events (Table 9.1). They include most of the more lethal and costly natural disasters. Geomorphic activity is a critical determinant of damages in many of these. There are serious concentrations of hazards in particular mountain regions that are rare elsewhere. They include areas with high explosivity volcanism, fire-susceptible mountain forests, and the aeolian hazards of arid, intermontane basins. Hazards that do not loom large on the global scene present singular dangers in certain mountains; for instance, glacier surges, jökulhlaups or crater lake outbursts.

One of the main observations emphasized here is that natural disasters identified with external forcing factors of climatic and geological processes generally involve, or become hazards through, extreme Earth surface processes that they initiate. In mountain storms, earthquakes and volcanic eruptions, for instance, it is often slope, glacial, fluvial, coastal or aeolian forces that are the main damaging agents. In fact, mass movements tend to be seriously under-represented in the risk and disasters literature because slope failure is a common 'secondary' hazard, triggered by 'primary' atmospheric, hydrologic and tectonic hazards. Yet, not only may slope failures cause much of the damage in these events, a majority of the more catastrophic landslides and sets of landslides occur in them. Sometimes this is well known, as in the great debris avalanche from Mt Huascaran in the 1972 Peruvian earthquake. Often it is not made clear in disaster reports.

Of course, virtually all types of subaerial process occur somewhere in the mountains, and contribute to mountain land risks. Some, not especially associated with rugged terrain, may be the more serious hazards. This applies quite generally to severe storms, including snowstorms and tropical

Table 9.1 – Reported numbers of natural disasters in mountain lands according to initiating agent, 1900–1988, ranked by largest impact in the countries involved

Initiating agent	Largest impact			
	Death toll	Made homeless	People affected	Economic losses
Earthquake	17	16	8	13
Volcanic eruptions	5	3	3	2
Drought with famine	2		5	3
Avalanche		2		1
Landslide	1			1
Flood	1	12	7	8
Fire	1			1
Tropical cyclone	1	1	2	2
Heatwave	1			
Storm			1	1

After United States Office of Foreign Disasters Assessment (1989).

cyclones, to droughts, floods and forest fires. The importance of these hazards must be recognized, but they are dealt with extensively in other literature and are of interest here mainly where mountain conditions adversely alter their frequency, intensity and patterns of occurrence.

To focus on geomorphic processes is to adopt a 'hazards perspective' on risks and disasters – 'hazards' being elements of risk resulting directly from dangerous agents and environmental conditions (Hewitt, 1997a). Nevertheless, it must be recognized that whether, and how far, these lead to harm depends at least equally on human activity. The map of geomorphic risk is a product of the interrelations between human geography and physiography. Indeed, the spatial and temporal pattern of endangerment depends primarily on *the exposure of people to dangerous conditions, the extent of their vulnerability, and provision or absence of possible protections.* These are matters influenced much more by society than nature. Of special concern is a range of modern developments that change, often adversely, these elements of risk, as well as having impacts on and through geomorphic conditions (Blaikie *et al.*, 1994; Lavell, 1994; Beniston, 2000). Some effort will be made, therefore, to balance a main emphasis on geomorphology, by indicating how human conditions influence the dangers it involves.

In these terms it is unlikely that mountain lands are *inherently* more dangerous than other regions. There is a common perception that disasters in the mountains reflect the severity and scale of natural processes, or a lack of development among their populations. However, these may well be environmental and cultural stereotypes based on outsiders' perspectives. Mountain societies have often benefited from a healthier environment, greater security, sometimes more, or more diverse, resources compared with surrounding lowlands. Studies of well-established mountain societies describe some uniquely effective 'risk averse' practices, land uses, settlement patterns and social arrangements. Historically among such peoples, land uses, permanent settlement and activity schedules have been designed to avoid sites at risk from severe Earth surface processes, or the seasons when they occur

(Waddell, 1983; MacDonald, 1994, 1996; Hewitt, 1997a). The point is that risk is as much about human adaptation and social organization as environmental conditions.

Nevertheless, the evidence suggests that mountain regions *have* become increasingly disaster-prone over the past century or two (Stone, 1992; Hewitt, 1992; Messerli and Ives, 1997). This is mainly a consequence of changing forms and impacts of human activities, and rapid, enforced social change. These can magnify or artificially release destructive forces through habitat abuse. People are more likely to be exposed to environmental extremes and unstable areas, including geomorphic processes that pose exceptional risks for human activities. Perhaps more important is how modern transformations have tied the fate of mountain lands to decisions made outside them, often with little or no regard to their conditions. Studies from many parts of the world show mountain peoples exposed and made more vulnerable to natural extremes by social upheaval, including vast numbers uprooted and resettled in unfamiliar or more dangerous settings (Stone, 1992; Blaikie et al., 1994; Hewitt, 1997a). While focusing on the geomorphic conditions that cause harm, it is essential to be aware that other, mainly social, conditions are what expose persons or activities to them, and fail to provide adequate protection.

This chapter looks first at the geophysical conditions governing the incidence, distribution and severity of geomorphic processes in mountains. Second, profiles of some distinctive and very destructive geomorphic processes or events are examined. A third section addresses regional risks, problems relating to the mix and relative significance of the range of hazards in given mountain landscapes. This will be illustrated with a case study of the Karakoram. Volcanic hazards are not covered in this chapter as they are dealt with by Thouret (this volume).

2 Effects of orogeny and orography

Some of the more distinctive natural hazards in the mountains arise from orogenic or geotectonic forces; from the orographic intensification of atmospheric and hydrological conditions; and in distinctive alpine, montane and mountain foot environments. These are discussed in detail elsewhere in the volume. Here we only need identify how they affect the risk environment, notably by concentrating and magnifying Earth surface processes.

Orogeny, or mountain building, is of general importance in determining the preconditions for dangerous erosion events. It generates high elevations and, in concert with erosion, maintains high relief and rugged terrain. Mountains are largely concentrated along lithospheric plate margins where distinctive, usually complex, bedrock formations and structures develop (Owen, this volume). Ongoing or stored tectonic stresses are introduced that enter risk as preconditions of slope instability. The relation of erosional development to geotectonics, past and present, is critical for the location and scope of dangerous sites. Dominant patterns of erosion, especially valleys and rockwalls, tend to follow structural lineaments, including faults, fold axes, regional joint systems and lithological boundaries. Scheidegger (1998) has explored this topic in terms of the 'geotectonic predesign' of landslides. As bedrock formations are unroofed and exposed by erosion, stress fields and unloading features develop in mountain masses and valley walls. They also influence slope safety, especially the risk of catastrophic rockwall failures (see below). These and other geomorphic hazards are intensified where stream incision and slope stability are affected by rapid uplift (Shroder, 1989; Burbank et al., 1996). Orogeny is also associated with the main concentrations of earthquakes and volcanoes. These, large and frequent weather fluctuations, glaciation and sensitivity to climate change, can further increase the incidence and magnitude of dangerous Earth surface processes.

Orographic effects involve the influence of mountainous topography on subaerial conditions. They act mainly through *topoclimate*, i.e., distinctive climatic zoning by elevation and the orientation of mountain slopes. The former, sometimes referred to as *verticality*, or the upslope variation in conditions with elevation, affects temperature, including the incidence of freeze–thaw, precipitation and its phase, cloudiness and intensity of sunshine. In risk terms, verticality serves to locate and intensify Earth processes and to complicate the adaptive context through the diversity of hazards, their spatial and temporal variability. Slope orientation modifies the extent and elevation range of altitudinal zones. This occurs in two ways: in relation to atmospheric circulation (windward versus lee slopes), and/or with aspect relative to the sun as a function of latitude. Shading by intervening mountain ridges adds further variability. Equally important is the way seasonality of conditions and processes varies in each elevation zone and by slope orientation. Many geomorphic processes, notably avalanches, debris flows and floods, have a different seasonal incidence or severity in different altitudinal zones (Hewitt, 1993b). The zones and mosaics of habitat, or ecotypes, that topoclimate helps to create are intimately involved in human settlement and economic activity. They prefigure critical zones in which environmental abuse can generate or aggravate hazardous processes (Messerli and Ives, 1997).

However, some of the most important and distinctive dangers arise from a third consequence of topoclimatic variability; the more or less rapid movements and exchanges between different zones, especially in a down-slope direction and along drainage lines. The latter, sometimes referred to as the 'cascade effect', is of singular importance in the power and reach of dangerous geomorphic processes. The down-slope moisture and sediment cascades coincide with major hazards. Glaciers, snow avalanches, debris avalanches, debris flows, floods and valley winds connect different subenvironments. Processes generated in one zone are swiftly and repeatedly introduced into others where they are not generated or seasonally absent; for example, avalanching below the snowline, or dust storms from arid valley floors or glacier moraines carried above the snowline or into well-vegetated areas. For the same reasons, the deposition and constructional landforms at the base of slope and down-valley sites can reflect different, larger and more episodic erosion or sediment entrainment, in higher zones. The cascade effect is also associated with processes generated in the mountains that can become hazards in surrounding lands, of which flood hazards and related sedimentation are usually the most significant.

Risk geography in and around mountain lands often depends upon human occupancy of distinctive but common landscape features. They include alluvial fans, and stream terraces, and dangerous sites on these associated with the cascade effect. Some of the worst calamities have been concentrated at erosional bluffs, bedrock gorges or depositional landforms around and in intermontane basins or piedmont zones, and along mountain lake or coast range shores.

In general, the higher magnitude and frequency of geomorphic processes in mountain lands gives rise to a greater number of large and potentially dangerous events. This is aggravated by the way mountain areas tend to be in more or less unstable transition from different past or changing present conditions (Hewitt et al., 2002). Legacies of orogeny and past climate change, notably glaciation and deglaciation, have left unstable slopes and bodies of sediment. These increase the variety, intensity and reach of hazardous processes. 'Paraglacial (re)sedimentation' is an example (Church and Ryder, 1972).

Again, one must emphasize that humans may well adapt to, or find singular advantages in, these active and diverse environments. However, they become unusually problematic in times of rapid social change and habitat abuse.

3 Earthquake country: geomorphic ingredients of seismic risk

The vast majority of earthquake disasters occur partly or wholly in mountain lands. They comprise more than one-third of mountain land disasters reported worldwide, and the single largest natural hazard. Clearly, the overriding factor is the occurrence of a majority of large earthquakes in or near orogenic belts. However, geomorphology enters earthquake risk in two important respects: through the Earth surface processes associated with seismic events, and landforms as a key to the mapping and interpretation of seismic risk. In a large fraction of 'earthquake' disasters, casualties and devastation are directly caused by Earth surface processes. More generally, landscape settings and landform characteristics are useful keys to the distribution of human vulnerability to earthquake, and the pattern of damage in disasters.

A survey of the 242 largest earthquake disasters between 1950 and 1990 showed 95% occurred partly or wholly in mountain lands (Hewitt, 1997a: figure 8.1). Severe damage from geomorphic processes was reported in more than half and was the largest cause of death and destruction in about one-third of the events (Table 9.2). These data probably underestimate the true extent and impacts of Earth surface processes, since many reports give few or no details about the mechanisms of damage.

Extremely destructive ('secondary') geomorphic hazards such as large landslides may be triggered by earthquake (Alexander and Formichi, 1993). These may lead to further ('tertiary') hazards such as damming of rivers by landslides or flood damage if the dam fails catastrophically (Hadley, 1964). Buildings sited on or below unstable slopes may be destroyed by slope failure. Those beside bodies of water may be destroyed by landslide as well as by seismically generated water waves.

Earthquake itself is primarily a threat to buildings or the built environment. Structural strength and design are critical, and therefore the building culture. But the choices, or lack of choice, over land use and building sites inscribes the social order upon the landscape (Figure 9.1). Often this is clearly a matter of poor-quality construction reflecting poverty, neglect or ill-considered development, with the poorer structures and people occupying the least stable land (Figure 9.2). But this is why geomorphic parameters help determine how slopes and built structures will perform under seismic shaking. Almost always, the buildings and land uses, whose safety ratings are indistinguishable, are destroyed in one part of a disaster zone but survive in another. This mainly reflects such variable conditions as bedrock, soils, watertables, slopes, vegetation cover, etc., rather than, say, distance from the epicentre or purely building quality alone.

Unsafe sites include those where sediments or topography tend to magnify seismic waves and the ground accelerations they cause (Figure 9.1(b)). They are often places where soils or bedrock are weak, susceptible to sudden collapse, or with a high watertable. In general, safety and damage patterns are closely related to slope steepness and geometry; the composition, strength or susceptibility to deformation of bedrock, near-surface sediments and soils; to surface and groundwater hydrology; vegetation cover, and to anthropogenic modifications of the landscape. Such factors are often referred to as microseismic conditions. They govern the local behaviour of seismic waves in the surface and near-surface environment (Reiter, 1990). Clearly, however, they also govern geomorphic responses to earthquakes. Thus, while social controls over building and siting are fundamental, patterns of structural damage and collapse often relate most directly to geomorphologically unfavourable sites. These, in turn, involve the relations of settlement patterns and developments to mountain landscapes.

3.1 Geomorphic hazards of piedmont or 'basal zone' ecotones

While earthquake disasters are overwhelmingly associated with mountain lands, most loss of life and destruction occurs along valley floors, in intermontane basins and, especially, mountain foot settlements (Figure 9.3). This applied to at least 75% of the earthquake disasters identified in the survey mentioned above and, in most cases, to the more severe devastation. One-third had damage concentrated in sea and lakeshore settings along mountainous coasts (Figures 9.4 and 9.5). A glance at topographical maps for, say, Agadir, Skopje, Gemona del Friuli, Erzurum, Spivak, Quetta, Kobe, Conception (Chile), Cuzco, Caracas, Managua, Los Angeles, Oakland and Anchorage, shows these

Table 9.2 – Examples of recent earthquake disasters in mountain lands with significant damage caused by geomorphic processes responding to seismic activity

Year	Location	Disaster
Strength loss or liquefaction of susceptible regolith		
1999	Western Turkey	17000 deaths, 600000 made homeless, US$5 billion in property damage mainly from collapse of recently constructed buildings, especially on susceptible alluvium around the Sea of Marmora.
1995	Japan, Kobe	Collapse of numerous modern structures on alluvium and artificial fill in Kobe Bay, main cause of some US$50 billion property damage. Death toll of 5502, mainly in housing collapses and fires in poorer districts.
1989	USA, Lomo Prieta, CA	Structural damage concentrated on soft mud and artificial fill of Oakland, 50 km from epicentre.
Large individual or great numbers of landslides		
2001	Indonesia	Three landslides caused most damage and 50 deaths.
1990	Iran, El-Burz Mts	Landslides caused significant part of 40000 deaths, >105000 injured, 500000 made homeless.
1989	Tajikistan, Pamirs	Villages buried by mudslides, and most of 274 deaths attributed to them.
Earthquake-triggered natural dams and dam-break floods		
1994	Colombia, Cauca	Landslide dams the Paez River causing severe river flooding and most of damage.
1993	Papua New Guinea, Markham Valley	Landslides block the Ume River, causing many casualties.
1960	Chile, Valdivia	An outburst flood from a landslide-dammed lake did most of the damage in the city. Casualties averted by warning system.
Tsunami damage along mountainous coastlines		
1998	Papua New Guinea	Tsunami resulting from submarine landslide triggered by earthquake 20 km offshore, killed 2182 of whom some 70% were children. At the shore, the wave reached 10 m on average and up to 30 m in places.
1994	Indonesia, Java	Tsunami drowned some 200 people and US$2.2 million in losses, mainly homes and fishing boats.
1992	Nicaragua	Tsunami caused most of some 116 deaths, loss of US$20–39 million, mainly homes and fishing boats.

After Hewitt (1997a).

places to be piedmont towns; a small sample of those are identified with and used to name catastrophes.

Mountain-foot landforms and Earth surface processes involve rapid transitions and a mosaic-like diversity. There is a transition from the rugged terrain and topoclimatic variety of the mountainous interior and, usually, to distinctive geomorphic environments of adjacent lowlands, plateaux or coasts. In particular, the piedmont is identified with marked hydrophysical transitions or thresholds.

(a)

(b)

Figure 9.1 *Mountain-foot settlements in earthquake disaster zones. (a) Distant view of the setting of the town of Montella, in the middle-scale mountains of Campagnia, central Italy, where concentrated pockets of damage during the 1980 earthquake disaster, associated partly with poor construction, were mainly located on base-of-slope alluvial sediments with high watertables. Poorer buildings on more solid foundations often survived intact. (b) Buildings in Montella damaged in the 1980 earthquake. This is at the lowest part of the settlement around a small stream and the collapse of buildings records both poorly designed and maintained masonry structures, and foundations rendered unstable by soft sediments and a high watertable*
Photos: K. Hewitt

Geomorphic processes bringing moisture and sediment down steep slopes, or out of canyons and tributary valleys, undergo sudden loss of energy, leading to rapid and variable sedimentation, and shifting transport paths. Depending upon regional climate, there may be avalanche deposits, torrential cones, sediment fans, pediments or deltas and beaches. Sediments derived from torrential and mass movement processes of the mountain interior interfinger fluvial, lacustrine, coastal or aeolian deposits. The mountain foot is often defined by faults or fault line features, and by conditions due to rising or

Figure 9.2 *Collapsed farmer's home in the earthquake of December 1974, Indus Kohistan, Himalayan mountains, Pakistan. The rubble walls completely disintegrated, while the heavy, flat roof collapsed, crushing the residents within. Note, however, the absence of trees on mountain slopes behind and the lack of timber in the walls of the home. Both reflect deforestation, and loss of wood resources, to timber merchants from lowland cities, within living memory (Hewitt, 1976)*
Photo: K. Hewitt

Figure 9.3 *Coastal mountain-foot settlement. The town of Kotor, Dalmatian coast of Croatia, heavily damaged in the 1979 earthquake (cf. Figures 9.4 and 9.5)*
Photo: K. Hewitt

Figure 9.4 *An example of earthquake damage to a road and port facilities along the Dalmatian coast of Croatia during the 1979 earthquake. The fissuring of coastal alluvium and artificial fill is evident in the foreground*
Photo: K. Hewitt

Figure 9.5 *Poorly designed or maintained buildings in the old town of Kotor, Croatia, damaged in the 1979 earthquake*
Photo: K. Hewitt

falling sea and lake levels. Quaternary climate change, tectonic uplift or faulting, have left relict erosion and depositional forms, terraces and the trenching or shallow burial of earlier sediments. A further range of complications arises where there are volcanoes.

Human adaptation is further complicated by the differentiation in habitat conditions, reflected in the patchiness of plant and animal communities and ecotones, boundaries or zones of rapid transition. Schimper coined the term 'Basal Zone' ecotype to identify these distinctive ecological conditions of mountain foot areas (Hewitt, 1984). The ecotones are particularly well developed, ecologically and economically important, where mountains lie at the margins of arid lands, in areas of seasonal drought and coastal zones.

However, it is not argued that mountain foot vulnerability is something ultimately decided by the physical environment. On the one hand, mountain foot habitats can offer many advantages and adequately secure sites. They are often strategically important for economic, transportation and security reasons. Many examples of sustained, long-term human occupancy are known. On the other hand, the most common victims of earthquake disaster in these settings reflect the extent to which human activity is a key to endangerment. From Central America to Afghanistan, the main victims in natural disasters are people recently forced into the piedmont towns, or impoverished and powerless folk obliged to settle the more hazardous sites (Maskrey, 1989). Squatter and other relatively poor neighbourhoods not only define the most vulnerable inhabitants from Lima, Peru, to the Los Angeles Basin, from Agadir in Morocco to Kobe, Japan (Hardoy and Satterthwaite, 1989), but their homes are generally the least earthquake-proof and are located on the more unstable parts of the landscape.

Other geomorphic hazards are also involved in basal zone settlements and reflect similar problems of ill-considered uses of dangerous sites. A catalogue of events and damage related to the basal zone would include a liturgy of flood damage, wildfires, landslide hazards, storms and coastal zone hazards. The overriding consideration is, again, concentrations of vulnerable human settlements and activities, ill-considered modernization, and the failure to provide adequate protection.

4 Hazards of glacial and nival zones

There are mountains in all latitudes with seasonal or perennial snow cover and, usually where the latter applies, glaciers and, in the colder latitudes and elevations, permafrost. Even in cold regions, the presence of mountains alters and, usually, increases the extent and importance of snow, ice and freeze–thaw in Earth surface processes. The bulk of mountain ice is found in the mid-latitude and subpolar ranges, but regionally important glaciers, rock glaciers, permafrost and related hazards occur in tropical high mountains even of the equatorial zone (Reynolds, 1992).

The main hazards associated with snow cover are avalanching on steeper slopes; snow loads on buildings; interference with transportation by avalanche deposits, snowfall and wind-induced snow drifting; and slushflow episodes (Gude and Scherer, 1998; Hestness, 1998). Of even wider importance are floods or droughts related to excessive or low amounts and concentrations of snow meltwaters, respectively. Each may be important in given regions, and each may be modified in an adverse sense by climate change and human activity.

4.1 Snow avalanches

Avalanching is a geomorphic hazard uniquely identified with the mountains. In the highest mountains, and humid, cold region mountains generally, avalanches have huge geomorphic and hydrological roles (Figure 9.6). They present singular dangers wherever humans and their property are exposed to them. Nevertheless, of the millions of avalanches in any given year only a very few are identified with major damage. It is rare for as many as ten incidents to enter the global disaster lists. In this sense,

Figure 9.6 *Avalanche path and deposits in the Nanga Parbat Himalaya. The avalanches threaten forest cover here. The foreground shows the remains of another hazard of avalanching, a breached snow dam on the river. However, the terraced fields beside the avalanche track reflect its role in providing moisture for summer crops*
Photo: K. Hewitt

avalanches present a far greater potential than actual danger. Yet, the real costs and losses resulting from avalanches are commonly underestimated. The reasons relate to the scale of individual events, issues of reportage and the nature of modern responses.

Events large enough to be classed as disasters in national, let alone global, reports rarely involve a single avalanche. If they are, most often the event involves multiple destructive avalanches in major snowstorms, sometimes earthquakes. Then, damage is more likely to be identified with the initiating hazard. This was the case for eastern Turkey in the unprecedented snowfalls of February 1992, and Kashmir in January 1995. In both cases, avalanching was the main cause of death and road closures, but they were widely reported as snowstorm disasters.

While avalanche burial and deaths of skiers, trekkers or mountaineers hardly ever go unreported in national and international media, the numbers are rarely of disaster magnitude. For instance, of 26 lethal avalanche events in the USA for 2001–2002, the worst case involved four deaths, and 20 separate events involved one death (State of Colorado, 2002). North American avalanche deaths overwhelmingly involve recreational, mountaineering and winter sports users (Avalanche Canada, 2002). The same applies in the European Alps (Servizio Valanghe Italiano, 2002) and New Zealand (Dingwall *et al.*, 1989). However, where modern developments are concerned, the severity and costs of the avalanche hazard are hardly apparent in the damage caused. All-weather transportation, mineral and timber extraction, or winter and high mountain sport and recreation, involve relatively high-profile or

Table 9.3 – Examples of recent avalanche disasters in the mountains

Date	Location	Disaster
26 October 1998	Iceland, Flateyer	In major snowstorm, avalanche engulfed 19 homes and killed 20 in coastal fishing village.
11 November 1997	Nepal	Snowslides and mudflows killed 49, trapped hundreds and 500 rescued during heavy snowstorms.
16 January 1995	India, Kashmir	More than 600 people trapped on Jammu–Srinagar highway, over 200 killed and 500 rescued.
1 February 1992	Turkey	Avalanches during exceptional snowfalls isolate some 7000 villages, cut roads and power for days, and killed some 210 people, mainly soldiers.
1990	Iran	An avalanche claimed 21 lives and trapped 110.

wealthy users who expect, and generally obtain, considerable hazard protection. Hence, the costs of avalanche defences give a better measure or indication of the risk. They may involve protection for mountain highways and railways, or wealthy towns and resorts; the costs of avalanche research, monitoring and forecasting; the artificial releases, ski patrols and rescue arrangements. In countries such as Canada, Switzerland and Norway expenditures amount to tens of millions of dollars annually. Meanwhile, the statistics leave no doubt that fatalities as well as damage continue to creep upwards, and seem only likely to expand with the developments taking place in many mountain lands, and the worsening security situation in some (Table 9.3).

Geomorphology is also engaged through the important role that physiography plays in the location and scale of avalanching. While snowfall, thermal conditions and wind action are critical, just where avalanching occurs, its scale and reach are profoundly affected by terrain (Perla, 1980; Lied and Toppe, 1989). The steepness, shape and elevation range of slopes; the geometry and extent of chutes and gullies that funnel avalanches; vegetation cover, the roughness and erodibility of soil and bedrock and conditions in the run-out zones, are geomorphic variables directly influencing the risk (LaChapelle, 1985; Owens and Fitzharris, 1989).

The more destructive events tend to occur when large amounts of debris are incorporated into an avalanche; for example, the most lethal event in the most lethal earthquake disaster in the western hemisphere. In the May 1970 earthquake in Peru, a snow and ice avalanche, triggered on the peak of Nevados Huascaran, picked up vast quantities of debris in its rapid descent and became a debris avalanche. This buried the town of Yungay and killed some 18000 people. A similar but smaller avalanche-turned-debris-avalanche had followed almost the same path in January 1962, killing approximately 4000 people (Pflaker and Ericksen, 1978; Oliver-Smith, 1986).

4.2 Mountain glacier hazards

The main conditions associated with glaciers that may endanger humans are:

1 ice environment hazards that apply on and at the margins of ice masses;
2 glacier advances or retreats due to climate change;
3 glacier surging: the sudden, catastrophic movement or expansion of an ice mass;
4 glacial meltwater floods;

5 ponding by ice and moraine dams, and outburst floods from them; and

6 proglacial hazards, such as sedimentation and erosion problems, in which the effects of glacier activity are felt beyond the glacial zone (Tufnell, 1984).

Dangers of valley glaciers and small ice caps in mountains include those on the ice itself, where there are seracs, crevasses or moulins, avalanches, sudden slush flows and torrents of meltwater, supraglacial ponds and unstable debris, wind chill and white-outs. These are of concern to mountaineers, trekkers and hunters, but usually involve small numbers of people.

Glacier advances may engulf land, settlements or infrastructure, as happened almost worldwide during the Neoglacial ('Little Ice Age'), which culminated in the middle decades of the nineteenth century (Grove, 1988). Dangerous advances continue to occur, mainly of isolated glaciers, but glacier retreat has been occurring in most regions since early in the twentieth century. This can lead to greater flooding, or the loss of water resources from smaller ice masses. It can leave behind unstable dead ice and thermokarst, glacial sediments and moraine-dammed lakes (see below).

Among glacial hazards felt beyond the glacial zone, sometimes far beyond the mountains, the most serious and widespread are floods. Glacial flooding depends to a large extent on rates of ablation. In mountains these relate mainly to the vertical extent and seasonal migration of thermal belts by elevation. The flow of glacier-fed streams tends to be highly variable on diurnal, weather system-related, seasonal and inter-annual scales. It may vary in more or less the opposite way to rain-fed streams at lower elevations in the same mountain ranges. The storm system that brings rain to lower slopes can shut down the ablation systems on glaciers high up through reduced solar insolation, low or freezing temperatures, and snowfall (Hewitt, 1993a). Conversely, sunny and dry episodes bring the highest rates and extent of glacier ablation so that glacial floods coincide with dry periods lower down. This is of particular importance in the tropical and subtropical high mountains.

Meanwhile, glacier 'plumbing', especially in the early part of the melt season, can retard runoff and then, as it opens up, lead to sudden episodes of high flow and sedimentation downstream (Collins, 1989). The variability of glacial streams, the concentration of runoff in the ablation season, and high yields of sediment from glacierized basins (see Caine, this volume), can lead to severe sediment and erosion problems downstream. Of particular concern is the formation of ice- and moraine-dammed lakes and outburst floods from them, and this is discussed in detail by Björnsson (this volume).

4.3 Glacier surges

In certain regions there are glaciers that are subject to sudden, catastrophic expansion or surging. These relatively short-lived events involve a sudden, large transfer of ice from the upper to the lower part of a glacier. Ice movement generally increases by at least an order of magnitude, sometimes two orders, compared to pre- and post-surge behaviour. One or more pulses of sharply accelerating flow move down the glacier, accompanied by a rise and severe crevassing of the ice surface. Surge events may last from a few months to several years. Some advance the glacier terminus several kilometres in a few months (Clarke, 1991; Paterson, 1994; Hewitt, 1998b). Others dissipate before reaching the terminus, especially those of tributary glaciers, but may lead to glacier advance later.

Some valley glaciers are known to surge as frequently as every 20 or 30 years. Once or twice a century is more common (Hewitt, 1998b). Surges are often accompanied by, or set the stage for, other hazards including ice margin mass movements, ponding and sudden release of meltwater ('jökulhlaups'), and heavy proglacial sediment loads (Benn and Evans, 1998; Björnsson, this volume).

Surging behaviour is related mainly to the character of each ice mass and conditions in its basin, especially thermal and sediment conditions at the bed. The timing of surges is, therefore, largely independent of what is happening with climate-induced glacier changes. However, surging glaciers are concentrated in just a few mountain ranges, suggesting a climate connection. The vast majority occur in subtropical and subpolar, maritime regions. The former include the Karakoram and Pamir Ranges of inner Asia, and the northern Argentinian Andes. Notable among the latter are the Alaska-Yukon ranges, Iceland and Svalbard glaciers. Meanwhile, surge magnitudes may vary according to climate trends.

Most known surges have caused little or no harm to humans. The regions or elevations where they occur have few or no settlements and small numbers of visitors. Nevertheless, they pose serious threats in certain regions, engulfing occupied land, generating sudden floods, disrupting local communications below and across glaciers, sometimes damming streams and triggering major landslide episodes (Figure 9.7).

5 Landslide hazards in the mountains

The whole range of known mass movements occurs somewhere in the mountains. Of special concern are large debris flows, catastrophic rockwall failures and the rock avalanches they can generate (Shroder and Bishop, 1998). The following sections concentrate on the latter.

5.1 Massive rockwall failure and catastrophic (excess) run-out events

Large rockslides followed by catastrophic run-out of debris are confined to regions of high relief and steep slopes. They involve some of the most lethal and destructive disasters directly resulting from geomorphic processes (Table 9.4). Others have dammed rivers causing highly destructive dam-break floods (see below). They are most likely to occur where rockwalls have been oversteepened by rapid stream incision and, especially, by glaciation and then exposed or 'debuttressed' by deglaciation (Hewitt, 2001). A linked series of factors govern the scale and scope of these risks. They begin with the causes of slope instability and failure, continue in a series of conditions governing run-out of debris, and may persist long afterwards in landscapes highly disturbed by rock avalanche emplacement.

The detachment zones of rockslides are usually identified with specific geological structures, bedding or foliation planes, faults or major joint systems. Additional sources of instability in active mountain belts are residual tectonic stresses as well as unloading stresses as the bedrock is uncovered

Figure 9.7 *Surge of the Bualtar Glacier, Karakoram, 1986. Shepherds stand where the former track from summer pastures gave easy access onto the glacier. Now its margin is a 20-m high, unstable ice wall, beyond which is a maze of crevasses and seracs generated by the surge*
Photo: K. Hewitt

Table 9.4 – Dimensions of some catastrophic landslides and associated mountain land disasters

Location	Type	Volume ($m^3 \times 10^6$)	Fall (m)	Run-out (km)	Deposit area (km^2)	Fatalities
Mayunmarca, Peru, 1974	Rock avalanche and mudslide	1000	1900	8	10	451
Mt Huascaran, Peru, 1970	Debris avalanche	100	3000	12	22.5	18000
Hope, B.C., Canada, 1965	Rockslide–rock avalanche	130	1100	3	3.5	4
Vaiont, Italy, 1963	Rockslide and flood	240	650	2	3	1925
Frank, Alberta, Canada, 1903	Rockslide–rock avalanche	40	1000	3	2.7	70
Elm, Switzerland, 1881	Rockslide–rock avalanche	10	600	2.2	1.5	115

After Voight (1978)

and exposed. Erosional history is critical, as it affects the height and geometry of slopes, the stress fields in mountain masses and rockwalls, albeit interdependent with pre-existing structural, lithological and stress conditions. There is widespread evidence of gravitational rock creep at the head or in the detachment zones of catastrophic rockslides (Radbruch-Hall, 1978; Bovis and Evans, 1996).

In most cases, the actual moment of catastrophic failure involves sudden loss of strength, usually induced by a high-magnitude triggering mechanism, most often an earthquake (Pflaker and Ericksen, 1978; Keefer, 1984). Many have also been associated with severe storms, some with rapid melting of snow and ice, and occasionally with lightning strikes (Hewitt, 1988).

Once massive rockwall failure occurs, the mass may remain relatively intact, as a translational slide, moving over one, or a limited number of, slide surfaces. This is more likely if the transport slope is short or not very steep, or if movement is down a well-defined plane of failure. The Vaiont catastrophe of 1963 in the Italian Alps was essentially of this sort. The rock mass slid down one wall into the man-made reservoir below, and stalled against the opposite wall. The loss of some 1925 lives was due to the water wave, forced out of the reservoir, descending upon towns in the valleys below (Selli and Trevisan, 1964; Muller, 1968).

Alternatively, the rock may be thoroughly crushed and pulverized in the descent. If there is substantial incorporation of moisture from snow, ice or entrained wet sediment, then it may transform into a debris avalanche such as the Yungay, Peru, case described above. However, if the mass remains essentially dry, a rock avalanche or 'sturzstrom' will result.

5.2 Rock avalanche hazards

Large magnitude, sudden occurrence and high-speed movement place rock avalanches among the most catastrophic subaerial processes. The conditions required for their occurrence mean they are largely confined to the world's high mountain areas (Voight and Pariseau, 1978). Classic studies have dealt with examples in the European Alps, the Western Cordilleras of the USA, Canada and Alaska, the Peruvian Andes and New Zealand (Heim, 1932; Voight, 1978). The Karakoram, with the highest known incidence of these events, is described below.

The rock avalanche originates from catastrophic failure of bedrock slopes, but refers to the exceptional, high-velocity run-out of debris (Hsu, 1975). It is a product of catastrophic crushing and grinding of bedrock – unlike, say, debris flows, which are formed from pre-existing, unconsolidated material. Rock avalanches generally involve at least one million cubic metres of bedrock and some are cubic kilometres in volume (Hewitt, 2002b). They generally travel many kilometres beyond the source slope, and at velocities exceeding 100 km h^{-1}, often >250 km h^{-1}. When movement falls below such high speeds the debris stops suddenly by frictional 'freezing'. The run-out of highly mobile material produces a large sheet of bouldery debris, sometimes tens of square kilometres in extent. The deposit is usually thin in comparison with area, lobate in plan, often with distinctive ridges and furrows.

In the most rugged mountains, morphology, thickness and deposit characteristics are complicated by topographic interference (Heim, 1932; Strom, 1996; Hewitt, 2002b). Distinctive run-up forms occur when the fast-moving debris stream encounters steep, opposing slopes. Where the bulk of the debris impacts a slope at a high angle, it will split into separate streams, one or more turning up and down valley – the 'deformed-T shape' of Nicoletti and Sorriso-Valvo (1991).

The rock avalanche rarely lasts more than two or three minutes. Yet, they can continue to influence landscape development over millennia, or tens of millennia. A great many have formed large and long-lived natural dams (Schuster, 1986). A modern example is Lake Sarez, more than 60 km long, and impounded by the 1911, Usoy rockslide–rock avalanche in the Pamirs of Tajikistan (International Strategy for Disaster Reduction, 2000). Rock avalanche dams often resist catastrophic collapse or rapid breaching when overtopped, but some have failed suddenly with catastrophic results (see below). Even when the dams have been filled or gradually breached, the deposits can remain sites of active and dangerous processes, or of potential instability if disturbed (Whitehouse and Griffiths, 1983; Hewitt, 1998a, 2001).

5.3 Natural dams and dam-break floods

Natural dam-burst floods arise from failure of landslide dams, ice- or moraine-dammed lakes in glacial terrain and, though rarely as large in the mountains, the bursting of log jams and river ice jams. Factors of risk in the natural dam-burst problem involve the size of the reservoir, the stability and mode of failure of the dam, and the behaviour of the outburst flood wave (Costa and Schuster, 1987). Where a relatively large dam forms and is filled, it can supply a flood vastly greater than rainfall or snowmelt events. A sudden, rapid breach may produce a singularly destructive, steep-fronted dynamic wave. This has much greater erosive power than the more usual, spread out kinematic waves, even when these are of equal or greater volume.

Dam-break flood waves will often entrain and deposit much larger sediment volumes and much larger boulders than other floods. They can reach and remove debris that has resisted erosion for long periods. The scope and scale of danger can also increase where the floods enter narrow mountain canyons, where the flood wave can grow in height and volume as it is frictionally retarded and may again become a dynamic roll wave. Sometimes these rebuilt flood waves are larger than the original outburst through incorporation of flows in the river, creating grave dangers for settlements or infrastructure situated not far above normal flood heights. In the 1926 and 1929 glacier-dam outbursts on the upper Indus, described below, this occurred repeatedly over the 1000-km travel of the flood waves through the Karakoram and western Himalaya (Hewitt, 1982). Some of the worst consequences have been where these rejuvenated flood waves emerge from a canyon into a broad, settled valley. This was seen in the disastrous Johnstown flood, Pennsylvania, in 1889, in that case after a rain-fall-induced collapse of an artificial dam, and in several of the Khurdopin Glacier outburst floods in the Karakoram, at Pasu where they leave the Shimshal gorge to enter the Hunza valley (see below).

6 Anthropogeomorphic hazards in the mountains

There are important landforms and processes in high mountains showing little or no evidence of human impacts: the vast areas of rockwall, the great gorges, glaciers and snow fields, a few remaining wildernesses and steeper forest lands. However, these are rarely the largest areas of mountain lands, nor those whose risks are of greatest significance for humans. Rather, of foremost concern are areas that support mountain agriculture and permanent settlements, where primary resource extraction occurs, the alpine pastures, and main lines of communication. In all these cases the landscape is strongly modified by human agency (Hewitt, 1989b).

Humans influence the map of geomorphic risk in mountains in two main ways. Settlement and land use influence danger from given hazards by the way they expose or protect people and property. Alternatively, human activity, especially using modern tools and energy sources, can drastically alter landscapes or the magnitude and incidence of dangerous processes. Examples in mountains include mass movements magnified or induced by deforestation, by highway and dam construction, or erosion and sedimentation associated with urbanization and mining, and ecological abuse along trails and on ski slopes (Hewitt, 1989b). These change the risks even from large earthquakes or torrential rains, whose incidence humans in no way influence.

The most widespread anthropogeomorphic hazards in mountains are associated with resource extraction, notably forestry, mining and water-resource schemes. The more frequent and costly of reported damages are associated with transportation routes. It would be difficult to overestimate the impact of modern road construction on erosion processes in mountains. Because highways involve high-profile and influential users, their dangers have received much attention of geomorphologists in recent years (Brunsden et al., 1975; Haigh, 1984; Lasa and Jingfang, 1992).

Another major concern is with artificial dam-break floods and mismanagement of flows into and from artificial reservoirs, given the ever-growing numbers of new dams in the mountains which include the very largest ones, and the aging of old ones. Most artificial dam-break disasters occur, or originate, in mountain lands (Hewitt, 1997b). Structural safety of dams and regulation of reservoir flows raise distinctive technical concerns (Jansen, 1980). However, geomorphology is engaged because, once released, the flood behaves like a natural dam-burst, being controlled by the same fluvial system features.

Modern influences not only affect geomorphic hazards through new technologies and managed landscapes, they can also adversely impact the long-term, often ancient, role of human societies in modifying and controlling settled landscapes. In a host of ways, humans have altered geomorphic and ecological conditions. Terracing of slopes is, perhaps, the most important, distinctive and widespread of these interventions in the mountains (Spencer and Hale, 1961; Price, 1981). It has tended to improve slope stability and reduce erosion rates. Some quite ancient examples have proved very resistant to damage in large earthquakes, notably some incredibly steep and long slopes of the high Andes terraced by the Inca and others. However, modern technologies and economic developments often marginalize the kind of small-plot, labour-intensive cultivation practices established in such contexts (Allan, 1995; Rieder and Wyder, 1997).

In general, one must emphasize how risks from geomorphic hazards in mountains are never only about natural processes. Moreover, there are few landscapes, and none permanently settled, where humans have not altered the pace and incidence of Earth surface processes (Banskota and Karki, 1994; Hewitt, 1997a).

7 Landscapes of risk

There are some technical and practical reasons for concentrating on specific natural processes that cause harm. It is in accord with the main thrust of modern geomorphology, which has tended to focus on process, and particular types of process such as fluvial or glacial. However, actual mountain areas or communities involve a range of different processes and attendant risk factors. At any given time, and over time, safety depends on the range of risks, their relative significance or impacts.

One way to think of this is in terms of the other and, for some, the ultimate concern of geomorphology, the landscape. The set of landforms in an area, and conditions affecting them, offer a framework for general hazard assessment. This, in turn, has two complementary aspects: hazards mapping and regional risk assessment. The former is an attempt to classify, and identify the forms and distributions of geomorphic hazards in a map area. This has been widely attempted in the mountains (Ives and Messerli, 1981; Fort, 1987) and can provide useful baseline information, especially in relation to development projects and activities. While hazards mapping lies outside the scope of this chapter, a few words about it seem in order.

Unfortunately, much of geomorphic hazards mapping has suffered from two severe constraints. First, there is the problem of magnitude–frequency relations of geomorphic processes, always at least as important for risk as geographical distribution. Second, the moment one begins to map risks, the large, if not overriding, role of the evaluation of human land use, exposure over time, vulnerability and protection assume paramount importance. Likewise, the mapping can be quickly rendered obsolete by the rapid changes these are undergoing in the mountains. In most cases neither of these considerations has been built into geomorphic risk mapping.

Geomorphic hazards maps have mostly shown morphological features, recording where dangerous processes are known to have occurred. There is rarely much or any sense of their magnitude and frequency distributions, or of human activities and property relations. The problem is often compounded by a tendency to rely on descriptions of the highest-magnitude or 'worst-case' events, often distant in time or location. In a general or global sense these are of compelling interest, but they are rarely representative of the places where the given process may occur or a good guide to adaptive measures. In most mountain areas, lack of adequate historical time series for hazardous processes is a major problem, as is limited knowledge of environmental and land-use change. An important development here is the use of historical images and archival materials to compare land-use changes over time (Ives, 1987; Ehlers, 1995; Byers, 2000).

Geographical Information Systems (GIS) offer ways to build temporal considerations and interactive complexities into spatial datasets and are an important area of innovation in risk 'mapping'. Here too, however, lack of necessary or commensurable data in most, if not all, mountain areas limits what can be done in practice rather than in principle. Finally, the state of research into some of the geomorphic hazards noted above is another constraint – for example, the field recognition, understanding and dating of catastrophic rockslides, or the relations of glacier surging to glacier damming or climate change.

GIS combining satellite and ground-based mapping and temporal monitoring data offer promising avenues for improved risk assessment. However, to the present time, if not always, extended experience with more conventional geomorphological field work in the given area is an essential basis for identifying risks, and compiling adequate profiles of geomorphic hazards. Local knowledge, cultural sensitivity and historical background are also essential to risk assessments in international and cross-cultural contexts. These relate to the broader problems of regional geomorphology and landscape interpretation, if highlighting the risk parameters. To indicate something of what is involved here, and to flesh out a sense of the range and contexts of the individual hazards described above, a case study will complete this chapter.

8 Regional risks: geomorphic hazards of the Karakoram

The Karakoram is a high, rugged mountain region of inner Asia (Figure 9.8). Valley slopes adjacent to human settlements often rise more than 3000 m, in places as much as 6000 m, and have been described as the steepest on Earth (Miller, 1984). The region has some of the highest rates of tectonic

(a)

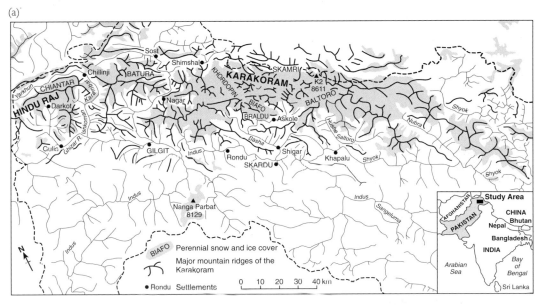

(b)

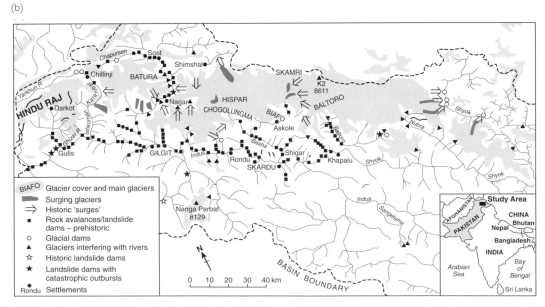

Figure 9.8 (a) Location map of the Karakoram showing the heavily glacierized zone, major mountain ridges, rivers and selected towns. (b) Examples and distribution of selected geomorphic hazards associated with major risks in the region

uplift and the pace of erosion is comparably high (see Owen, Caine, both this volume). Geomorphological studies attest to a great variety and incidence of large, potentially destructive processes (Brunsden and Jones, 1984; Goudie *et al.*, 1984; Kalvoda, 1992; Shroder, 1993). Some of the more important geomorphic hazards will be reviewed, but first it is essential to have a sense of the nature of human settlement and activities in the region (Figure 9.8).

8.1 The human ecology of the Karakoram

The inhabitants live in hundreds of villages scattered along the valleys up to elevations of about 3000 m asl, and in a few small, but rapidly growing, towns (Figure 9.8). For most people and settlements the economy is still, primarily, one of subsistence based on irrigation agriculture and animal herding. The villages are situated in dry, rainshadowed valleys, their crops dependent upon meltwaters descending from the higher, more humid elevations. Appropriately, they have been called 'mountain oases' (Whiteman, 1985). Much vaster areas of the mountains serve the pastoral economy. Herds of goats, sheep, cattle and yaks graze the valley floors and less extreme slopes between the villages in the winter half of the year. They follow the retreat of the snow to high 'alpine' pastures through the summer season, where shepherds live in temporary villages some days or even weeks away from their permanent abodes (Butz, 1996). Most of the routes now used for trekking, and by mountaineers heading to the high peaks, pass through the pastures and were originally developed for the pastoral economy.

Almost everywhere, firewood and dung are used for heat and cooking, small-scale water mills grind grain, sun and wind dry crops. Families generally build their own homes and other structures, although larger settlements may have men who specialize in carpentry and construction. The region is home to more than a dozen, widely different, linguistic groups with distinctive histories and cultures, but all are Muslim (Dani, 1989). There is a gendered division of labour which tends to place men and women in differing risk positions, women being most affected by hazards in and around their homes and villages, men in the high pastures and through involvement in modern developments (Azhar-Hewitt, 1989; Ives, 1997).

Since 1947 most of the region has been administered by Pakistan, the eastern Karakoram by India. Despite war and civil unrest in surrounding areas, notably Afghanistan and Kashmir, prosperity has improved over several decades, and most valleys have fast-growing populations. Development programmes, technological change and a large and growing tourist industry increasingly affect local people.

8.2 Earthquake disasters

There is a long history of destructive mass movements triggered by earthquakes, and earthquake damage in villages from the collapse of structures and fires in the predominantly wood and wattle walls and close-packed buildings (Coburn *et al.*, 1984). In the Karakoram itself, only relatively minor, if widespread, seismic activity has been recorded in recent decades, but the area has been affected by large earthquakes in surrounding regions (Miller, 1984).

Among several damaging earthquakes in recent decades, the worst was centred in the Northwest Himalaya just south of the Karakoram: the Pattan earthquake of December 1974 (Hewitt, 1976). It destroyed many villages and some small towns, and perhaps as many as 7000 people were killed. The shaking led to collapse of large numbers of houses, but landslides caused the greater part of the damage. However, in my view it was more of a 'deforestation' than an earthquake disaster. On the one hand, most of the destructive landslides came from slopes denuded in recent decades. On the other, a local shortage of timber meant that newer buildings tended to use rubble and mud walls rather than wooden framing, while many traditional homes had been neglected so that rotten timbers led to collapse (also see Figure 9.2). Perhaps the most threatening question relating to an earthquake involves the many prehistoric, catastrophic rockslides in the region, and the fact that most settlements, including the larger towns, lie on or beside the rubble of past rock avalanches, or where they have formed large landslide lakes. Where the trigger of such events is known, it is often an earthquake (see below).

8.3 Snow avalanches

These occur at all elevations. In the zone of permanent habitation they are generally confined to the winter and, especially, spring. Most long-time settlements have been located clear of avalanche run-out areas. Hence, disasters affecting village lands and communities are more likely to be associated with recently occupied agricultural land, sometimes housing, notably where new roads have been built. For example, in December 1988 an avalanche wiped out the village of Arandu-Gund, a satellite of Arandu at the head of the Basha Valley in Baltistan. It killed 42 residents and almost all their livestock, and injured 25 persons. The rest were forced to seek refuge in surrounding communities, where most remain to this day, although a few have returned to Arandu-Gund. The site was first occupied about 50 years earlier when arable land became scarce in Arandu. It lies on the outer part of a large sediment fan at about 3000 m asl, fed by steep gullies with small ice masses at their heads up to 1500 m higher. Normally the dangers are from debris flows or torrents, which the village location took into account. But the great avalanche of 1988 swept over most of the fan.

Avalanche risks affect shepherds and animals travelling to and residing in the summer pastures, and there are reports of losses in most years. Some old as well as newer settlements have suffered their worst impacts indirectly, through temporary damming of streams by avalanches and subsequent floods. A disastrous example occurred at Ratul, one of the Hopar villages of Nagyr, in the spring of 1978. The stream coming from a hanging valley suddenly dried up, blocked by an avalanche. Men of the village went to try to clear a channel before the lake became too large, but the dam burst before they were finished, drowning many of them. The flood descended with little warning upon the village below, killing more people and animals, destroying homes and much of the arable land. About half the surviving families were relocated on reclaimed land on the Bualtar ('Hopar') Glacier moraines.

Many blockages and damage by avalanches occur along the Karakoram and Indus Highways, and the many roads that join them from tributary valleys. The problems usually occur between March and June and may close the roads for several days. At this time many of the resources and personnel for road maintenance are fully occupied in clearing avalanches.

At high elevations avalanching is a year-round danger (Figure 9.9). A large number of the mountaineers killed on the Karakoram have been swept away by avalanches. High-altitude porters from the villages, an elite and well-paid but high-risk group, have a steady toll of losses too. In part, this reflects the heavy snowfalls and all-season incidence of severe snowstorms at high altitudes. In part, it reflects the attraction of the largest concentration of peaks over 8000 m, including K2 (8590 m) the second highest in the world, and countless other peaks. The competitive nature and aggressive risk-taking of modern mountaineering plays no small part.

8.4 Landslides

The region is notorious for large and potentially destructive debris flows and mudflows. They are a particular hazard during spring snowmelt and summer rainstorms at lower elevations. The Karakoram Highway and other roads are also repeatedly blocked by these mass movements, and at dozens of places in a given year. Debris flows occur at higher elevations throughout the summer as the snowline retreats upwards and melting snow saturates debris in chutes and gullies. There is a close relation between snowfall, or avalanche-deposited snow, and the incidence of larger debris flows. Occasionally,

Figure 9.9 *Large, late-summer avalanche from the north flank of Gannissh Chissh or 'Spantik' peak (7030 m), in the Karakoram. The avalanche is of intermediate size for the area, and descended 2.5 km over a front of 2 km, and travelled 6 km horizontally. There is all-year avalanching at these elevations, although avalanche character changes with season and storm events. This is a significant hazard for summer grazing lands, and the main cause of death for mountaineers and trekkers in the region*
Photo: K. Hewitt

as a result of an episode of rapid melting of heavy snowpack or avalanche deposits, high-magnitude debris flows reach and destroy settled areas and agricultural land. They may also result from the bursting of small glacier lakes, or exceptional rainstorms (Figure 9.10). A number of these occur somewhere in the region in most years (Kreutzmann, 1994).

Rockwalls make up the single most prevalent landform, and countless small and moderate-sized rock falls occur at all altitudes, their incidence, again, tending to reflect the vertical migration of temperatures with season, including the zone of frequent freeze–thaw cycles. Catastrophic rockslides are not so common as large debris flows but are a significant hazard (Figure 9.8). Recent surveys have identified a total of 186 rockslide–rock avalanche events in the region (Hewitt, 2001). Seven events, including the only twentieth- century examples, descended onto glaciers and it seems likely that more rock avalanches have and will occur in the extensive glacierized areas of the Karakoram than elsewhere. However, the inventory so far is dominated by prehistoric events, recognized from their deposits in the inhabited river valleys. The rock avalanche deposits all exceeded 10×10^6 m³ in volume, many over 200×10^6 m³, and a few more than 1×10^9 m³. Vertical displacements, from the head of the detachment zone to the farthest run-out of debris, were at least 1000 m, in some cases more than 2000 m. Source slopes are in excess of 40°, often over 60°. Maximum horizontal displacements exceeded 6 km, in some cases more than 10 km (Hewitt, 1998a, 2001). Rockslide–rock

Figure 9.10 *Path of the 1986 debris flow through Haldi, a village at the mouth of the Saltoro River in Baltistan. Initiated by an intense, local thunderstorm in summer, the typical flow of mud with large boulders buried several square kilometres of precious arable land, destroyed about 50 houses and many trees supplying fruit or timber Photo: K. Hewitt*

avalanches were found in every elevation zone between 1800 and 7200 m asl and in all geological terranes (Hewitt, 1999). They are divided almost equally among igneous, metamorphic and metasedimentary rock types (Hewitt, 2002b).

These events have serious but hitherto unrecognized implications for natural hazards assessment. The numbers now known come from surveys of barely 20% of the Karakoram, but involve a much larger fraction of the zone of permanent settlement. In fact, dozens of settlements, including some of the main towns and visitor destinations, lie on or beside rock avalanche deposits. Almost all settlements and most of the roads and other infrastructure lie on, or across, sedimentary features controlled by cross-valley rock avalanche barriers (Hewitt, 1998a). Many of the worst sites of slope instability and recurring debris flows along the highways are where they cross rock avalanche deposits. Only two historical events are known to have engulfed inhabited areas in the past 200 years, but four episodes were identified in the glacier zone in the last 15 years. Without dates for a large fraction of the cases now discovered, it is impossible to determine the contemporary risk in the zone of permanent settlement. Important questions are whether the rock avalanche deposits represent a declining post-glacial risk, or one dependent mainly on rare, extreme triggering events such as large earthquakes, great summer storms or exceptional quantities of snow and glacier melt waters. Each of these has triggered catastrophic rockslides in the past.

8.5 Landslide dams

Floods and unseasonal low flows on the Himalayan streams are the most serious problems originating in the mountains and affecting Pakistan and India (Hewitt, 1989a; Ives and Messerli, 1989). Here, however, only certain special problems affecting people in the Karakoram will be considered. Small blockages resulting from spring and summer debris flows can occur in most valleys. These may cause local inundations and hardships, and hundreds of examples are known from the valleys (Kreutzmann, 1994). Debris flows may impound lakes of moderate size on the main rivers. A very large debris flow dammed the Ghizar River in July 1980 just above its junction with the Gilgit at Gupis (Nash and Hughes, 1986). The dam survives to the present time (2001), with slight and slow incision of the outlet stream. The worst damage was from inundation of village land upstream, but the lake is also filling rapidly with sediment, promising new land to occupy in the future. In the late 1990s three other debris flows blocked the same river, causing similar damage from inundation of village lands.

The largest flood disaster known on the Indus occurred in 1841 as the result of a rock avalanche triggered by an earthquake. The debris blocked the Indus River below Nanga Parbat (8126 m) in the western Himalaya (Shroder, 1993). The dam survived for six months before failing catastrophically and releasing a huge flood wave. This caused damage and deaths all the way to the Indian Ocean. The second largest flood wave on record came from a landslide dam in the Hunza valley in 1858. It had similar, if less destructive, results.

Meanwhile, more than 100 of the rock avalanches identified above formed relatively long-lived landslide dams, as indicated by extensive lacustrine deposits (Hewitt, 2002a). Early or catastrophic failure seems to be rare, but not unknown. In a few cases the lakes were hundreds of metres deep at the barrier and more than 100 km long when full. A repeat of some of the larger examples in the Karakoram would not only drown hundreds of villages, but cause severe water-supply problems for most of Pakistan while the reservoir filled up.

8.6 Glacier hazards

On the higher parts of the Karakoram heavy snowfall supports the largest contiguous area of glaciers outside polar regions (Hewitt et al., 1989). There are several thousand individual glaciers, but some 30 large ice masses dominate the glacial zone. The more serious dangers involve outburst floods and surges.

Every major tributary of the Indus draining the high Karakoram ranges has a history of glacier damming, and rarer but catastrophic outburst floods (Table 9.5). There have been at least 35 large examples over the last 200 years, generating destructive flood waves that travelled hundreds of kilometres through the mountain canyons. Mason (1935) argued that dams were always the result of exceptional glacier advances, or surges, but this is not true of many of the examples in Table 9.5. For example, the Chatteboi–Karambar dams reform in almost every year, and there is no reason to suppose the large dams at the Biafo and Batura glaciers have been the result of surges (Hewitt, 1969).

In the last 100 years, 26 sudden, rapid advances have been reported involving 17 glaciers (Hewitt, 1998b). At least 12 other glaciers have features associated with surging. Several surges caused devastation in, or other losses for, mountain settlements. Locally they bring increased erosion, sudden, local floods and prevent movement across the glaciers. Where this involves paths leading to high pastures, firewood sources, hunting areas and trekking routes for visitors, villagers may suffer loss of livelihood. All of these problems were associated with the surge of the Bualtar Glacier in 1986 (see Figure 9.7).

The greatest danger from surges is where they form an ice dam on a significant stream. The last of many glacial lake outburst floods, large enough to cause damage as far as the Indus Plains, occurred in 1926 and 1929 (Mason, 1929; Gunn, 1930; Lyall-Grant and Mason, 1940). They derived from ice dams of the Chong Khumdan Glacier on the Upper Shyok. Dams were observed from 1925 to 1931, resulting in three catastrophic outbursts and several minor ones. The glacier is particularly hazardous because it may reseal dams over several years, even after one or more catastrophic failures. In fact, it was the most dangerous glacier in the historical records of the Indus Basin; a threat compounded by two nearby glaciers, the Kitchik Khumdan and Sultan Chusskku, that have also surged and dammed the river. Their advances are out of phase with the Chong Khumdan (Visser, 1932). The present status of these glaciers is uncertain. Mason (1929) estimated the surging interval of the Chong Khumdan at around 90 years, suggesting it should be watched carefully over the next few years.

Another significant threat involves the Karambar Glacier, on a tributary of the Gilgit River. It surged

Table 9.5 – Glaciers interfering with Upper Indus streams in the present or Neoglacial advances, including known ice and moraine barriers, and glacial lake outburst floods (GLOFs)

River basin	Total	Present day	Modern dam[a]	GLOFs
Chitral				
Yarkhun	9	5	6	–
Laspur	3	3	I	–
Turikho	5	5	4	
Gilgit				
Ghizar	I	I	I	–
Yasin	I	I	I	
Ishkoman-Karambar	9	8	5	9[b]
Bagrot	2	I	I	–
Hunza				
Chapursan	3	2	3	2
Shimshall	3	3	2	5
Hispar	I	–	I?	
Main Hunza	4	I	I	?
Indus (Baltistan)				
Khumdan (Shyok)	4	4	4	19
Nubra	3	2	2	–
Saltoro	I	–	⌐	–
Hushe	2	–	I	2
Braldu	2	–	I	I
Basha	I	–	I	I

Notes:

[a]Number of glaciers with reported dams. Some have formed more than one dam.

[b]Countless small ones of local impact, especially from the perennial Chatteboi dam on Karambar.

three times in the twentieth century and in 1905 formed a large ice dam on the Karambar River. The dam failed suddenly giving rise to the largest and most destructive flood recorded on the Gilgit River, some damage continuing beyond its junction with the Indus. The Karambar surges of the mid-1950s and 1993 advanced the glacier some 3–4 km, interfering with, but not damming, the river.

The latest threat from a surge-related ice dam was in 1999 and 2000, from the Khurdopin Glacier in the Shimshal valley where it crosses the mouth of its Virjerab tributary. This is thought to be the source of some nine historic floods recorded over the past 200 years on the Upper Indus (Mason, 1930). Satellite imagery showed deformation and rapid displacement of ice 1–2 km down-valley between 1998 and 1999, while the lower glacier thickened and broke up into severely crevassed ice. All of this suggests a surge, though we have no detailed measurements. The middle and lower reaches of the Khurdopin display the 'meandering' or looped ice streams and medial moraines associated with surging glaciers elsewhere. Local residents say the glacier is subject to sudden advances and surface changes, and long periods when the lower tongue stagnates. Like the Chong Khumdan, impound-ments have been breached and resealed over as many as four successive years.

The last great flood in 1961 destroyed half of Shimshal's cropland, many houses and animals, and one-third of the productive land of Pasu in Hunza. Fortunately, the lakes of 1999 and 2000 did not reach a large size and drained relatively slowly in early summer, when existing flows in the rivers were still not high. There was some damage to agricultural land and irrigation channels around Shimshal village. Preparations in case of large floods caused more disruption of life than the damage suggests. The larger losses were because the heavily crevassed ice prevented Shimshal shepherds from taking their animals to high pastures along the Khurdopin and its tributary valleys.

8.7 Severe storms

Severe rainstorms occur in parts of the region in almost every summer. An increase in their occur-rence and severity has been the worst problem of climatic change in the past three decades, damag-ing crops and houses designed for snow and dry conditions. The main sources of destruction are the landslides that heavy rain can trigger. The worst are large debris flows such as those on the Ghizar mentioned above. These are processes that may devastate areas of long-term permanent settlement on sediment fans and valley floors (cf. Figure 9.8). Heavy rains also result in many debris slides and rock falls from unstable cuts along the roads, which may be blocked for days or weeks (Figure 9.11). Karakoram rains hardly ever cause floods on the main rivers. The catchments below the snow line are too small in winter rains, and summer storms shut down melting in the more extensive high-altitude snowfields and glacier ablation zones. The latter dominate the flow of the Indus streams in the Karakoram.

One of the most damaging recent disasters was in September 1992, when an influx of monsoon air, remarkably late in the year, brought severe rains and heavy damage to settlements and roads throughout the region (Hewitt, 1993a). For several days after the storm ended, practically everyone in the villages, men and women, old and young, could be observed working together to rebuild homes, and clear and rebuild clogged or damaged irrigation ditches. Shepherds who had been in the high pastures left the animals in the care of a few older children and raced down to help the injured and homeless, and with the rebuilding. Roads throughout the region were cut and blocked by land-

Figure 9.11 *Section of the Karakoram Highway, linking Pakistan and China, blocked in May 1987 by debris flows triggered by an intense local thunderstorm*
Photo: K. Hewitt

slides, rockfalls and debris flows. There were innumerable blockages to the Karakoram Highway and flights from Gilgit and Skardu were cancelled, leading to some major problems for tourists trapped in remote areas. The problems would have been worse had this not been at the end of the tourist season. Typically, however, the flow of the main Indus streams in the Karakoram declined rapidly as snow and low temperatures shut down glacier ablation.

However, conditions in the Karakoram received little and delayed attention from the media and officials, because of developments to the south. The same system had brought torrential rains in the front ranges of the Himalayas producing huge floods there. These and, especially, a sudden 'planned' release of impounded waters from the large Mangla Dam drowned several thousand people and wrought havoc far down the Indus Plains.

8.8 Changing risks and priorities

The significance of geomorphic hazards in this region, as in most mountains, has been changing swiftly in recent decades. Perhaps the most important transformations came from the building of an all-weather highway in the 1970s, linking the plains of Pakistan to China: the Karakoram Highway or (Kreutzmann, 1991). Many unsurfaced 'jeepable' roads have been built since, linking all but a few villages to the Karakoram Highway and, ultimately, the main centres of lowland Pakistan. With the tourism boom of the 1980s and 1990s, many hotels were built and other services provided in towns, and arrangements for large numbers of trekking and mountaineering expeditions to the high peaks and glaciers. There is still little mechanization in the villages, but electricity from small hydroelectric plants is becoming quite widespread. More and more villages on the roads are hiring machines for agricultural processes, notably threshing of grain, from itinerant contractors. Each year more land is devoted to cash crops such as tree fruits and seed potatoes, destined for urban markets.

Formally, it seems that only men's lives and risks are affected by these changes. Growing numbers of them spend their time in, or looking for, wage work, ways to trade in village products, or establishing small businesses. Large numbers act as porters and guides for tourists and mountaineers, serve in the army, with road crews, or on major construction works. Many head down to the lowland cities and even overseas to earn 'hard currency'. These activities expose them to different, sometimes increased risks, compared with village life (Kreutzmann, 1995; MacDonald and Butz, 1998). Deaths of soldiers and other government servants in recent conflicts and terrorism have been disproportionately men from the northern mountains of Pakistan and India. Hundreds from the Karakoram valleys died in the conflict with India around Kargil in 1999–2000. However, death and injury from avalanches, landslides, altitude sickness and frostbite have probably affected more soldiers, as well as those working on the highways and for mountaineering expeditions (Dixit, 1992). Village men also find themselves, or their investments, directly at risk through increased dependency on the tourist trade when there is a sharp drop in visitors. This has happened in the new millennium because of nuclear weapons development, open conflict and fear of terrorism.

It is, however, an illusion to imagine that these changes do not affect women and their risks. Most women may remain secluded in village roles, tending crops and animals, taking care of children; but risks are changing there too. With men away in paid work, women's workloads and responsibilities in the traditional subsistence economy have increased. Children and the elderly also play greater roles in the domestic and subsistence economy. At the same time, many development projects at the village level tend to increase women's work or to open up land or areas of settlement subject to other, sometimes greater, geomorphic hazards than traditional lands and activities.

In sum, the Karakoram may well seem to be one of the most rugged and 'remote' of high mountain regions, its people among the last 'romantic' and 'traditional' societies. But the stereotypes hardly define the scope of contemporary risks and disasters, nor where they are headed. Of course, the region has an ancient history connected to, and influenced by, developments in surrounding regions of inner and south Asia. However, the scope and pace of recent changes, the forms and impacts of modernization, herald accelerated outside influences and accommodations to them. With this, the significance of geomorphic hazards for residents, local and national governments, and visitors alike, have changed. Environmental risks can hardly be understood without reference to how these society-driven changes affect land use, economic and social activity, apart from developments with manifestly adverse impacts on the habitat. Hence, important as the understanding and monitoring of dangerous geomorphic processes are, they offer only a prologue to understanding risk. This also requires an assessment of human exposure and vulnerability, social and physical protection, and what happens to these in a changing world.

References

Alexander, D. and Formichi, R., 1993. Tectonic causes of landslides. *Earth Surface Processes and Landforms*, **18**: 311–38.

Allan, N.J.R.L. (ed.), 1995. *Mountains at risk: current issues in environmental studies*. New Delhi: Manohar.

Avalanche Canada, 2002. Avalanche accidents in Canada, volume 4: 1984–1996, 1–17. http://www.avalanche.ca/accident/trends

Azhar-Hewitt, F., 1989. Women's work, women's place: the gendered life-world of a high mountain community in Northern Pakistan. *Mountain Research and Development*, **9**: 335–52.

Banskota, M. and Karki, A.S. (eds), 1994. *Sustainable development of fragile mountain areas of Asia: regional conference report, 13–15 December*. Kathmandu, Nepal: International Centre for Integrated Mountain Development (ICIMOD).

Beniston, M., 2000. *Environmental change in mountains and uplands*. London: Arnold.

Benn, D.I. and Evans, D.J.A., 1998. *Glaciers and glaciation*. London: Arnold.

Blaikie, P., Cannon, T., Davis, I. and Wisner, B., 1994. *At risk: natural hazards, people's vulnerability, and disasters*. London: Routledge.

Bovis, M.J. and Evans, S.G., 1996. Extensive deformations of rock slopes in the southern Coast Mountains, British Columbia, Canada. *Engineering Geology*, **44**: 163–82.

Brunsden, D. and Jones, D.K.C., 1984. The geomorphology of high magnitude–low frequency events in the Karakoram Mountains. In Miller, K.J. (ed.), *The international Karakoram project*. Cambridge: Cambridge University Press, 345–88.

Brunsden, D., Doornkamp, J.C., Fookes, P.G., Jones, D.K.C. and Kelly, J.H.M., 1975. Large-scale geomorphological mapping and highway engineering design. *Quarterly Journal of Engineering Geology*, **8**: 227–53.

Burbank, D.W., Leland, J., Fielding, E., Anderson, R.S., Brozovic, N., Reid, M.R. and Duncan, C., 1996. Bedrock incision, rock uplift and threshold hillslopes in the northwestern Himalayas. *Nature*, **379**: 505–10.

Butz, D., 1996. Sustaining indigenous communities: symbolic and instrumental dimensions of pastoral resource use in Shimshal, Northern Pakistan. *The Canadian Geographer*, **40**: 36–53.

Byers, A.C., 2000. Contemporary landscape change in the Huascaran National Park and Buffer Zone, Cordillera Blanca, Peru. *Mountain Research and Development*, **20**: 52–63.

Church, M. and Ryder, J., 1972. Paraglacial sedimentation: a consideration of fluvial processes conditioned by glaciation. *Geological Society of America Bulletin*, **83**: 3072–95.

Clarke, G.K.C., 1991. Length, width, slope influences on glacier surging. *Journal of Glaciology*, **37**: 236–46.

Coburn, A., Hughes, R., Illi, D., Nash, D. and Spence, R., 1984. The construction and vulnerability to earthquakes of some building types in the Northern Areas of Pakistan. In Miller, K.J. (ed.), *The international Karakoram project*. Cambridge: Cambridge University Press, 226–52.

Collins, D.N., 1989. Seasonal development of subglacial drainage and suspended sediment delivery to melt waters beneath an alpine glacier. *Annals of Glaciology*, **13**: 45–50.

Costa, J.E. and Schuster, R.L., 1987. *The formation and failure of natural dams*. US Geological Survey, Open-File Report 87–392. Department of the Interior. Vancouver, WA: Government Printing Office.

Dani, A.H., 1989. *History of northern areas of Pakistan*. Islamabad: National Institute of Historical and Cultural Research.

Dingwall, P.R., Fitzharris, B.B. and Owens, I.F., 1989. Natural hazards and visitor safety in New Zealand's National Parks. *New Zealand Geographer*, **45**: 68–79.

Dixit, K.M., 1992. The porter's burden. *Himal*, **8**: 32–39.

Ehlers, E., 1995. Dies Organisation von Raum und Zeit – Bevolkerungswachstum, Ressourcemanagement und angeplasste Landnutzung im Bagrot/Karakorum. *Petermanns Geographische Mitteilungen*, **139**: 105–20.

Fort, M., 1987. Geomorphic and hazards mapping in the dry, continental Himalaya: 1:50,000 maps of Mustang District, Nepal. *Mountain Research and Development*, **7**: 222–38.

Goudie, A.S., Brunsden, D., Collins, D.N., Derbyshire, E., Ferguson, R.I., Hasmet, Z., Jones, D.K.C., Perrot, F.A., Said, M., Waters, R.S. and Whalley, W.B., 1984. The geomorphology of the Hunza Valley, Karakoram mountains, Pakistan. In Miller, K.J. (ed.), *The international Karakoram project*. Cambridge: Cambridge University Press, 359–410.

Grove, J.M., 1988. *The Little Ice Age*. London: Routledge.

Gude, M. and Scherer, D., 1998. Snowmelt and slushflows: hydrological and hazard implications. *Annals of Glaciology*, **26**: 381–84.

Gunn, J.P., 1930. *Report on the Khumdan Dam and Shyok Flood of 1929*. Lahore: Government of the Punjab Publication.

Hadley, J.B., 1964. *Landslides and related phenomena accompanying the Hebgen Lake earthquake of August 17th, 1959*. US Geological Survey Professional Paper 435-K. Washington, DC: Government Printing Office, 107–38.

Haigh, M.J., 1984. Landslide prediction and highway maintenance in the Lesser Himalaya, India. *Zeitschrift für Geomorphologie*, N.S., **51**: 17–37.

Hardoy, J.E. and Satterthwaite, D., 1989. *Squatter citizen: life in the urban third world*. London: Earthscan.

Heim, A., 1932. *Bergsturz und Menschenleben*. Zürich: Fretz and Wasmuth.

Hestnes, E., 1998. Slushflow hazard – where, why and when? 25 years of experience with slushflow consulting and research. *Annals of Glaciology*, **26**: 370–77.

Hewitt, K., 1969. Glacier surges in the Karakoram Himalaya (Central Asia). *Canadian Journal of Earth Sciences*, **6**: 1009–18.

Hewitt, K., 1976. Earthquake risk in the mountains. *Natural History*, **85**: 30–37.

Hewitt, K., 1982. Natural dams and outburst floods of the Karakoram Himalaya. In Glen, J. (ed.), *Hydrological aspects of high mountain areas*. IAHS Publication 138. Wallingford: IAHS Press, 259–69.

Hewitt, K., 1988. Catastrophic landslide deposits in the Karakoram Himalaya. *Science*, **242**: 64–67.

Hewitt, K., 1989a. Hazards to water resources development in high mountains: the Himalayan sources of the Indus. In Starosolszky, O. and Melder, O.M. (eds), *Hydrology of disasters: proceedings of the technical conference in Geneva, November 1988*. London: World Meteorological Organisation/James and James, 294–312.

Hewitt, K., 1989b. Human society as a geological agent. In Fulton, R.J. (ed.), *Quaternary geology of Canada and Greenland: geology of North America*. No. K-7. Ottawa: Geological Survey of Canada/Geological Society of America, 619–34.

Hewitt, K., 1992. Mountain hazards. *GeoJournal*, **27**: 47–60.

Hewitt, K., 1993a. Torrential rains in the Central Karakoram, 9–10 September 1992. *Mountain Research and Development*, **13**: 371–75.

Hewitt, K., 1993b. The altitudinal organization of Karakoram geomorphic processes and depositional environments. In Shroder, J.F. Jr (ed.), *Himalaya to the sea: geology, geomorphology and the Quaternary*. New York: Routledge, 159–83.

Hewitt, K., 1997a. *Regions of risk: a geographical introduction to disasters*. London: Addison Wesley Longman.

Hewitt, K., 1997b. Risk and disasters in mountain lands. In Messerli, B. and Ives, J.D. (eds), *Mountains of the world: a global priority*. New York: Parthenon Publishing Group, 371–408.

Hewitt, K., 1998a. Catastrophic landslides and their effects on the Upper Indus streams, Karakoram Himalaya, northern Pakistan. *Geomorphology*, **26**: 47–80.

Hewitt, K., 1998b. *Recent glacier surges in the Karakoram Himalaya, south central Asia*. American Geophysical Union. http:www.agu.org/eos_elec/97016e.html

Hewitt, K., 1999. Quaternary moraines vs catastrophic rock avalanches in the Karakoram Himalaya, Northern Pakistan. *Quaternary Research*, **51**: 220–37.

Hewitt, K., 2001. Catastrophic rockslides and the geomorphology of the Hunza and Gilgit River valleys, Karakoram Himalaya. *Erdkunde*, **55**: 72–93.

Hewitt, K., 2002a. Postglacial landform associations in a landslide-fragmented river system: the transHimalayan Indus streams, Central Asia. In Hewitt, K. *et al.* (eds), *Landscapes of transition: landform assemblages and transformations in cold regions*. Dordrecht: Kluwer Academic, 63–92.

Hewitt, K., 2002b. Styles of rock-avalanche depositional complexes conditioned by very rugged terrain, Karakoram Himalaya, Pakistan. In Evans, S.G. and DeGraff, J.V. (eds), *Catastrophic landslides: effects, occurrence, and mechanisms*. Reviews in Engineering Geology, 15. Boulder, CO: Geological Society of America, 1–33.

Hewitt, K., Byrne, M.L., English, M., and Young, G. (eds.) 2002. *Landscapes of transition: landform assemblages and transformations in Cold Regions*. Dordrecht: Kluwer Academic.

Hewitt, K., Wake, C.P., Young, G.J. and David, C., 1989. Hydrological investigations at Biafo Glacier, Karakoram Range, Himalaya: an important source of water for the Indus River. *Annals of Glaciology*, **13**: 103–108.

Hsu, K.J., 1975. Catastrophic debris streams (struzstroms) generated by rock falls. *Geological Society of America Bulletin*, **86**: 129–40.

International Strategy for Disaster Reduction, 2000. *Usoi landslide dam and Lake Sarez: an assessment of hazard and risk in the Pamir Mountains, Tajikistan*. ISDR Prevention Series, No. 1. New York and Geneva: United Nations Publication.

Ives, J.D., 1987. Repeat photography of debris flows and agricultural terraces in the Middle Mountains, Nepal. *Mountain Research and Development*, **7**: 82–86.

Ives, J.D., 1997. Comparative inequalities – mountain communities and mountain families. In Messerli, B. and Ives, J.D. (eds), *Mountains of the world: a global priority*. New York: Parthenon Publishing Group, 61–84.

Ives, J.D. and Messerli, B., 1981. Mountain hazards mapping in Nepal: an overview. *Mountain Research and Development*, **1**: 223–30.

Ives, J.D. and Messerli, B., 1989. *The Himalayan dilemma: reconciling development and conservation*. London: Routledge.

Jansen, R.B., 1980. *Dams and public safety*. Washington, DC: Government Printing Office.

Kalvoda, J., 1992. *Geomorphological record of the Quaternary orogeny in the Himalaya and Karakoram*. New York: Elsevier.

Keefer, D.K., 1984. Landslides caused by earthquakes. *Geological Society of America Bulletin*, **95**: 406–21.

Kreutzmann, H., 1991. The Karakoram Highway: the impact of road construction on mountain societies. *Modern Asian Studies*, **25**: 711–36.

Kreutzmann, H., 1994. Habitat conditions and settlement processes in the Hindu Kush–Karakoram. *Petermanns Geographische Mitteilungen*, **138**: 337–56.

Kreutzmann, H., 1995. Globalisation, spatial integration, and sustainable development in Northern Pakistan. *Mountain Research and Development*, **15**: 213–27.

LaChapelle, E.R., 1985. *The ABC of avalanche safety*. Seattle, WA: The Mountaineers.

Lasa, J. and Jingfang, N., 1992. A tentative appraisal of the environmental impact of railway construction in mountain areas. In Walling, D.E., Davies, T.R. and Hasholt, B. (eds), *Erosion, debris flows and environment in mountain regions*. IAHS Publication No. 209. Wallingford: IAHS Press, 413–18.

Lavell, A. (ed.), 1994. *Viviendo in Riesgo: communiades vulnerable y prevencion de desastres en America latina*. Santafe de Bogota, Colombia: La Red, Tercer mundo Editores.

Lied, K. and Toppe, R., 1989. Calculation of maximum snow-avalanche run-out distance by use of digital terrain models. *Annals of Glaciology*, **13**: 164–69.

Lyall-Grant, I.H. and Mason, K., 1940. The Upper Shyok glaciers in 1939. *Himalayan Journal*, **12**: 52–63.

MacDonald, K.I., 1994. The mediation of risk: ecology, society and authroity in a Karakoram mountain community. Unpublished Ph.D. Thesis, Waterloo: University of Waterloo, Ontario.

MacDonald, K.I., 1996. Indigenous labour arrangements and household security in Northern Pakistan. *Himalayan Research Bulletin*, **16**: 29–35.

MacDonald, K.I. and Butz, D., 1998. Investigating portering relations as a locus from transcultural interaction in the Karakoram region of Northern Pakistan. *Mountain Research and Development*, **18**: 333–43.

Maskrey, A., 1989. *Disaster mitigation: a community based approach*. Oxford: Oxfam.

Mason, K., 1929. Indus floods and Shyok glaciers. *Himalayan Journal*, **1**: 10–29.

Mason, K., 1930. The glaciers of the Karakoram and neighbourhood. *Records of the Geological Survey of India*, **63**: 216–78.

Mason, K., 1935. The study of threatening glaciers. *Geographical Journal*, **8**: 24–35.

Messerli, B. and Ives, J.D. (eds), 1997. *Mountains of the world: a global priority*. New York: Parthenon Publishing Group.

Miller, K.J. (ed.), 1984. *The international Karakoram project*. Cambridge: Cambridge University Press.

Muller, L., 1968. New considerations on the Vaiont slide. *Rock Mechanics and Engineering Geology*, **6**: 1–91.

Nash, D. and Hughes, R., 1986. The Gupis debris flow and natural dam. *Disasters*, **10**: 8–14.

Nicoletti, P.G. and Sorriso-Valvo, M., 1991. Geomorphic controls on the shape and mobility of rock avalanches. *Geological Society of America Bulletin*, **103**: 1365–73.

Oliver-Smith, A.S., 1986. *The martyred city: death and rebirth in the Andes*. Prospect Heights Ill: Waveland Press.

Owens, I.F. and Fitzharris, B.B., 1989. Assessing avalanche-risk levels on walking tracks in Fiordland, New Zealand. *Annals of Glaciology*, **13**: 231–36.

Paterson, W.S.B., 1994. *The physics of glaciers* (third edition). Oxford: Pergamon/Elsevier Science.

Perla, R.I., 1980. Avalanche release, motion and impact. In Colbeck, S.C. (ed.), *Dynamics of snow and ice masses*. New York: Academic Press, 397–462.

Pflaker, G. and Ericksen, G.E., 1978. Nevados Huascaran avalanches, Peru. In Voight, B. (ed.), *Rockslides and avalanches I. Natural phenomena*. New York: Elsevier, 277–314.

Price, L.W., 1981. *Mountains and man: a study of process and environment*. Berkeley, CA: University of California Press.

Radbruch-Hall, D.H., 1978. Rock creep, scale effects and related problems. In Voight, B. (ed.), *Rockslides and avalanches I. Natural phenomena.* New York: Elsevier, 607–58.

Reiter, L., 1990. *Earthquake hazard analysis: issues and insights.* New York: Columbia University Press.

Reynolds, J.M., 1992. The identification and mitigation of glacier-related hazards: examples from the Cordillera Blanca, Peru. In McCall, G.J.H., Laming, D.J.C. and Scott, S.C. (eds), *Geohazards: natural and man-made.* London: Chapman and Hall, 143–58.

Rieder, P. and Wyder, J., 1997. Economic and political framework for sustainability of mountain areas. In Messerli, B. and Ives, J.D. (eds), *Mountains of the world: a global priority.* New York: Parthenon Publishing Group, 85–102.

Scheidegger, A.E. 1998. Tectonic predesign of mass movements, with examples from the Chinese Himalaya. *Geomorphology,* **26**: 37–46.

Schuster, R.L. (ed.), 1986. *Landslide dams: processes, risk, and mitigation.* Geotechnical Special Publication No. 3. New York: American Society of Civil Engineers.

Selli, R. and Trevisan, L., 1964. Caratteri e interpretazione della frana del Vaijont. *Giornale do Geologia,* **32**: 7–123.

Servizio Valanghe Italiano, 2002. *Statistiche Degli Incidenti in Valanga.* Commissione Technica del C.A.I. www.cai-svi.it/ital5anni.php

Shroder, J.F., Jr, 1989. Hazards of the Himalaya. *American Scientist,* **77**: 565–73.

Shroder, J.F., Jr (ed.), 1993. *Himalaya to the sea: geology, geomorphology and the Quaternary.* London: Routledge.

Shroder, J.F., Jr and Bishop, M.P., 1998. Mass movement in the Himalaya: new insights and research directions. *Geomorphology,* **26**: 13–35.

Spencer, J.E. and Hale, S.A., 1961. The origin, nature and distribution of agricultural terracing. *Pacific Viewpoint,* **2**: 1–40.

State of Colorado, 2002. *US & world avalanche accident statistics 2000–2001.* http://geosurvey.state.co.us/html.

Stone, P.B. (ed.), 1992. *The state of the world's mountains: a global report.* London: Zed Books.

Strom, A.L., 1996. Some morphological types of long-runout rockslides: effects of the relief on their mechanism and of rockslide deposits distribution. In Senneset, C. (ed.), *Landslides.* Rotterdam: Balkema, 1977–1982.

Tufnell, L., 1984. *Glacier hazards.* London: Longman.

United States Office of Foreign Disasters Assessment, 1989. *Disaster history: significant data on major disasters worldwide, 1900–present.* Washington, DC: Agency for International Development.

Visser, Ph.C., 1932. Gletscheruberschiebungen im Nubra-, und-Shyock-Gebiet des Karakorum. *Zeitschrift für Gletscherkunde,* **21**: 29–44.

Voight, B. (ed.), 1978. *Rockslides and avalanches I. Natural phenomena.* New York: Elsevier.

Voight, B. and Pariseau, W.G., 1978. Rockslides and avalanches: an introduction. In Voight, B. (ed.), *Rockslides and avalanches I. Natural phenomena.* New York: Elsevier, 1–70.

Waddell, E., 1983. Coping with frosts, governments and disaster experts: some reflections based on a New Guinea experience and a perusal of the relevant literature. In Hewitt, K. (ed.), *Interpretations of calamity: from the viewpoint of human ecology.* London: Allen and Unwin, 33–43.

Whitehouse, I.E. and Griffiths, G.A., 1983. Frequency and hazard of large rock avalanches in the central Southern Alps, New Zealand. *Geology,* **11**: 331–34.

Whiteman, P.T.S., 1985. *Mountain oases: a technical report on agricultural studies (1982–84) in Gilgit District, Northern Areas, Pakistan.* Gilgit, Pakistan: FAOUNDP, PAK/80/009.

10
Mountain hazards in China

Tianchi Li

1 Introduction

China is a particularly mountainous country. Mountains cover about two-thirds of its total land area of 9.6 million km^2 (Wang et al., 1986). China has a population of approximately 1.3 billion people, and about one-third of this, and two-fifths of the total cultivated land area, are distributed in the mountainous part of the country. Furthermore, there are rich water, mineral and forest resources in the mountainous areas, and these areas are also important for tourism. For these reasons, mountains are important for the economic development of the country and have great productive potential and vast development prospects. On the other hand, the mountains have complex geologic structures, intense geotectonic and earthquake activity and abundant water supply, as a result of the monsoonal climate, which combine to make China one of the countries most susceptible to mountain hazards. Since human economic activities have expanded to the mountain regions, exploitation of mountains has intensified these hazards. This chapter examines the types and distribution of mountain hazards, their impacts on socio-economic development and the techniques used in China to reduce the impacts of mountain hazards.

2 The main mountain systems in China

Figure 10.1 shows the location of the main mountain ranges in China. On the basis of their absolute relief, these mountains can be classified into four categories:

1 Extremely high, folding-faulting mountains with elevation >5000 m asl. These mountains are generally covered with continuous snow and well-developed features of modern glaciation. They are distributed mostly in western China, and include the mountains of Altay, Tianshan, Kunlun and Altun in northwest China, and the mountains of Tanggula, Gangdise, Nyaingentanglha and the Himalaya on the Tibet (Qinghai–Xizang) Plateau.

2 Strongly uplifted, faulting, high mountains, 3000–5000 m asl in elevation. Generally, they extend above the treeline but below the snowline. Nivation processes dominate, with well-developed periglacial landforms and ancient glaciation features. They are distributed mostly in central-west China, for example the mountains of Hengduan, Bayan and Qilian.

3 Moderately uplifted, folded and faulted middle mountains, 1500–3000 m asl in elevation. These mountains have a humid or semi-humid climate and have undergone severe erosion during the Quaternary period. This category includes most of the mountains located in central and eastern China.

4 Low rolling mountains and hills, generally between 500 and 1500 m asl in elevation, with a few peaks above 2000 m asl, under a humid or semi-humid climate. This category includes the

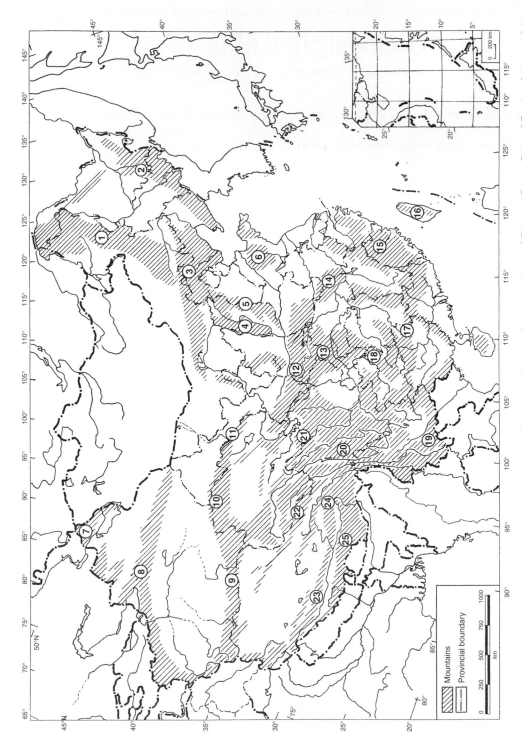

Figure 10.1 *Map of China showing the main mountain systems: 1, Great Khingan; 2, Changbai and Qianshan; 3, Yanshan; 4, Luliang; 5, Taihang; 6, Taishan; 7, Altay; 8, Tianshan; 9, Kunlun; 10, Altun; 11, Qilian; 12, Qinling; 13, Daba; 14, Dabie; 15, Tianmu, Wuyi and Donggan; 16, Chungyang; 17, Nanling; 18, Dalou; 19, Ailao; 20, Hengduan; 21, Bayan Har; 22, Tanggula; 23, Gangdise; 24, Nyainqentanglha; and 25, Himalaya.*

mountains of Tianmu, Taishan and Nanling in eastern and southern China, and the mountains of Yanshan, Great Khingan, Changbai and Qianshan, in north and northeastern China.

The types and occurrence of hazards in the mountain areas of China are affected by physiographic factors which are specific to each of these larger mountain systems. Therefore, the distribution of mountain hazards is fairly complicated and reflects the zonation of different physiographic conditions.

3 The type, causes and distribution of mountain hazards in China

3.1 The main types of geomorphic hazards

Landslides, rock falls, debris flows and outburst floods from the failure of natural dams and glacial lakes are the main types of natural hazards in the mountains of China. This chapter will mainly focus on these types of geomorphic hazard. Other, less frequent and/or damaging, hazards in the mountainous areas of China include excessive soil erosion, snow avalanches and the collapse of karst features.

A rock fall is defined as the rapid movement of weathered rock (sometimes including soil) from steep slopes or cliffs. A landslide is defined as a continuous slope movement that has a clear sliding surface, with small or large dimensions, occurring on gentle slopes and moving slowly or rapidly. Landslides are often influenced by groundwater flows (see Onda, this volume). Landslide development is a long-term process and can be divided into five stages: elastoplastic deformation; microfissure development; uniform displacement; accelerated displacement; and rapid sliding.

A debris flow is defined as a flow of a mixture of water, soil and air, with a unit weight of >1.2 t m^{-3}. The behaviour and characteristics of a debris flow lie between those of general sediment-laden flow and a landslide. In most cases, a debris flow is similar to a landslide. The evolution from landslide to debris flow is one of increasing water content and decreasing sediment content. According to viscosity, debris flows can be classified as (i) low-viscous debris flow or fluid debris flow (unit weight from 1.3 to 1.8 t m^{-3}), (ii) viscous debris flow (unit weight from 1.8 to 2.3 t m^{-3}), and (iii) plastic debris flow (unit weight >2.3 t m^{-3}) (Wu and Li, 2001). Debris flows usually have large impact forces, and vibrating and abrasive effects (Figures 10.2 and 10.3).

Outburst floods can be defined as flash floods that cause significant damage in the river valleys of mountains and the adjacent lowlands. They are often caused by glacial lake outburst in the high mountain areas (also see Björnsson, this volume), and the failure of landslide and debris flow dams in the high, rugged mountain areas (also see Hewitt, this volume).

3.2 The main causes of mountain hazards

The main hazards, such as rock falls, landslides and debris flows, are the result of the interactions between natural environmental and human conditions. Rainstorms, strong earthquakes and human activities constitute three of the most important triggering factors.

3.2.1 Heavy rainstorms

Monsoon rainstorms initiate many landslides and debris flows each year in China. Rainfall thresholds necessary to cause landslides and debris flows have been studied by many scientists (Li and Li, 1985; Zhong et al., 1993; Tan et al., 1994). The studies carried out in China show that: (i) if cumulative precipitation is between 50 and 100 mm day^{-1}, small-scale and shallow landslides and debris flows will occur; (ii) if cumulative precipitation over two days is 150–200 mm, and daily precipitation is approximately 100 mm, the number of landslides and debris flows has a tendency to increase proportionally with precipitation; and (iii) if cumulative precipitation exceeds 250 mm in two days, and has an average

Figure 10.2 *A view of a viscous, turbulent debris flow that occurred in the Jiangjia ravine,*
Dongchuan District, Yunnan Province. The width of the ravine channel is approximately 60 m.
The debris flow was triggered by a heavy rainstorm. The solid load is made up of rock
fragments, boulders up to 3-m diameter, and fine-grained sediment (>5% of the total solid
load). The unit weight of the debris flow was approximately 2 t m^{-3}, the height of the
turbulent surge was >5 m and the velocity was 10 m s^{-1}
Source: *reproduced with the permission of ICIMOD, Nepal.*

Figure 10.3 *A view of a viscous debris flow in the Jiangjia ravine, Dongchuan District,*
Yunnan Province. The 7-m-high building is the old debris flow observation station that was
constructed in the 1980s. The distance between the channel bed and the building
foundations is about 15 m. Heavy rainstorms may trigger more than debris flows each year.
The content of the solid load was similar to that for Figure 10.2. The fast-flowing inertia and
the shape of the banks caused the debris to rise c.10 m, with some stones rising >15 m
Source: *reproduced with the permission of ICIMOD, Nepal.*

intensity of more than 8 mm h^{-1} in one day, the number of large and vast landslides and debris flows increases dramatically. Therefore, it may be suggested that 50 mm precipitation in one day is a threshold necessary to create landslides and debris flows.

These studies also show that, under the same rainfall conditions, the main geomorphic hazards in mountains differ in their quantity, size and density owing to different geological and topographical conditions. Therefore, geomorphic hazards have obvious regional characteristics and patterns. For a given region, geological and geomorphological conditions are the decisive factors under which landslides and debris flows can be induced (Liu et al., 1992; Liu and Tang, 1995).

3.2.2 Earthquakes

It is well recognized that an earthquake can trigger landslides if the magnitude (m) of the earthquake is >4.0 M. China experiences many earthquakes, and at least 656 earthquakes with magnitude greater than 6.0 M took place between 1767 BC and AD 1976. Of the total earthquakes, 33.5% triggered landslides, except those in marine areas, Tibet and Taiwan (Feng and Guo, 1985). The earliest record of landslides triggered by an earthquake was found in the Chronicle on Bamboo slip in 1767 BC, which said that landslides induced by an earthquake blocked the Yi and Luo Rivers in Shaanxi Province. Other examples include the 59 huge landslides induced by the Tianshai earthquake ($M=$ 8.0) in 1657, 337 landslides by the Tongwei earthquake ($M=7.5$) in 1718, 675 landslides by the Haiyuan earthquake ($M=8.5$) in 1920, and 170 landslides triggered by the Songpan earthquake ($M=$ 7.2) in 1976.

Several studies on landslides (e.g., Li, 1979a, b; Feng and Guo, 1985) show that: (i) an earthquake of $M=4.0$ can trigger small-scale landslides and rock falls; (ii) an earthquake of $M=6.0$ can trigger many landslides and rock falls in mountains covered by loess and other loose materials; and (iii) for an earthquake of $M=7.0$ or greater, larger and disastrous landslides and debris flows can be induced over a large area.

The influence of geology is reflected in both geologic structure and lithologic characteristics. The Songpan earthquake of 16 August 1976 induced >170 slumps and slides, which occurred predominantly along the active tectonic faults in the seismic region. On the slopes consisting of loosened limestone and igneous rocks, rock falls occurred readily, but on slopes consisting of claystone, shale and phyllite the rock falls were fewer. The types of slope and slope angle also have a great influence on the occurrence of landslides and rock falls. Landslides and rock falls are rare on straight slopes, but they are common on convex, concave and complex slopes. Statistical data for the five most earthquake-prone areas for the period 1973 to 1976 show that landslides generally do not occur on slopes <25°, and that 90% of all landslides occur on slopes with slope angles ranging from 30° to 50°. In most cases, rock falls happen on slopes with angles ranging from 60° to 75°. Such figures are of great help in selecting a relatively safe zone in an area subject to earthquake activity. The safest zone in the mountain areas in China will be one with slope angles <25° (Li, 1979a, b).

3.2.3 Human activities

In addition to the main natural triggering factors, such as earthquakes and heavy rainstorms, mountain hazards can also be created by human activities, such as road and irrigation canal construction, deforestation, overgrazing and overexploitation of mineral resources. During the past 50 years, large-scale deforestation, unplanned urban growth and badly engineered mountain roads have been major contributing factors to geomorphic hazards in the mountain areas of China.

3.3 Regional distribution of mountain hazards in China

Mountain hazards have a wide spatial distribution and this is affected by geology, geomorphology and earthquake activity, as well as climatic and meteorological conditions. The following is a brief summary of the regional distribution of geomorphic hazards in the main mountain areas.

In the mountains of northeastern China, hazards are dominated by debris flows and landslides, which are caused by rainstorms during the monsoon season. Most of these hazards are distributed in the southwestern part of the Changbai Mountains and the Qianshen Mountains in southern Liaoning Province (Song and Li, 1998). For most of the Great Khingan Mountains in Heilongjiang Province, there are only a few small slumps and rock falls locally where sedimentary rocks form steep fault scarps, and along river banks where the rocks and soils undergo freeze–thaw and also channel erosion (Li, 1989).

In northern China, small- to medium-scale debris flows and landslides are found in the Yashan, Lulian and Taihang Mountains, and most of these are caused by very heavy rainstorms. They are usually very destructive, such as the debris flows and landslide at Fengcheng in 1985, and the debris flow in the Zuochao ravine, Quyan, in 1987.

In the Altay, Tianshan, Kunlun and Qilian mountains in northwestern China, landslides are not common because of dry climatic conditions, although some rock falls have occurred on the higher peaks and canyon walls. A few debris flows have been triggered by snow melting and heavy rainstorms in some small catchments with easily eroded glaciofluvial deposits (Chen, 1992). There are also other destructive mass movements in these mountain ranges.

The Qinling and Daba Mountains in central China, and the adjacent mountains on the Loess Plateau area, comprise one of the regions most prone to landslides and mudflows triggered by rainstorms and earthquakes. The landslides and mudflows are characterized by their large scale, wide distribution and strong activity. More than 2000 landslides have occurred along both banks of the Beilong River in the west of this region. In the mountain areas of the Loess Plateau, most of the landslides occur in the southern region, which is crossed by the Weihe, Jinhe and Taohe rivers. Famous landslides in this area include the bedrock landslide near Putouyan, and the large deep-seated landslide of Saleshan which occurred on 7 March 1983. The volume of the Saleshan landslide has been estimated at $>40 \times 10^6$ m³, and it was the largest slide in this region in recent years. Near Baoji city, in the middle reaches of the Weihe River, there are 43 large loess landslides occupying nearly 90% of the total 97 km length of the bank (Li and Wang, 1992). These landslides were induced mainly by rainfall, earthquakes and artificial irrigation. The mudflows in this region, such as those in the Biechuan and Tonghe River basins, are characterized by their large scale, intense activity and destructiveness (Li and Cheng, 1998).

The mountainous areas in the south Yangtze River basin and the southeast coastal mountains and hills experience monsoon climatic conditions. Thus, heavy rainfall from typhoons is the main triggering mechanism of debris flows and slumps on the mountain slopes, which are composed of reddish clay soils. In June 1982, a rain event of about 600 mm during seven days induced 9579 small and large landslides in the three counties in central Jiangxi Province (Li, 1989). Comparatively, these slumps and debris flows were small in size and less destructive.

The Three Gorges Reservoir area of the Yangtze River valley is one of the most hazard-prone areas in China. In this area, 2490 landslides and 90 debris flow ravines have been found. Since 1982, more than 70 landslides and debris flows, including 40 large landslides, have been reported and have killed more than 400 people. In the 1999 rainy season, 420 slope failures, landslides and debris flows occurred in the Three Gorges Reservoir area, causing death and injury to 512 people and total direct economic losses of US$62 million.

The Hengduan Mountains are located in the transition zone from the Tibetan Plateau to the

Sichuan Basin. The combination of steep slopes, sheared rocks, frequent earthquakes, heavy rainstorms and snow melting makes this region particularly hazard-prone. In this region, the catastrophic debris flow ravines are concentrated along the large rivers that flow along the major active fault lines, such as the Daying River and Xiaojiang River faults in Yunnan Province, and the Xiangshu River and Anning River faults in Sichuan Province. All of these fault lines are associated with strong earthquake activity (Wu and Li, 2001). In the last 50 years, the incidence of debris flows in this region has greatly increased because of serious environmental degradation. For example, the catchment of the Xiaojiang River in Yunnan Province has lost more than half of its forest area, and the vegetation cover has been reduced to between 2.5% and 8%. As a result, the number of debris flow ravines has increased from 38 at the beginning of the 1950s to 101 at present. The Yangtze River and its tributaries drain much of these mountains, and large landslides have frequently dammed the Yangtze River and its tributaries, causing great floods both upstream and downstream of the landslide dams. Table 10.1 presents examples of well-documented large landslide dams.

Landslides, snow avalanches and debris flows associated with glaciers, earthquakes and rainstorms are common in the mountain areas of south and southeastern Tibet. Large landslides occur frequently, creating serious problems every year for road transportation on the China–Nepal and the Sichuan–Tibet highways. In 1983, 36 vehicles and >100 people were buried by a huge rock avalanche from the Peilong ravine. Debris flows in this region are the result of high precipitation, glacier melting, snow avalanches and slope movements in till and colluvium or strongly weathered bedrock. Large landslides and debris flows often dam rivers, creating flooding upstream of the dam and possibly catastrophic flash floods downstream. The sites of landslide- and debris flow-dammed rivers are concentrated in the river valleys of the Bodoi, Yigong, Parlung, Dongjug and Niyanqu rivers, surrounding the great bend of the Yarlung River (upstream of the Brahmaputra River). During the past 50 years, at least 21 large debris flow events have had a big impact on highways, farmland and local communities. The Guxiang debris flow ravine is the most well-known site in China in terms of frequency and magnitude (Du *et al.*, 1992; Lu *et al.*, 1999). The largest landslide in China in recent years is the Yigong landslide (volume *c.*300×10^6 m^3) and occurred on 9 April 2000 (see Section 5.3.3 for further details).

In the Himalaya of southern Tibet, geomorphic hazards are also dominated by landslides, debris flows and glacial lake outburst floods (GLOFs). The flash floods caused by glacial lake outbursts at higher altitudes have resulted in huge problems for the people in the counties of Diangi, Jilong and Jiangzi, as well as to the neighbouring countries of Bhutan and Nepal. During the past 50 years, 15 GLOFs have been recorded and recognized as a common type of catastrophic mountain hazard in southern Tibet (Xu, 1987). Some GLOFs are reported to have created long-term secondary environmental degradation and socio-economic problems, both locally and in neighbouring downstream countries. The enormity of GLOF phenomena is demonstrated by the 1954 Jiangzi flood (Sangwang–Cho GLOF), during which 300×10^6 m^3 of water was released from the glacial lake with a 40-m-high surge flood on the Nyang Qiu River, leaving in its wake up to 5 m of debris in some areas and flood damage to cities as far downstream as Gyangzi (120 km) and Xigazi (200 km). It is estimated that the peak discharge from this outburst flood was >10000 m^3 s^{-1}.

The types and distribution of mountain hazards in China are illustrated in Figure 10.4. It shows that mountain hazards are widespread throughout the vast mountain areas of China. It is well known that landslides and debris flows are of special significance in southern mountain areas, the Three Gorges area of the Yangtze River, the mountain areas of the Loess Plateau area, including the provinces of: Sichuan, Yunnan, Guizhou and Tibet in southwest China; Gansu, Qinghai in northwest China; Hubei, Hunan and Shaanxi in central China; and Fujian, Jiangxi and Taiwan in southeast China. Owing to the

Table 10.1 – Examples of well-documented landslide dams in China

Name of landslide	Year	River dammed	County (Province)	Landslide volume (m^3)	Blockage dimension			Lake dimension		Dam failed?
					Height (m)	Length (m)	Width (m)	Length (km)	Volume (m^3)	
Daluba	1956	Qianjiang	Qianjiang (Sichuan)	40×10^6	70	1170	1040	5	70×10^6	No
Shigaodi	1881	Jinsha	Qianjia (Yunnan)	530×10^6	110	1000	–	50	270×10^6	Yes
Diexi	1933	Minjiang	Maowen (Sichuan)	150×10^6	255	400	1300	17	400×10^6	Yes
Luchedu	1935	Jinsha	Huili (Sichuan)	90×10^6	50	250	500	–	–	Yes
Bitang	1961	Bitang	Xunhua (Qinghai)	80×10^6	65	1000	–	1.5	4.2×10^6	No
Zepozhu	1965	Donghe	Xichang (Sichuan)	7.2×10^6	51	–	650	1	2.7×10^6	Yes
Pufu	1965	Pufugou	Luguan (Yunnan)	450×10^6	179	1100	800	1.8	5×10^6	Yes
Tanggudong	1867	Yalong	Yajiang (Sichuan)	68×10^6	175	650	3000	53	680×10^6	Yes
Guanjiayuan	1981	Changgou	Mianxian (Shaanxi)	2×10^6	30	250	200	–	10×10^6	Yes
Liangjiazhuang	1983	Gancha	Zhenan (Sichuan)	4.12×10^6	68	80	350	1.5	1.5×10^6	Yes
Diaobanya	1988	Baisha	Wanyuan (Sichuan)	100×10^4	38	75	100	0.5	–	No
Zhongyangchun	1988	Xixi	Suxi (Sichuan)	765×10^4	30	150	600	–	–	Yes
Zhamulongba	2000	Yigong	Bomi (Tibet)	300×10^6	130	1500	2600	17	500×10^6	Yes

Adapted from various technical reports.

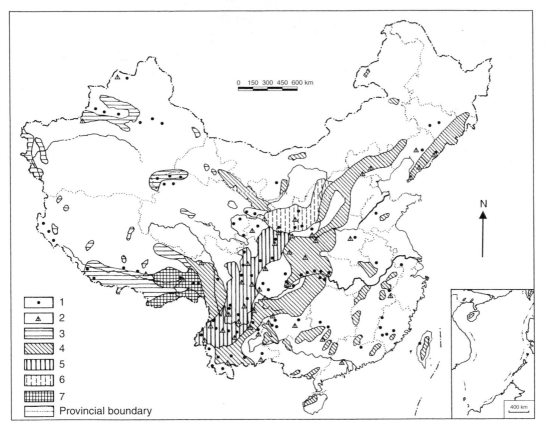

Figure 10.4 *Map of China showing the main types and distribution of mountain hazards: 1, large individual landslide site; 2, large individual debris flow site; 3, landslides and fluid debris flows and glacier lake outburst floods (GLOFs) induced by melting snow and ice; 4, landslides and low-viscosity debris flows or fluid debris flows triggered by heavy rainstorms; 5, landslides and viscous debris flows triggered by heavy rainstorms; 6, loess landslides and mudflows triggered by heavy rainstorms; and 7, landslides, viscous debris flows and GLOFs triggered by heavy rainstorms and melting snow*

lack of detailed and precise hazards statistics for much of the country, especially for very sparsely populated areas in the far western parts of China, the distribution of hazards in far western China is partly subjective.

4 Social and economic impacts of mountain hazards

4.1 Fatalities resulting from mountain hazards

China has suffered more fatalities from mountain hazards than any other nation. The recorded history of such disasters goes back nearly 4000 years to 1767 BC. According to historic records and technical reports, the total deaths from mountain hazards exceeds 250000 during the period from 186 BC to AD 2000. The large number of such deaths in mountain areas is related to individual rapid landslides and debris flows, and flooding from landslide dams and GLOFs. Some of the best documented large disasters derived from historical and technical records, and examples of the author's experiences, are presented in Table 10.2, each of which caused >100 deaths.

According to incomplete statistics, the total number of people reported killed by mountain hazards was 8798 during the period 1990–2000, and ranged from 200 deaths in 1992 to 1573 deaths in 1998.

Table 10.2 – Major mountain hazards in China that have killed at least 100 people

Year	Province (Autonomous region)	Affected area	Type of mountain hazard	Number of deaths
186 BC	Gansu	Wudu	Rock and debris avalanche	760
100 AD	Hubei	Zigui	Rockslide and avalanche	>100
689	Shaanxi	Huaxian	Loess and rockslide	>100
1072	Shaanxi	Huaxian	Rockslide and avalanche	>900
1310	Hubei	Zigui	Rockslide and avalanche	3466
1558	Hubei	Zigui	Rockslide and avalanche	>300
1561	Hubei	Zigui	Rockslide and avalanche	>1000
1680	Shaanxi	Tonghe	Debris flow	2385
1718	Gansu	Tongwei	Earthquake-induced landslide	40000
1786	Sichuan	Luding	Flood resulting from landslide dam failure	100000
1847	Qinghai	Beichuan	Loess and rockslide	>100
1856	Sichuan	Qianjiang	Rockslide induced by earthquake	>1000
1870	Sichuan	Batang	Rockslide induced by earthquake	>2000
1888	Beijing	Fangshan	Debris flow	>1000
1891	Sichuan	Xichang	Debris flow	>1000
1897	Gansu	Ningyuan	Loess and rockslide	>100
1917	Yunnan	Daguan	Rockslide	1800
1920	Nigxia	Haiyuan	Loess landslide induced by earthquake	100000
1926	Sichuan	Ganlu	Debris flow	230
1933	Sichuan	Maowen	Flood resulting from landslide dam failure	2429
1935	Sichuan	Huili	Rock and debris slide	250
1943	Henan	Lushan	Debris flow	>100
1943	Qinghai	Gonghe	Loess and mudstone slide	123
1951	Taiwan	Tsao-Ling	Flood caused by landslide dam failure	154
1954	Tibet	Jiangzhi	Flood caused by glacier dam failure	450
1956	Henan	Lushan	Debris flow	290
1964	Gansu	Lanzhou	Landslide, debris flow and mud flow	137
1965	Yunnan	Laquan	Rockslide	444
1965	Sichuan	Huidong	Rockslide	400
1966	Gansu	Lanzhou	Landslide and debris flow	134
1968	Sichuan	Yuexi	Debris flow	120
1970	Sichuan	Mianning	Debris flow	104

Table 10.2 – (continued)

Year	Province (Autonomous region)	Affected area	Type of mountain hazard	Number of deaths
1971	Sichuan	Ya'an	Debris flow	159
1972	Sichuan	Lugu	Debris flow	123
1974	Sichuan	Nanjiang	Landslide	193
1975	Gansu	Zhuang-lang	Loess landslide caused flooding along the shores of the reservoir and downstream	>500
1980	Hubei	Yunnan	Rockslide and avalanche	284
1981	Sichuan	Ganlu	Debris flow	360
1983	Gansu	Dong-xiang	Loess landslide	277
1984	Yunnan	Yinmin	Debris flow	121
1984	Sichuan	Ganlu	Debris flow	>300
1987	Sichuan	Wuxi	Rock avalanche	102
1989	Sichuan	Xikou	Rockslide and debris flow	221
1991	Yunnan	Zhaotong	Landslide	216
1994	Henan	Xiyu, Lingbao	Debris flow	>2000
1996	Yunnan	Yunnan	Landslide	156
1997	Sichuan	Meigu	Landslide	150
2002	Yunnan	Yuangyang	Landslides	107

Adapted from various historical and technical records.

Thus, on average, about 800 people were killed by mountain hazards in the 1990s. This accounts for 60% of the total casualties from all natural hazards for the same period. More than 90% of casualties were in rural mountain areas.

4.2 Social and economic impacts of mountain hazards

Mountain hazards cause much damage to economic development, properties and the quality of human life. There are approximately 100 000 landslide sites and 50 000 debris ravines, which are distributed in 950 counties/cities of the 31 provinces. The area of landslide, rock fall and debris flow activities covers 44.8% of the total land area of China. A total of 250 cities, 36 railway lines and their branch lines, 44 large-scale mines, 1.4×10^6 ha of farmland, a large number of highways, and hydro-electric facilities are endangered by mountain hazards. The economic loss caused by mountain hazards was US$1364 million in the 1950s, US$1509 million in the 1960s, US$1693 million in the 1970s, US$1809 million in the 1980s and US$2597 million in the 1990s. In recent years, the *annual* loss as a result of mountain hazards reached US$300–500 million (Luo, 1994).

In the past, the impact of mountain hazards has generally been localized but today damage to transportation facilities, especially railways and highways, has caused significant social and economic impacts to large areas. One report, for the period 1974–1976, recorded more than 1000 landslides of medium or large scale along China's 20 main railways. These landslides interrupted the railway network 44 times, or more than 800 h per year. The losses caused by these landslides was about US$49 million and the estimated cost of stabilizing them was US$2 billion. Landslides along a 154-km section of the Baoji to Tianshui railway alone interrupted traffic for almost 4700 h and involved repair costs of some US$675 million in the three decades to 1989. In addition, there are 1386 debris flows ravines where debris flow may damage 3000 km of railway, and the cost to control these ravines was estimated to be US$70 million. In the summer of 1985, debris flows buried pavement and several tunnels of the Dongchuan railway, which was closed for half a year. The direct economic losses were about US$8 million. In the summer of 1981, a debris flow from Liziyida ravine destroyed a bridge of the Chengdu–Kunming railway and derailed a passenger train. As a result, 360 passengers were killed and the economic losses were about US$3 million. It was the worst railway accident in China's history.

The cost of mountain hazards for highways may be even greater, given their greater extent and volume of traffic. Much of the 2413-km Sichuan–Tibet highway, from Chengdu to Lhasa, passes through the Hengduan mountains. Each year the highway is seriously affected by landslides, avalanches and debris flows.

Outburst floods from the failure of natural dams also greatly impact on human life and properties along river valley's. Casualties from individual landslide dam failures can be many thousands. The world's worst recorded landslide dam disaster occurred when the 1786 Kangding–Luding earthquake ($M = 7.5$) in Sichuan Province triggered a huge landslide that dammed the Dadu River. After 10 days the landslide dam was overtopped and breached, and the resulting flood extended 1400 km downstream and drowned approximately 100 000 people (Li et al., 1986). Similarly, the earthquake on 25 August 1933 ($M = 7.5$), centred near Diexi in northwestern Sichuan Province, induced many landslides, one of which formed a natural dam 250 m high across the valley in the upper reaches of the Minjiang River, creating a lake. The dam was overtopped 45 days later and a great flood rushed down the valley for a distance of 250 km, killing at least 2500 people (Li et al., 1986).

River traffic is also important for China's economy. For large rivers such as the Yangtze, Jinsha and Yalong that pass through mountainous terrain, landslides and debris flows which enter and block, or partly block, their flow can be serious hazards for river traffic. Part of a large landslide in the Yunyang district of Sichuan Province in 1982 slid into the Yangtze River. It disrupted traffic on the river for four years. Costs of stabilization and dredging to restore navigation were US$32 million. At least five major landslides in the Three Gorges Reservoir area of the Yangtze River are known to have blocked navigation on this important river during the past 1700 years. Some that also created lakes obstructed movement for several decades. The worst, a great landslide in 1542 in Hubei Province, obstructed navigation for 82 years and prevented the river from flowing entirely in the dry season. In 1985, the Xintan landslide, with a volume of 20×10^6 m^3, slid down into the Yangtze River, creating a 70 m high wave, which capsized 90 ships in the 10-km reach immediately up- and downstream of the landslide site. In the Jinsha River, in the upper reaches of the Yangtze River basin, landslides and debris flows created more than 400 dangerous shoals in the middle and lower reaches of the river, significantly impacting navigation.

5 Mountain-hazard mitigation

Since the 1960s, great efforts have been made to reduce the losses from mountain hazards, including activities that reduce the likelihood of the occurrence of damaging hazards and activities that minimize the effects of the hazards that do occur.

5.1 Regional mountain-hazard studies and mapping

The purpose of regional hazard studies and mapping programmes is to identify the areas where mountain hazards are either statistically likely or immediately imminent, to represent these hazardous locations on maps, and to disseminate such information to planners, engineers and policy makers. Therefore, regional studies and mapping of mountain hazards are often considered as the first steps in the mountain-hazard mitigation process.

In the last 20 years, regional hazard investigations at various levels from national, to provincial to small catchment areas have been carried out. As a consequence, mountain-hazard distribution and risk zonation maps for China, at various scales and using a variety of factors, have been prepared and published by a variety of academic institutions and governmental organizations. These include:

1 Landslide Distribution Map along Chinese Railway (1:7 500 000), published in 1978;
2 Landslide Distribution Map of the Reservoir Area of the Longyangxia Hydroelectric Power Station on the Yellow River (1:100 000), published in 1984;
3 Debris Flow and Landslide Map of the Xiaojiang Watershed (1:200 000), published in 1987;
4 Landslide Distribution and Hazard Zonation Map of the Three Gorges Reservoir Area (1:200 000), published in 1988;
5 Manual of Debris Flow Distribution and Zonation Map in China (1:6 000 000), published in 1991;
6 Distribution Map of Landslide Hazards in China (1:6 000 000), published in 1991;
7 Debris Flow and Landslide Distribution and Risk Zonation Map of the Panxi Area of Sichuan Province (1:300 000), published in 1993;
8 Landslide Distribution Map of Shaanxi Province (1:750 000), published in 1995;
9 Landslide Distribution and Risk Zonation Map of Sichuan Province, China (1:1 000 000), published in 1996;
10 Debris Flow Distribution and Risk Zonation Map of Sichuan and Chongqing (1:1 000 000), published in 1997; and
11 Landslide and Debris Flow Distribution and Risk Zonation Map of Yunnan Province (1:500 000) published in 1998.

More recently, geo-hazards zonation and preparation plans for 414 counties have been completed under the coordination of the Geo-environment Monitoring Bureau of the Ministry of Land and Resources.

On the basis of landslide identification and hazard zoning maps, governments, project teams and threatened communities can develop hazard preparation plans accordingly. For instance, in the Three Gorges hydropower project area, from the dam site to the city of Chongqing, 2490 landslides have been identified and hazard zonation maps prepared. For mitigating landslide hazards to the settlement and to systems of communication and lines of essential supplies, a Master Plan of Geo-hazards Mitigation and Management has been prepared by the Leading Group of Geo-hazard Mitigation and Management of the Three Gorges Project. According to the Master Plan, it has been decided to treat 957 landslides, of which 198 landslides will be stabilized by various engineering methods, such as drainage systems, piles and retaining walls, 232 landslides should be avoided for any construction, and 151 landslides will be carefully monitored by instrumentation. In addition, 79 km of unstable bank slopes and 1428 high-cut slope sites will be treated by engineering methods to prevent new landslides. The total cost of landslide hazard mitigation measures has been estimated to be US$488 million. All of these activities will be completed by June 2003, before the water level behind the dam rises to 135 m asl, as stated by the Leading Group of Geo-hazard Mitigation and Management of the Three Gorges Project (Ministry of Land and Resources, 2001).

5.2 Monitoring, forecasting and early-warning systems

The monitoring, forecasting and warning systems are expected to provide early warning of incipient hazards so that the local people can be alerted and, if necessary, evacuated. Since the 1970s, some landslide and debris-flow monitoring systems have been established. Examples include the landslide monitoring system at Xintan, Hubei Province, operated by the provincial government of Hubei and the Yangtze Valley Planning Office, and the landslide monitoring system at Jing Long Shan in the Ertan hydro-power station, operated by the Institute of Mountain Hazards and Environment of the Chinese Academy of Sciences (Li, 1991). The most common methods of monitoring are field observations, and surface and subsurface instrumentation.

Successful predictions of large-scale landslides, based on measurements of surface displacement amounts and rates, have been presented by Cheng and Sun (1988), Luo (1988) and Wang (1989). For instance, the Xintan landslide was accurately predicted based on measurement data and other indications of slope failure. The landslide, of 20×10^6 m^3, occurred on 12 June 1985 on the upper slopes of Xintan town. The warning was given one day before the event so that all 1371 inhabitants of the town were safely evacuated.

Other examples of successful hazard warning include the Jimingshi landslide (Wang, 1992) and the Huagci landslide (Liao and Xu, 1997), and these were based on measurements of surface deformation for over a year. The warning was given one day before large-scale sliding took place and more than 2000 inhabitants from each of the two landslide areas were safely evacuated. Also, the loss of property was reduced to a minimum.

In predicting the time of landslide occurrence, the biggest difficulty is determining the critical rate of displacement. The problems of obtaining such data include: (i) the variation of characteristics and type of movement of each landslide; and (ii) insufficient available data on which to base warning criteria. To solve this problem, the critical rate of displacement can be established based on previous displacement records and the data on the sliding velocity of some rapid landslides. According to the recorded rate of displacement of rapid landslide motion in China, probably 10 mm day^{-1} could be regarded as the initial criteria rate for early warning. The initial criteria can be checked and re-established during the monitoring period. Based on studies of plastic deformation power and creep rupture, and the basic principles of slope failure, Liao and Xu (1997) proposed a new method to predict the failure time of a slope and successfully used this in the prediction of several large-scale landslides.

The monitoring and warning system of the Dongchuan Debris Flow Observation Station consists of radio-transmission rain gauges, vibration meters, ground sound meters, and an indicator of remote ultrasonic flow level. Similarly, debris-flow monitoring and warning systems have been installed in some of the dangerous debris-flow ravines along the Chengdu–Kunming railway. The relationship of antecedent precipitation and rainfall intensity has also been used to predict debris flows in the mountains of China (Cui, et al., 1994; Chen et al., 1996). The accuracy of forecasting debris flows is as high as 85% in the Dongchuan Debris Flow Observation Station.

The largest landslide and debris-flow monitoring and warning system was established between 1990 and 1994 in the upper reaches of the Yangtze River by the Yangtze Valley Planning Commission. As one of the most comprehensive monitoring systems, it consists of three first-class working stations located in the cities of Yibin, Longnan and Wanxian, nine second-class working stations located in the counties of Huili, Zhaotong, Bijie, Leibo, Fulin, Zigui, Zhouqu, Lixian and Luoyong, and 59 field monitoring stations located in the critical landslide and debris-flow hotspots, covering 11×10^4 km^2 of hazard-prone areas. By the end of 2000, 158 landslide and debris-flow events were successfully predicted (based on displacement and indications of forerunning slope failure), and a total of 33100 local inhabitants were safely evacuated from the hazard areas and the direct economic loss was reduced by US$12 million (Zhang, 2001).

In addition to the above-mentioned monitoring/warning systems implemented by technical institutions, community-based monitoring and early-warning systems have been developed and established in many provinces. In Sichuan Province, a geo-hazard monitoring and prevention network has been established at the provincial level for the first time in China. A typhoon and heavy rain-based forecast system was also successfully used to predict landslides and debris flows at the community level with support from professional institutions in many hazard-prone areas. For instance, in the year 2001, 231 hazard events were successfully predicted in the provinces of Sichuan, Fujian, Yunnan, Henan, Hunan and Jiangxi, saving the lives of 4200 people and reducing direct economic losses by more than US$10 million (Ministry of Land and Resources, 2001).

5.3 Prevention and control works of mountain hazards

The prevention and control works carried out are based on the following principles: (i) the saving of human life has top priority; and (ii) the safety of public structures, buildings and road traffic is of secondary importance. Also, if a landslide or debris flow dams a river, the prevention of flooding becomes a major concern.

5.3.1 Landslide prevention and control

The landslide prevention and control works to be undertaken must be carefully selected, taking the mechanisms of the landslide in question into full consideration. The prevention and control works used in China fall into seven main categories and these are shown in Table 10.3.

1 *Avoiding landslide sites.* When deciding upon the location of mountain railways, highways and other public works, large-scale landslide-prone areas should be avoided as much as possible. Where large landslide-prone areas cannot be avoided, prevention measures should be taken to stabilize them before construction.

Table 10.3 – Summary of landslide prevention and control works used in China

Category	Landslide prevention and control works
(1) Problem avoidance	Avoid existing landslides, relocation
(2) Surface-water drainage works	Channels or ditches, seepage-water prevention works
(3) Subsurface drainage	Tunnels, subsurface trenches, deep-seated counterfort drains, drill vertical drainage holes, horizontal boreholes, slope-seepage ditches, drainage wells of ferroconcrete, drainage wells with liner plates, water catchment well works
(4) Support structures	Retaining walls, anchored retaining walls, crib works, gabions, stabilization trenches, piling works (driven piles), caisson pile works
(5) Excavation	Removal, flattening and benching
(6) River structure works	Erosion control dams, consolidated dams, revetment groins, spur dikes, groundsel works, groyne works
(7) Other methods	Bio-engineering and planting vegetation, blasting and hardening, etc.

Adapted from various technical records.

2 *Surface drainage.* Draining runoff water and preventing it from entering a landslide-prone area is often carried out to control landslides, and this is the least expensive technique. The methods of surface drainage include reshaping of slopes, construction of ditches, and sealing all tension cracks so that rainwater cannot infiltrate and build up porewater pressure. Experience indicates that surface drainage is successful in controlling shallow debris and earth landslides. The ditches/drainage system should be inspected after any heavy rainfall/runoff and cleared if necessary, as relatively minor blockages can lead to spectacular erosion and new landslides.

3 *Subsurface drainage.* The main purpose of subsurface drainage is to reduce water pressure. The methods used to drain subsurface water include tunnels, subsurface trenches, deep-seated counterfort drains, vertical and horizontal boreholes and water catchment well works. The deep-seated counterfort drain is the main measure used to treat medium-scale landslides. It is also an important subsidiary measure in treating large-scale landslides because it not only drains subsurface water but also has a strong supporting force against sliding. The best method for draining subsurface water from a rock landslide is to drill horizontal holes. If the water pressure in the tension cracks and the sliding plane is reduced to zero, the safety factor of the landslide will be improved significantly.

4 *Support structures.* The natural and easiest ways to stabilize a landslide are either to place a heavy counterweight at the toe of the landslide or by building piling works and retaining walls at suitable sections of the landslide. Retaining walls were used extensively as the main method for controlling landslides along railways during the 1960s in mountain areas. Since the 1970s, a vertically anchored and chair-type retaining wall has been widely used to supersede the gravity retaining wall. It can reduce masonry use by about 20% and is especially suitable for conditions in which landslide outlets are higher than the base of the slope (Wang, 1985). In recent years, concrete piles (deep foundation piping with broad diameters) have been used to control landslides. Most anti-skid piles are driven piles and have large rectangular sections of 1×1 m^2 to 2×3 m^2. The depths of the piles are from 10 m to 30 m depending upon the thickness of the sliding body. The interval between piles is normally 2.4 times the pile width (Pan, 1988). During the 1980s, this kind of pile had been used extensively for controlling large landslides in China because of its great capacity to resist slides, the small amount of masonry needed, convenient construction, and the fact that it can easily be constructed manually using simple instruments. More recently, anchored piles have been widely used to control large landslides, and they can reduce masonry use by about 30% compared with driven piles (Zhong *et al.*, 1994; Wang *et al.*, 1996).

5 *Excavation and fill.* Excavation is the removal of rock and soil from the head of a landslide in order to reduce the driving force and thereby improve the stability of the landslide. This method is only suitable for rotational landslides. It is ineffective on planar failures on 'infinite' slopes or on flow-type landslides. In large complex landslides with more than one potential sliding surface, toe loading will protect against all failures, but a cut may destabilize some sliding surfaces. In general, a combination of soil removal from the head and fill at the toe of the landslide is most suitable for controlling medium-sized rotational slides.

6 *River structure works.* In China, many railways and highways run along river valleys. Landslides are frequently caused by river erosion, which damages the roads and interrupts traffic. The construction of river structure works, including revetment groins, spur dikes and groundsel works, can prevent river erosion and increase the resistance forces of landslides. The best method for controlling landslides on river banks is to construct a riprap at the toe of a landslide. A riprap is relatively easy to construct and is effective on many types of eroding banks and landslides. Heavy ripraping built into the slope acts as a permeable toe buttress, increasing resistance to failure. The minimum riprap size may be estimated from the largest boulders in the river bed. Where boulders of the correct size are not available, gabions or wire-mesh baskets can be constructed and filled with boulders.

7 *Bio-engineering methods.* Bio-engineering methods for landslide stabilization include planting trees and reseeding grass. The roots of the plants tend to reinforce the loose deposits and also enter into the joints of the bedrock. This living reinforcement will increase the cohesion of the loose material, and thus the slope may develop more factors of safety against sliding. Under certain natural conditions, shallow landslides can be stabilized by planting trees, but these methods cannot replace structural methods of landslide control. A combination of structural and bio-technical methods is often the best way to stabilize landslides (Li, 1999).

5.3.2 Debris flows

A debris flow can be generally divided into three zones: formation zone, transportation zone and deposition zone. In most cases, landslides located in the upper and middle reaches of a watershed are in turn the formation zone of a debris flow. Thus, landslide stabilization is one of the most important aspects of debris-flow control (Li, 1998).

In the past 40 years, many structural methods have also been developed and widely used to mitigate debris-flow hazards (Kang, 1996). These methods include flood-regulating reservoirs in the upper parts of a debris-flow ravine, silt-trap dams, check dams and large-sized mesh dams, diversion flumes and vegetative measures. Many debris flows in the Yunnan and Sichuan Provinces have been controlled by integrated measures (Figure 10.5), which are a combination of structural and vegetative measures used in the upper, middle and lower reaches of the debris-flow watershed, with an overall management of soil, water, forest and farmland (Wu et al., 2001). In many debris-flow control projects, check dams are key structures with the following functions (Li and Yang, 1994):

1 storing sediment and reducing sediment discharge by stopping debris and soils from, landslide or erosion areas;

2 stabilizing landslides and potential slope failure by causing siltation behind the check dam;

3 preventing down-cutting of the channel bed by trapping sediment; and

4 dissipating the energy of fast-flowing water by creating small waterfalls.

5.3.3 Outburst floods from landslide dam failure

The longevity of landslide dams ranges from several hours to hundreds of years, but most landslide dams last several days to a few months. Usually landslide dams fail because of overtopping as a result of the lack of protected spillways followed by breaching from erosion by the overflowing stream. Spillways are the most simple and common methods for preventing dam failure and subsequent flooding. Pipes, tunnels, outlets and diversions have also been used to prevent dam failure and control discharges from landslide dam lakes in many places. In a few cases, extensive blasting measures have been used to excavate new river channels through landslide dams.

It is usually possible to estimate accurately the extent and rate of upstream flooding from landslide dams. Such estimates require knowledge of the height of the dam crest, rates of stream flow into the dam lake, rates of seepage through or beneath the dam, and information on the topography upstream from the dam.

The peak discharge of downstream flooding resulting from landslide dam failure depends on the process of dam failure. The failure process can be classified into three types: (1) failure caused by erosion of overtopping; (2) instantaneous failure by sliding; and (3) progressive failure by piping. For

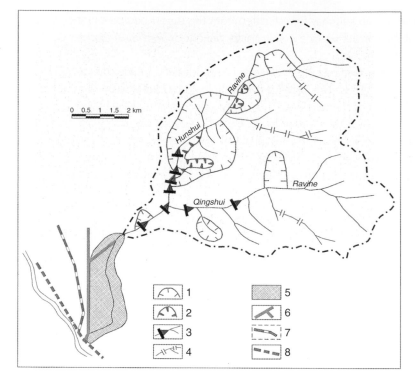

Figure 10.5 *Map of the Dachao River catchment in Dongchuan District, Yunnan Province, showing the location of engineering structures to stabilize landslides and debris flows: 1, relatively stable landslide; 2, active landslide; 3, large check dam; 4, group of small check dams; 5, debris flow alluvial fan; 6, artificial channel; 7, highway; and 8, railway*

the rapid assessment of downstream flood potential, the peak discharge of downstream flooding can be estimated by the regression equation given by Costa and Schuster (1988):

$$Q = 0.063 \, P, \qquad (1)$$

where Q is peak discharge (m^3 s^{-1}) and P is potential energy (J). The potential energy is the energy of lake water behind the dam prior to failure and can be computed as the product of dam height (m), volume (m^3), and specific weight of water (defined as the product of density and gravitational acceleration).

When assessing downstream floods, the following should be taken into account: (i) type and characteristics of the landslide dam; (ii) size of the lake (length, width, depth and volume); (iii) size of the dam (length, width, height and volume); (iv) physical characteristics of the geological materials making up the dam; (v) determination of the mechanism(s) of the dam failure; (vi) the nature of the valley downstream; and (vii) the rates of sediment and water flow into the newly formed lake and rates of seepage through the dam.

To illustrate the effect of mitigation works in reducing the damage caused by an outburst flood from a landslide dam, a large flood in eastern Tibet in June 2000 is used as an example. Owing to the sudden increase in temperature, and snow and ice melting in eastern Tibet, a huge and complex landslide occurred at 20:00 h on 9 April 2000 in the upper part of the Zhamulongba watershed, in which about 300 × 10^6 m^3 of displaced debris, soil and ice dammed the Yigong River in eastern Tibet. This

Figure 10.6 *View (looking northeast) of the landslide area along the Zhamulongba stream, eastern Tibet. The landslide originated in the southwest of the upper part of the catchment (upper background) at an elevation of 5520–5610 m asl, and moved 8 km downslope along the main channel and dammed the Yigong River (see Figure 10.7)*
Photo: T. Li

natural landslide dam was approximately 130 m thick, 1.5 km long (cross river) and 2.6 km wide (along the river) and was created in only eight minutes (Figure 10.6). After 9 April 2000, the discharge of the stream flowing into the dammed lake was about 100 m³ s⁻¹. The water level in the lake rose about 0.5 m day⁻¹ in April and 1 m day⁻¹ in May because of an increase in the inflow. According to the rate by which the water level rose, as well as historical records of hydrological data for the Yigong River, the dam would have overtopped two months later if no control measures were taken (Zhu and Li, 2000). In the two months following the landslide, the Tibetan Government spent US$7.2 million to mitigate the potential hazard from the failure of the dam, including the construction of a 150-m long large spillway channel through the dam to release water from the lake soon after the landslide occurred.

In early June 2000, the water level in the impounded lake was rising by about 2 m day⁻¹, and the lake attained an estimated volume of >3000×10⁶ m³ of water. On 8 June, water started to flow through the spillway channel with a velocity of 1 m s⁻¹ and open flow discharge of 1.2 m³ s⁻¹. However, because inflow to the lake remained greater than outflow, the lake level continued to rise by more than 5 m on the 9 June 2000 and the volume of water in the impounded lake was estimated to be 4000×10⁶ m³. As a result, the dam partially failed on 10 June 2000 because of the collapse of both banks of the spillway channel owing to rapid erosion. The dam was breached to a depth of 57 m and a width of 336 m over six hours and approximately 3000×10⁶ m³ of water was released by the evening of 11 June 2000 (Figure 10.7). The maximum flood discharge was estimated at 126000 m³ s⁻¹, 36 times higher than the normal flood discharge of the Yigong River. It was also estimated that the maximum flood discharge would have been 240000 m³ s⁻¹, and the volume of released water would have been 4000×10⁶ m³, if the channel had not been constructed (Wan, 2000).

The resulting flash flood, which was more than 50 m deep, flowed into the Palong River (a tributary of the Yarlung River) near the town of Tongmei, from which point the velocity became even greater because of the steep gradient of the Palong River. The flood reached the town of Medog, located on the left side of the Great Bend Gorge of the Yarlung River, about 200 km downstream from the site of the landslide dam, in 3.5 hours. The average velocity of the flood was 16 m s⁻¹ (60 km h⁻¹). The flash

Figure 10.7 *View of the remains of the breached Yigong landslide dam, looking upstream along the Yigong River, eastern Tibet. The dam covered 5 km². The remains of the landslide dam (central background) rise about 120 m and 80 m on the left and right side of the river bed, respectively*
Photo: T. Li

flood was estimated to be more than 200 m deep, and more than 300 m deep in the narrow Great Bend Gorge of the Yarlung River because of the narrow width of the gorge. The high-velocity flood triggered numerous new landslides and avalanches along both sides of the Yigong, Palong (Figure 10.8) and Yarlung (Brahmaputra) rivers. The heavy sediment loads carried by the flood waters and the sediments contributed by new landslides changed the landscape and hydrological regimes in many sections of the rivers.

The flash flooding was on a scale seldom seen in Tibet, China and the Arunachal Pradesh State of northeastern India, and resulted in economic losses estimated at more than US$22.9 million in northeastern India as reported by Indian Newspaper (*The Hindustan Times*, 13, 14 June 2000, New Delhi). In the Himalayan region, many rivers flow down from high mountains to countries in the lowlands. For mitigating such outburst flood hazards, inter-country flood-warning systems should be set up in river valleys where more than one country is involved. There is also a need to develop a mechanism to share the costs and benefits of flash-flood mitigation works.

6 Conclusions

Mountains comprise approximately two-thirds of the total land area of China. Mountain hazards, such as landslides, rock falls, debris flows and outburst floods, have a wide distribution. The occurrence and distribution of such hazards are affected by physiographic factors, such as geology and geomorphology, as well as by climatic and meteorological conditions. Heavy rainstorms and strong earthquakes are the main natural triggering factors. Human activities, such as road and reservoir construction, the exploitation of mineral resources and deforestation, have accelerated mountain hazard processes and increased damage.

The annual economic losses caused by mountain hazards has now reached US$300–500 million. The total number of people killed by mountain hazards during the period from 1990 to 2000 was 8798. Thus, on average, approximately 800 people were killed each year by mountain hazards in the 1990s, accounting for 60% of the total casualties for the same period. It has been found: (1) that 80% of mountain hazards are distributed in the rural areas of the mountains and that only 20% of such

Figure 10.8 *A view of erosion and landslides along the Palong River created by an outburst flood (>100 m high) from the landslide dam failure on 10 June 2000 in the Yigong River valley, eastern Tibet*
Source: *photo by Gary McCue, reproduced with permission of ICIMOD, Nepal.*

hazards are distributed in the urban and mining areas of mountains; and (2) that >90% of casualties were in rural areas. The impacts of mountain hazards on development are great, and apparently growing. Sustainable mountain development, therefore, cannot be achieved without mitigation of mountain hazards.

The damage caused by mountain hazards can be effectively reduced by avoiding hazards or by reducing the likelihood of their occurrence. Identification and zoning of mountain hazards are essential for assessing potential damage and quantifying risks and can be used by planners, decision-makers and communities. Mountain-hazard forecasting and early-warning systems also depend greatly on hazard zonation maps. Therefore, the preparation of mountain-hazard zoning maps is of high priority. This requires the combined efforts of geomorphologists and geologists.

Many monitoring and warning systems have been set up at critical danger sites, and successful prediction of landslides and debris flows has been made based on information on the displacement of slope movement and indications of forerunning slope failure, as well as heavy-rain forecasting. The losses caused by mountain hazards have also been effectively reduced by using physical measures to prevent or control existing landslides and debris flows and by preventing potential hazard problems. The natural damming of rivers by landslides, rock avalanches and debris flows is a significant hazard in many mountain river valleys, especially in the high, rugged mountains of southwestern China. Spillways are the most simple and common method for preventing natural dam failure and subsequent flooding.

To achieve a dramatic reduction of mountain hazards there is a need to help local governments and communities to prepare comprehensive hazard mitigation programmes and integrate them with local development plans. Geomorphologists can make great contributions to mountain-hazard mitigation.

References

Chen, J.W., Wang, K. and Zhu, P.Y., 1996. The relation between rainfall and debris flow in upper reaches of Yangtze River. In Du Ronghuan (ed.), *Debris flow observation and research*. Beijing: Science Press, 116–20.

Chen, Y.N., 1992. *Debris flow survey at Ala Ravine, Tianshan Mountain*. Urumqi: Xinjiang Science and Hygienic Press.

Cheng, Y. and Sun, Z., 1988. The deformation failure and the prevention of landslide, Daye Open Iran Mine's Xiangbishan Northern Wall. In *Editions of the committee of typical landslide in China*. Beijing: Science Press, 167–71.

Costa, J.E. and Schuster, R.L., 1988. The formation and failure of natural dams. *Geological Society of America Bulletin*, **100**: 1054–68.

Cui, P., Liu, S.J., Tang, W.P. and Chen, Y.Y., 1994. Debris flow monitoring and forecasting in China: progress and prospect. In Shen, J.Z., Sun, H. and Schaefer, T.L. (eds), *Natural disaster mitigation and reduction in China and United States – development and perspective*. Beijing: China Ocean Press, 357–63.

Du, R.H., Liu, X.L. and Duan, J.F., 1992. Basic characteristic of debris flows in China. In Li, T. and Shang X.C. (eds), *Mountain hazards and environment in China*. Chengdu: Southwest Jiaotong University Press, 1–5.

Feng, X.C. and Guo, A.N., 1985. Earthquake landslide in China. In *Proceedings of the IVth international conference and field workshop on landslide, 1985*. Tokyo: The Japan Landslide Society, 339–44.

Kang, Z.C., 1996. *Debris flow hazards and their control in China*. Beijing: Science Press.

Li, M. and Li, T., 1985. An investigation of landslides induced by heavy rainfall during July 1985 in the eastern part of Sichuan Province, China. In *Proceedings of the PRC-US-JPN trilateral symposium on engineering for multiple natural hazard mitigation*. Beijing: L.5.1–L.5.14.

Li, P., 1991. Research progresses of landslide integrated observation in Jinlongshan Region. *Mountain Research*, **9**: 205–209.

Li, T., 1979a. A study on the relationship between earthquake and landslide and the prediction of seismogenic landslide areas. In *Collected works of landslides, no. 2*. Beijing: China Railway Publishing House, 127–32.

Li, T., 1979b. Some problems on seismogenic landslide in Songpan–Pingwu area. In *Collected works on landslides, no. 2*. Beijing: Railway Publishing House, 133–44.

Li, T., 1989. Landslides – extent and economic significance in China. In Brabb, E.E. and Harrod, B.L. (eds), *Proceedings of the 28th international geological congress, symposium on landslides, Washington, D.C., 27 July*. Rotterdam: Balkema, 271–87.

Li, T., 1998. Assessment of failure and success of preventing damages of debris flows caused by landslides in Laogan Ravine, Yunnan, China. In Sassa, K. (ed.), *Environmental forest science*. Dordrecht: Kluwer Academic Publishers, 519–27.

Li, T., 1999. Bio-engineering for landslide control in upland watershed. In Bhatta, B.R., Chalise, S.R., Myint, A.K. and Sharma, P.N. (eds), *Recent concepts, knowledge, practices and new skills in participatory integrated watershed management*. Kathmandu: ICIMOD, 81–86.

Li., T. and Wang, S., 1992. *Landslide hazards and their mitigation in China*. Beijing: Science Press.

Li, T. and Yang, W.K., 1994. Poverty alleviation through watershed management in mountain areas: the case of Xiaojiang watershed, Yunnan Province. In Banskota, M. and Sharma, P. (eds), *Development of poor mountain areas: proceedings of an international forum, Beijing, 22–27 March 1993*. Kathmandu: ICIMOD, 238–43.

Li, T., Schuster, R.L. and Wu, J.S., 1986. Landslide dams in south-central China. In Schuster, R.L. (ed.), *Landslide dams – processes, risk, and mitigation*. American Society of Civil Engineers Geotechnical Special Publication No. 3, Reston, VA: American Society of Civil Engineers Press, 146–62.

Li, Z.S. and Cheng, Y.H., 1998. The debris flow calamities in Qinling and Bashan mountain areas and the solution to them. In Cui, P., He, Z.W., Wang, C.H., Chen, S.C. and Liu, S.J. (eds), *Research on mountain disasters and environmental protection across Taiwan Strait*. Chengdu: Sichuan Scientific and Technical Press, 439–43.

Liao, X.P. and Xu, J.L., 1997. Theory and practice of time prediction for landslides. In *Proceedings of the international symposium on landslide hazard assessment, Xian, 1997*: 279–90.

Liu, X.L. and Tang, C., 1995. *Risk zone assessment of debris flow*. Beijing: Science Press.

Liu, X.L., Wang, S.G. and Zhang, X.B., 1992. Influence of geologic factors and landslides in Zhaotong, Yunnan Province, China. *Environmental Geology and Water Science*, **19**: 17–20.

Lu, R.R., Tang, B.X. and Zhu, P.Y., 1999. *Debris flow and environment of Tibet*. Chengdu: Publishing House of Chengdu Science and Technology University.

Luo, P., 1988. Xintan landslide and urgent forecast before sliding. In *Editions of the committee of typical landslides in China*. Beijing: Science Press, 191–99.

Luo, Y.H., 1994. Distribution and economic loss evaluation of landslides, debris flow and rock-falls in China. *Development and Management of Land*, **4**: 49–55.

Ministry of Land and Resources, 2001. *Bulletin of land and resources of China 2001*. Beijing.

Pan, Y., 1988. The studies and control of Zhajiatong landslide. In *Editions of the committee of typical landslide in China*. Beijing: Science Press, 361–66.

Song, D.R. and Li, Y., 1998. Debris flow disaster in Northeast China. In Cui, P., He, Z.W., Wang, C.H., Chen, S.C. and Liu, S.J. (eds), *Research on mountain disasters and environmental protection across Taiwan Strait*. Chengdu: Sichuan Press of Science and Technology, 218–20.

Tan, W.P., Wang, C.H., Yao, L.K., Jin, Y.T. and Yang, W., 1994. *Area predicting and forecasting of rainfall-induced debris flow and landslide with example of Paxi Area*. Chengdu: Sichuan Press of Science and Technology.

Wan, H.B., 2000. A brief of risk reduction of the huge Yigong landslide in Tibet, 2000. *Disaster Reduction in China*, **10**: 28–31.

Wang, C., 1989. Prediction of large landslides near the Longyan Dam on the Langyan Gorge. In *Selected papers on landslides*. Chengdu: Sichuan Press of Science and Technology, 190–97.

Wang, F.D., 1992. A specific example of monitor and prealarm of landslide in Jimingshi of Ziqui County. *Bulletin of Soil and Water Conservation*, **12**: 39–45.

Wang, G., 1985. The measures for controlling landslides on the railways in China. In *Proceedings of the fourth international conference and field workshop on landslides*. Tokyo: The Japan Landslide Society, 34–40.

Wang, H.Q., Liao, Z.H. and Zhou, X., 1996. Antislide pile with pre-strain anchor cable – new antisliding structure. In Editorial Committee (eds), *Landslide research and control (i)*. Chengdu: Sichuan Press of Science and Technology, 48–68.

Wang, M.Y, Zhu, G.J., He, Z.D. and Zheng, L., 1986. The mountains and mountain systems in China. *Mountain Research*, **4**: 67–74.

Wu, J.S. and Li, T., 2001. Behaviour and characteristics of debris flows. In Li, T., Chalise, S.R. and Upreti, B.N. (eds), *Landslide hazard mitigation in the Hindu Kush – Himalayas*. Kathmandu: ICIMOD, 203–14.

Wu, J.S., Li, T. and Yin, C.Q., 2001. Debris flow control and management: case studies from the Sichuan and Yunnan Provinces of China. In Li, T., Chalise, S.R. and Upreti, B.N. (eds), *Landslide hazard mitigation in the Hindu Kush – Himalayas*. Kathmandu: ICIMOD, 291–312.

Xu, D.M., 1987. Characteristics of debris flow caused by outburst of glacier lakes on the Poiqu River in Xizang (Tibet), China. *Journal of Glaciology and Geocryology*, **9**: 24–26.

Zhang, X.L., 2001. Landslide and debris flow monitoring and warning system in the upper reaches of the Yangtze River. In *Landslide and debris flow mitigation training manual prepared by water conservation human resource development project of China and Japan*. Beijing: Ministry of Water Conservation, 88–103.

Zhong, D.L., Xie, H., Cheng, Z.L. and Wang, B.Z., 1993. *A comprehensive prevention study on mountain disasters in low mountain and hill areas in Xiuyan Man Autonomous County*. Chengdu: Sichuan Press of Science and Technology.

Zhong, D.L., Xie, H. and Wei, F.Q., 1994. Establishment of forecasting and warning system of debris flow and observation station in the upper reaches of the Yangtze River. *Mountain Research*, **12**: 119–23.

Zhu, P. and Li, T., 2000. Flash flooding caused by landslide dam failure. *ICIMOD Newsletter*, **38**: 4–5.

11
Geomorphic processes and hazards on volcanic mountains

Jean-Claude Thouret

1 Introduction

Volcanic mountains are distinctive in shape and composition, and in terms of the interaction between volcanic activity and geomorphic processes. The extent and rate of geomorphic processes of denudation are exacerbated on volcanic mountains because of the thickness of unconsolidated materials and earthquakes. Volcanoes are also subjected to specific processes, such as ash falls, pyroclastic flows and volcanic mudflows or lahars. Unlike ordinary mountains, which are formed by tectonic uplift and erosion, volcanoes are *constructed* and eroded. Volcanic mountains have a distinctive impact on human activity. It is commonly thought that, owing to the fertility of volcanic soils, volcanoes are favoured areas for settlement, but these soils are only potentially fertile under regimes of careful land management (James *et al.*, 2000). Volcanic mountains possess all the geomorphic processes and hazards of mountains worldwide (see Hewitt, and Li, both this volume) but have additional hazards posed by eruptive activity.

Since AD 1783 eruption-related deaths have totalled 220000 (Tanguy *et al.*, 1998). Most deaths resulted from post-eruption famine and epidemic disease (30%), pyroclastic flows and surges (27%), lahars (17%) and volcanogenic tsunamis (17%). Volcanic fatalities are small compared with those of floods and earthquakes but the potential threat from a massive eruption is greater today than ever before, because of large concentrations of populations living around volcanoes.

There are subtle differences in the terms 'hazard' and 'risk'. *Hazard* refers to an event that is potentially dangerous. Potentially destructive tephra falls, pyroclastic flows and lahars are volcanic hazards occurring at a particular site within a given period of time. *Risk* is the probability of loss or the degree of harm caused when the dangerous event occurs. *Vulnerability* is the proportion of the elements exposed to a hazard (e.g., humans, buildings, agriculture and other economic activity) and likely to be lost if the dangerous event occurs. *Value* may include the number of people at a particular site, monetary value of property, and projected productive capacity.

In the case of volcanic hazards, the relation between hazard and risk can be defined by the qualitative 'risk equation' (Fournier d'Albe, 1979):

volcanic risk = hazard × vulnerability × value.

A *geomorphic hazard* results from any landform change that adversely affects the geomorphic stability of a site and that intersects the human use system with adverse socio-economic impacts (Slaymaker, 1996). Geomorphic hazards are characterized by magnitude, frequency and areal extent. On volcanic mountains, geomorphic hazards can be endogenous (i.e., caused by eruptive and/or seismic processes) and exogenous (i.e., caused by erosion).

This chapter first describes the distinctiveness of volcanic mountains whose geomorphic characteristics promote rapid erosion processes. Second, the main geomorphic hazards acting on volcanic mountains, either direct, mixed or delayed, are examined. Finally, geomorphic methods used to assess volcanic hazards are described.

2 Distinctiveness of volcanic mountains

2.1 What is a volcanic mountain?

A volcanic mountain grows around a vent through eruption of magma, as lava or pyroclasts. This combination of material provides greater strength and explains the steepness of the slopes of most composite volcanoes. In some eruptions, so much lava is erupted that the volcano is no longer supported, and it collapses, creating a caldera whose diameter may range from 2 km to 60 km across. Pyroclastic flows spread out rapidly from the volcanic centre in all directions and cover large areas. The largest ash- or pumice-flow deposits, known as ignimbrites, have covered areas of thousands of square kilometres, with volumes up to 3000 km^3. Repeated pyroclastic eruptions can build up an ignimbritic plateau (e.g., Cappadocia, Turkey).

The composition and the effusion rate of magma help to distinguish three principal types of volcanic mountains. (1) Countless eruptions of fluid basaltic lava flows, poor in silica; build up a gently sloped mountain called a shield volcanoe, such as Mauna Loa in Hawaii. (2) In contrast, viscous lava forms thicker, shorter flows, which build up a steep-sided stratovolcano. Most viscous lava, being gas-rich, erupts explosively, forming tephra falls (clouds of fine volcanic debris), pyroclastic flows (hot avalanches of ash and larger fragments) or pyroclastic surges (a hot blast of fine-grained ash). The pyroclastic deposits form a pile around a crater, building a cone, which, if large enough, is a volcanic mountain. (3) Silicic volcanoes usually take the form of a dome complex, and although they are highly explosive they usually achieve only modest topographic expression.

2.2 Geomorphic types and regional tectonic setting

Volcanoes can be classified into groups. The *monogenetic* group contains small mountains, such as cinder cones and domes, whereas the *polygenetic* group contains voluminous edifices such as shield volcanoes and stratovolcanoes. It is possible to distinguish six main types of volcanic constructional and erosional landforms (Table 11.1), based on geomorphic parameters (shape, size, scale, mono/polygenesis, constructional/excavational origin), combined with volcanological criteria (types of activity and erupted products, magma composition and volume; Ollier, 1988; Francis, 1993). Based on the morphology of 699 volcanoes, Suzuki (1977) found six 'volcano series': stratovolcano (62%) , stratovolcano with caldera (10%), shield volcano (11%), shield volcano with caldera (3%), caldera volcano (7%) and monogenetic volcano (6%). Almost three-quarters of the edifices conform to the popular image of a volcano, i.e., a conical mountain. Large stratovolcanoes and shield volcanoes represent 86% of the volcanic edifices worldwide. Table 11.2 lists nine volcano groups whose geomorphic size and shape are combined regionally with their tectonic setting. The last two groups are not volcanic mountains.

2.3 Distinctiveness and characteristics of composite volcanic mountains

Composite volcanoes are the product of multiple eruptions spanning thousands or millions of years. A composite volcano can be divided into two parts: a composite cone and a surrounding ring plain. Key characteristics include:

Table 11.1 – Classification of volcanic landforms with emphasis on volcanic mountains

First group of monogenetic hills and small mountains (4 subgroups):

(1)   pyroclastic landforms: cinder or scoria cones; hydromagmatic tuff cones and tuff rings, maars and diatremes;

(2)   lava-made landforms: endogenous and exogenous domes, crater row or fissure vent;

(3)   mixed landforms of intra- or subglacial volcanoes: tuyas (table mountains) and mobergs;

(4)   submarine landforms: seamounts and guyots.

Second group of medium to large, high composite volcanic mountains (3 subgroups):

(1)   stratovolcanoes: simple cone with summit crater; composite cone with sector collapse scar (summit amphitheatre), composite cone with a caldera or a somma;

(2)   twin cones; compound or multiple cones with elongate ridge; clusters of cones;

(3)   shield volcanoes with shallow-sloped flanks and caldera in summit region: Hawaiian shields and domes; Galapagos, Icelandic, and scutulum-type shields.

Third group of high plateaux with altitudinal range and dissected relief (3 subgroups):

(1)   continental flood basalt plateaux;

(2)   dissected plateaux of ash flows and ignimbrite sheets;

(3)   intermediate-silicic multivent centres that lack a central cone; rhyolithic centres; silicic volcanic lava field with multiple domes and calderas.

Fourth group of calderas on stratovolcanoes or on uplifted basement:

(1)   collapse and explosion somma;

(2)   collapse on Hawaiian shield volcano;

(3)   collapse in basement and resurgent caldera;

(4)   very large and complex resurgent calderas.

Fifth group of landforms resulting from a combination of eruptive and/or erosional processes on volcanic mountains:

(1)   horseshoe-shaped avalanche caldera from a flank failure of magmatic (Bezymianny-type), gravitational (Fugendake) or mixed origin (Bandaï-san, La Réunion cirques);

(2)   amphitheatre-shaped, erosional calderas (Haleakala, Maui).

Sixth group of landforms resulting from denudation and inversion of relief:

(1)   eroded cone; inverted small-scale landforms: necks, culots;

(2)   much eroded cone, inverted lava flow and planèze;

(3)   roots of palaeo-volcanic mountain: dissected cauldron and hypovolcanic complex.

Adapted from Thouret (1999).

Table 11.2 – Volcano type according to their geomorphic size and shape, and to their regional tectonic setting

Volcano type	Description	Tectonic setting	Examples
Scoria cone, dome and compound lava field	Single hill or small mountain 0.2–1 km across, 0.1–0.4 km high	Monogenetic field Near graben or rift in continental setting	Chaîne des Puys, France, 35 km long × 10 km (75 cones, domes and maars)
Volcanic island seamount, guyot	Single mountain of medium scale, 1–3 km high, 1–2 km across	Mid-ocean spreading ridge volcanism	Ritter Island Mid-Pacific ridge
Oceanic basaltic shields, compound massifs on islands	Huge mountains up to 9 km high, volume of about 40 000 km³	Hot spot in intra-oceanic plate	Piton de la Fournaise, La Réunion; Mauna Loa, Hawaii
Composite or compound cones	Large mountains tens of km across and 4–6 km high above plateaux	Continental rift	East Africa; Basin and Range, USA
Large stratovolcanoes	Huge mountains 10–20 km across, up to 6 km high	Micro-continental arc Young island arc	Japan; New Zealand; Indonesia; Tonga-Kermadec, Aleutians
Polygenetic volcanic range: stratovolcanoes, large domes, wide calderas	Arc hundreds to thousands km long, tens of volcanoes 20 km across, spacing 10–100 km	Continental arc, active margin	Cascades Range, USA; Kamtchatka, Russia
Composite volcanoes above high mountain range or plateau	Line(s) of tens of stratocones, compound lava-flow fields, domes and calderas	Continental marginal arc with oceanic trench subduction zone	Northern, central, southern and austral Andes
Large calderas and caldera complexes, resurgent calderas	On mountainous relief Modest topographic relief	Large swell of the crust over hundreds of km²	Long Valley, USA; Cerro Galán, Andes; San Juan, USA; Valles
Continental plateaux and traps	Widespread lava-flow fields and traps	Intraplate flood volcanism	Eastern Australia; Deccan, India; Columbia River Basalt Plateau, USA

Adapted from Cas and Wright (1987).

1 they commonly achieve great topographic relief, in the range of 2000–4000 m, providing a large potential energy for masses of rock and water;

2 they have steep slopes, commonly about 30°, which favour landslides that transform into mobile debris flows;

3 they have specific landforms such as cones and lava flows, and piles of pyroclastic debris, either as layers of tephra which mantle slopes or as ring plains surrounding the cone;

4 they occur in tectonic regions prone to earthquakes which may destabilize large masses of rocks;

5 they commonly have explosive eruptions. Phreatic or phreatomagmatic explosions can be triggered by magma interacting with groundwater while magmatic explosions are produced by rapid depressurization and vesiculation of rapidly ascending magma;

6 they have a high proportion of loose or poorly consolidated rock debris, which is highly permeable and structurally weak;

7 they are shaped by deep, radial drainage which commonly carves gullies;

8 the rapid denudation of high volcanic mountains supplies huge volumes of sediment to the surrounding lowlands.

Most composite volcanoes are characterized by a concave-upward profile. Mass transfer through erosion and deposition modifies the shape of a volcano from a cone with straight slopes to one with a concave-upward profile. Aggradation must exceed degradation to maintain a simple conical shape (Davidson and de Silva, 2000) but there is a limit to the size of composite cones. Edifice heights are in the range of 2000–2500 m but rarely exceed 3000 m. Maximum volumes are about 200 km³ for arc-related volcanoes, although intraplate composite volcanoes can be bigger (e.g., Kilimanjaro, Tanzania).

Volcanic mountains result from *opposing constructive and destructive forces* (Davidson and de Silva, 2000). Composite cones, in particular, result from short-term eruptive phenomena that construct volcanoes and the persistent erosive processes that wear them down. Peaks of erosion follow closely on from peaks of eruption as pyroclastic material is rapidly removed from the cone. The biggest single obstacle to the growth of a volcano is gravitational collapse, and the subsequent debris avalanches represent one of the deadliest geomorphic hazards on composite volcanoes. Gravitational collapses are now recognized as being part of the normal mode of activity of composite volcanoes. The most spectacular examples involve the failure of 10–30% of an edifice in a matter of minutes. The morphologic signatures of collapse are an amphitheatre on the cone and hummocky topography on the avalanche deposits. These deposits extend from the talus apron out to several tens of kilometres from the summit. Gravitational collapses require a trigger, and many are probably earthquake-triggered (e.g., Mt St Helens, USA, 1980).

A detailed chronology of composite volcanoes reveals four characteristics.

1 Constructive processes operate more rapidly than on ordinary mountains, typically days to years. Paricutin, a scoria cone in Mexico, began on 20 February 1943. After a year of activity it was 325 m high. About 2 km³ of lava and tephra were erupted during nine years of activity. Construction of an imposing cone may take only less than 5% of the active lifetime of the volcano at a maximum rate of 1–5 km³ ka⁻¹ (Hildreth and Lanphere, 1994), but the length and rate of cone-building episodes are quite variable.

2 Active composite volcanoes display a wide range of recurrence time intervals between eruptions. Sakurajima (Japan) and Sangay (Ecuador) have been erupting persistently, on an almost daily basis, for decades. More typically, composite volcanoes erupt on the 0.1 to 10-year timescale with clusters of minor outbursts of varying frequency. The next characteristic timescale

of many recently active volcanoes is 100 to hundreds of years, as shown by Mt St Helens, Mt Fuji (Japan) and Mt Pinatubo (Philippines).

3 Composite volcanoes can remain active for as long as half a million years and published examples of greater longevity (e.g., Etna, Italy) are rare. It is important to distinguish the active life of a volcano from its longevity as a topographic entity. The time span over which it remains recognizable is largely a function of degradation rate, which is a function of climate and lithology. Volcanic cones are better preserved in arid and cool climates than in humid equatorial climates. In the hyper-arid conditions of the Central Andes, it may take several million years to reach the initiation of 'flat iron-shaped' lava flows on a cone subjected to gully erosion; in humid tropical areas such as Indonesia, it may take only 20000 years.

4 Erosion starts to work on a volcano as soon as it starts growing. Large volcanoes are more erosion-prone than small volcanoes made of the same material, for they have a larger catchment area and the greater potential for runoff, whereas small volcanoes may remain entirely within the hydrological limits of 'no erosion' (Ollier, 1988).

3 Geomorphic hazards on volcanic mountains

Because volcanoes are rapidly and repeatedly constructed and worn down, geomorphic hazards on volcanoes may be more frequent than on most other mountain types. Table 11.3 and Figure 11.1 illustrate the main types of direct (syn-eruptive) and indirect (post-eruptive) volcanic hazards. First, geomorphic hazards may result from eruptive processes, such as lava flows, tephra falls, pyroclastic flows and surges. These phenomena are generally more hazardous on explosive stratovolcanoes than on basaltic volcanoes. Second, volcanogenic tsunamis, sediment-water flows, lahars and debris avalanches are mixed geomorphic hazards for they can be triggered by endogenous and/or exogenous processes. Third, geomorphic hazards result from erosion and sedimentation in catchments disturbed by eruptions, and from mass movements on extinct volcanic mountains.

The effects of hazardous eruptive phenomena can be ranked according to distance from the volcano vent (Figure 11.2). The effects can be either immediate or delayed and may last long after an eruption, such as the persistent lahars since the 1991 Mt Pinatubo event in central Luzon. Table 11.3 itemizes the characteristics of volcanic hazards which are pertinent to identify risks. Even though the average duration is 50 days, an eruption may last several months (e.g., Galunggung in Java, 1982–83), and quasi-continuous eruptive activity may last for decades (e.g., Sakurajima, Japan, since 1956). Recurrent hazardous activity may generate new hazards, such as secondary eruptions in pyroclastic-filled catchments, which can disturb the drainage network.

3.1 Geomorphic hazards related to eruptive activity

3.1.1 Lava flows

Most lava flows move sufficiently slowly for people to avoid them, but some are fast enough to cause loss of life. Two eruptions of high-discharge, low-viscosity flows of lava have occurred on Hawaii and one in Congo during historic times, with velocities of 40–100 km h^{-1}. As many as 600 people died in the extremely fluid, fast-moving lava flows in 1977, when the lava lake at Mt Nyiragongo, Congo, rapidly emptied. Again in January 2002, a flank fissure-fed lava poured down the slopes of Nyiragongo and through the eastern Congolese city of Goma, 18 km south of the crater, destroying the city, displacing hundreds of thousands of people, and killing about 50 people. Other examples of devastating lava flows in recent decades include those of Kilauea (Hawaii), Etna (Italy) and Paricutin (Mexico). The land covered by lava is generally rendered uninhabitable and unproductive, perhaps for centuries.

Table 11.3 – Principal types of hazardous volcanic processes and selected examples

Type	Characteristics pertinent to risk	Example
Direct hazards		
Fall processes:		
Tephra fall VF	Can extend 1000+ km downwind, and produces impenetrable darkness; surface crusting encourages runoff.	Vesuvius, 1631, 1906
Ballistic projectiles C	High-impact energies; fresh bombs above ignition temperatures of many materials	Soufrière Saint-Vincent, 1812
Lava flow:		
Lava flow F	Bury or crush objects in their path;	Kilauea, 1960, 1983–…
Domes R	Noxious haze from sustained eruptions.	Merapi, Soufrière Hills Montserrat, 1995–…
Pyroclastic flow:		
Pyroclastic flows VF	Small flows travel 5–10 km down topographic lows, but large flows travel 50–100 km; large flows climb topographic obstructions	Pinatubo, 1991, Unzen 1991–93,
Pyroclastic surges F		Mount Pelée, 1902, Taal, 1960
Laterally directed blast VR	Knock down all constructions	Bezymianny, 1956, Mount St Helens, 1980
Debris flow:		
Primary (eruption-triggered) debris flows VF	Velocities may exceed 10 m s^{-1}; rapid aggradation, incision or lateral migration; hazard may continue for months or years after eruption	Nevado del Ruiz, 1985, Kelud, 1919
Jökulhlaups C	Can occur with little or no warning.	Katla, 1918, Grimsvötn, 1996
Sector collapse and flank failure:		
Debris avalanche C	Emplacement velocities up to 100 m s^{-1}; create topography, pond lakes; can produce tsunamis in coastal areas	Mount St Helens, 1980,
magmatic origin R		Bezymianny, 1956,
phreatic origin R		Bandaï-san, 1883, Ontake, 1984,
no eruption; seismogenic C		Shimabara, 1792
Other eruptive processes:		
Phreatic explosions VF	Damage limited to proximal areas but can be lethal; corrosive, reactive; low pH in water; CO_2 in areas of low ground	Soufrière de Guadeloupe, 1976,
Volcanic gases and acid rains VF		Dieng plateau, 1979

Table 11.3 – (continued)

Type	Characteristics pertinent to risk	Example
Indirect hazards		
Earthquakes and ground deformation F	Limited damage; subsidence may affect hundreds of km².	Sakurajima, 1914, Usu, 2000
Tsunami R	Can travel great distances; exceptionally, waves to 30+ m.	Krakatoa, 1883
Secondary debris flows VF	Can continue for years.	Santa Maria, 1902–1924
Post-eruption erosion and sedimentation F	Can affect extensive areas for years after eruption.	Irazu, 1963–64, Pinatubo, 1991–2000
Atmospheric effects C Air shocks, lightning	Limited effects.	Mayon, 1814, Agung, 1960
Post-eruption famine and disease R	Limited effects at present.	Lakagigar, 1783

Notes: Processes having a direct impact on geomorphic hazards are italicized. Historical frequency of adverse effect, damage, and/or death are: VF, very frequent; F, frequent; C, common; R, rare; VR, very rare (from Blong, 1984).
Adapted from Tilling (1989). Characteristics pertinent to risk modified from Blong (2000).

3.1.2 Tephra falls

Tephra includes all airborne products of explosive eruptions except gas. Volcanic bombs, blocks and lapilli (fragments 2–64 mm across) which leave the vent with ballistic trajectories are called projectiles. Particles which fall from the eruption column or the plume are termed tephra fall. Tephra fall consists of ash and lapilli, particularly pumiceous lapilli with low specific density that can be transported large distances downwind.

It is impact energy rather than size or density that determines the degree of hazard posed by a projectile (Blong, 1984). Distances travelled by projectiles vary between 0.3 km (for projectiles weighing tonnes) and 3–6 km (for those weighing kilograms). The temperature on impact is significant because projectiles can set fire to vegetation and man-made objects several kilometres from the vent. As the tephra disperse downwind the concentration of particles declines, so that the thickness deposited also declines exponentially with distance. The areas experiencing tephra fall in a major eruption may be very large: the ash fall around Krakatoa (Java) in 1883 covered about 827 000 km², but much of this area received less than 1 cm thickness. The hazards produced by tephra fall result principally from the accumulation of the deposits and extreme darkness. Much concern is voiced over the respiratory effects of ash but the main causes of death are heat and gas.

The weight of ash deposited can bring down roofs and cause serious damage as well as injury to people. Most of the 300 deaths resulting from the eruption of Mt Pinatubo in 1991 occurred outside the 30 to 40-km radius of evacuation as a result of roofs collapsing under the weight of ash. The ash on the roofs had been made more dense by the effect of rain: a wet ash deposit of 15 cm may have a load of 200 kg m^{-2}. Fine-grained deposits of tephra may represent a greater hazard than coarse-grained tephra because they will be more easily eroded by wind and water.

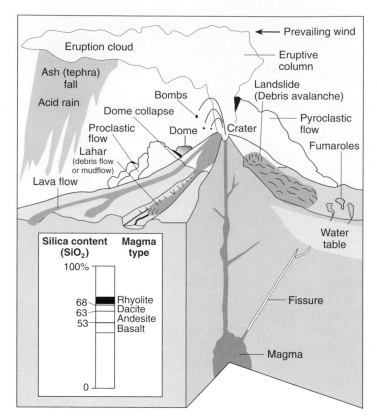

Figure 11.1 *Volcanic eruptive and noneruptive processes which lead to direct and indirect hazards on and around a volcanic mountain*
Source: *from Myers et al. (1998) reproduced with permission of the USGS.*

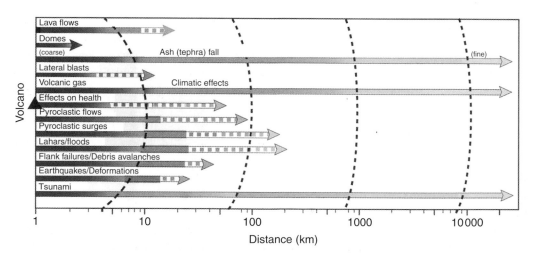

Figure 11.2 *The theoretical influence of distance (average, solid lines; exceptional, dashed lines) on the impact of destructive phenomena associated with volcanic eruptions. Except for volcanic gas, atmospheric effects and effects on health, all volcanic phenomena may lead to geomorphic hazards*
Source: *reprinted from Global Environmental Change, vol. 2, Chester, D.K., Degg, M., Duncan, A.M. and Guest, J.E., The increasing exposure of cities to the effects of volcanic eruptions: a global survey, pp. 89–103. © 2001, with permission from Elsevier.*

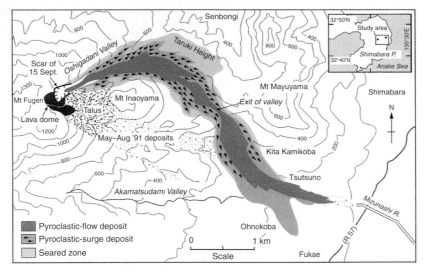

Figure 11.3 *Map showing deposit distribution of the 15 September 1991 pyroclastic flow at Unzen, Japan, and detachment of the ash-cloud pyroclastic surge from the main basal avalanche of pyroclastic flows*
Source: *reprinted from Nakada, S., Hazards from pyroclastic flows and surges. In Sigurdsson, H., Houghton, B., McNutt, S.R., Rymer, H. and Stix, J. (eds),* Encyclopedia of volcanoes, *pp. 945–55.* © 2000, *with permission from Elsevier.*

After eruptions, the slopes of volcanoes are mantled by loose tephra-fall deposits, which can be remobilized by heavy rains into lahars. In the well-documented Irazu eruption in Costa Rica in 1963 (Waldron, 1967), the ash destroyed the vegetation cover and then hardened to form an impervious crust which led to increased runoff and flash floods.

3.1.3 Pyroclastic flows and surges

Many of the largest volcanic disasters in history have involved pyroclastic flows and surges (e.g., 29000 people killed at Mt Pelée, Martinique, in May 1902). Pyroclastic flows contain mixtures of hot lava blocks, ash, pumice and volcanic gas, descending slopes at very high speeds (Nakada, 2000). They follow topographic lows and valleys as they move. Pyroclastic surges are more diluted flows, lacking a high concentration of particles and containing a lot of gas. Surges can be generated directly by column collapse, and also by separation from the moving, relatively dense, main body of the pyroclastic flow. Surges rarely have sufficient force to destroy buildings in their path and are short-lived compared with pyroclastic flows because of their lower momentum and density. The area impacted by surges is limited to a few kilometres from the source (Figure 11.3), although they can run over topographic barriers. 'Blasts' or laterally directed explosions are also a short-lived diluted turbulent current, but they may travel larger distances, such as 35 km from the Mt St Helens summit.

Pyroclastic flows can derive from gravitational collapse of Plinian (high and sustained) eruption columns. For example, at Pinatubo in 1991, more than 4 km³ of ash and pumice flowed from such a collapse within a few hours. Pyroclastic flows can also originate directly from the vent, as produced by the August 1980 eruptions of Mt St Helens. Lateral explosions directly from a summit dome may produce small but devastating pyroclastic flows such as those at Mt Pelée, Martinique, on 8 and 20 May 1902. Finally, pyroclastic flows may originate from the gravitational collapse of lava domes or lava flows: they were frequently observed during the 1991–95 eruption at Unzen (Japan) and the

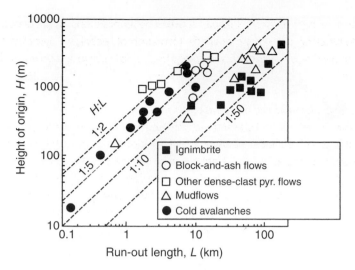

Figure 11.4 *Height of origin (H) versus run-out distance (L) of various pyroclastic and other flow deposits. Mudflows have lower H/L ratios than cold avalanches, and large ignimbrites have lower ratios than small pyroclastic-flow deposits, suggesting that both lubrication of the granular material and the mass of flow influence mobility. Note, however, that* H *values for pyroclastic flows derived from column collapse are uncertain*
Source: *reprinted from Freundt, A., Wilson, C.J.N. and Carey, S.N., Ignimbrites and block-and-ash flow deposits. In Sigurdsson, H., Houghton, B., McNutt, S.R., Rymer, H. and Stix, J. (eds), Encyclopedia of volcanoes, pp. 582–99. © 2000, with permission from Elsevier.*

1995–2002 eruption at Soufrière Hills, Monserrat. At Unzen, partial collapse of the lava domes generated frequent pyroclastic flows. On 3 June 1991, pyroclastic flows reached the Kita Kamikoba area (Figure 11.3), and the associated ash cloud surge killed 43 people in the evacuated zone (Nakada, 2000).

The relief of volcanic mountains can influence some physical properties of flows. Pyroclastic flows and surges are hot and very mobile and their speed is generally >10 m s^{-1} and sometimes >100 m s^{-1}. Their travel distance depends on the mass of tephra thrown out and the height at which the pyroclastic flow begins to descend. The ratio of the height (H) dropped over the travel distance (L) is as small as 0.2–0.29 for large-scale pyroclastic flows and as high as 0.33–0.39 for small block-and-ash flows (Figure 11.4). This shows that larger flows are more mobile than smaller flows. High-mobility pyroclastic flows may be fluidized by volcanic gases released from the interior of lava particles or by expansion of heated air trapped in the moving front. From the relationship between volume erupted and the H/L ratio, one can predict the travel distance from the starting point of pyroclastic flows, assuming a constant apparent ratio (H/L) for the flow. Areas impacted by pyroclastic flows are likely to be affected in future eruptions. The magnitude of pyroclastic flows increases generally with time over a period of days to weeks. Since dome eruptions commonly last a few years, the pyroclastic flow hazards change. As a lava dome increases in volume and height with time, pyroclastic flows tend to travel larger distances because of the increase in H, leading to larger-scale dome collapse. As the valleys around a dome become filled, later pyroclastic flows are likely to descend further and to spread out more (Nakada, 2000).

Observations of recent eruptions suggest that rugged topography and high relief may affect transport and deposition of pyroclastics (Nakada, 2000). The interactions of pyroclastic currents with topography include blocking, downslope drainage, formation of secondary pyroclastic flows in valleys, and development of dividing streamlines, as indicated by the flow-surge laid down by the 18 May 1980 blast at Mt St Helens.

3.1.4 Large-scale and caldera-forming explosive eruptions

Large-scale explosive eruptions (>10 km^3 in volume) producing widespread ignimbrites are historically rare. Ignimbrites travel distances of several tens of kilometres from the crater. These powerful eruptions are responsible for the highest death toll in recent history: Krakatoa in 1883 caused 36 000 deaths, mainly from tsunamis associated with the caldera formation. Large caldera-forming explosive eruptions (>100 km^3) probably represent the most catastrophic geologic events that ever affected the Earth's surface, other than large meteorite impacts. If an eruption comparable in size with that of Toba in Sumatra should occur in a heavily populated area such as Naples, the effects would be devastating (Lipman, 2000).

In recent years, intense seismic unrest has occurred at Long Valley in the USA (1980 to present), Campi Flegrei in Italy (1983 to present), and Rabaul in Papua New Guinea (1971–1994). The unrest at Rabaul culminated in 1994 when two volcanoes erupted simultaneously on opposite sides of an old caldera, but no large eruption occurred. Both Campi Flegrei and Rabaul have already experienced multiple caldera-subsidence events associated with large eruptions, so additional caldera-forming events are a significant long-term hazard. One of the future tasks for volcanologists will be to recognize precursors and determine the time intervals between precursors and future, potentially devastating, caldera-forming eruptions.

3.2 Geomorphic hazards of mixed origin

Mixed geomorphic hazards include volcanogenic tsunamis, lahars, jökulhlaups, and debris avalanches. Unlike other volcanic hazards, lahars, jökulhlaups, and debris avalanches do not require an eruption. They can be triggered by rainfall or by edifice failure long after an eruption.

3.2.1 Volcanogenic tsunamis

Historical records of volcanic eruptions show that approximately 17% of victims have died as a result of tsunamis. The lethal effect of volcanogenic tsunamis is the unexpected transfer of energy from isolated volcanoes to sea waves, which travel rapidly to densely populated shorelines. In the historical record, 92 examples of tsunamis of volcanic origin can be attributed to eight causal mechanisms (Neall, 1996). Earthquakes accompanying eruptions, pyroclastic flows, and submarine explosions each account for 20% of all volcanogenic tsunamis. Caldera collapse is responsible for about 10%, avalanches of cold and hot materials about 14%, with lahars, air waves and lava avalanching of minor extent.

The most calamitous volcanogenic tsunamis in recorded history occurred during the eruption of Krakatoa on 27 August 1883 (Neall, 1996). Two large waves swept along the Sunda Strait shorelines to be followed by a gigantic wave that inundated the nearby coasts. Approximately 36 000 lives were lost, mostly from drowning. The tsunamis probably originated by one of three mechanisms. The first is large-scale collapse of the northern part of Krakatoa island as part of a caldera-forming process. The second is the discharge of approximately 12 km^3 of subaerially generated pyroclastic flows which were violently emplaced into the sea. The third is a major debris avalanche into the sea north of Krakatoa, supported by a hummocky submarine morphology.

3.2.2 Lahars

Lahar is an Indonesian term that describes a flowing mixture of rock debris and water (other than normal stream flow) from a volcano, which encompasses a continuum from debris flows (sediment concentration >60% per volume) to hyperconcentrated flows (sediment concentration from 20% to 60% per volume). People in distal areas commonly neither expect the event nor anticipate the destructive power of lahars (Vallance, 2000). Lahars are more deadly and devastating than pyroclastic flows for several reasons. They flow farther down slopes to the more heavily populated plains. The rock fragments carried by lahars make them especially destructive, while abundant liquid allows them to flow over gentle gradients and inundate areas far distant from their source. Requiring only the sudden mixture of large amounts of water with abundant, loose and easily eroded debris on a volcano slope, they can be formed in a variety of ways. They occur more frequently and over longer periods of time than pyroclastic flows.

Primary lahars are contemporaneous with eruption. They are triggered by crater lake overflow, subglacial eruption, snowmelt, pyroclastic flow entering streams, landslides, avalanches into lakes and streams, and rain during eruption. Secondary lahars occur during volcanic quiescence and can be generated by avalanches, crater lake failure, heavy post-eruptive rain, and earthquake-induced avalanches. Crater or caldera lakes, and volcanically debris-dammed lakes can break out months to years after eruptions, producing post-eruptive lahars. Lahars resulting from intense rainfall often occur after explosive eruptions, such as the large 9 December 1963 lahar at Irazu, Costa Rica, which generated economic losses estimated at US$15 million.

Because clay-rich sediment is uncommon on active volcanoes, lahars induced by sudden water release are clay-poor. In contrast, clay-rich mudflows begin as landslides from a volcanic mountain that is saturated by hydrothermal fluids and weakened by hydrothermal alteration (e.g., the Osceola mudflow at Mt Rainier, USA: Vallance and Scott, 1997). Clay-rich, collapse-induced lahars appear to be more common at ice-clad volcanoes than at edifices that are free of ice.

3.2.3 Geomorphic hazards from interactions of volcanic activity with snow and ice

3.2.3.1 Jökulhlaups

Many volcanoes lie beneath ice caps and, owing to prolonged geothermal activity or to sudden explosive activity, large quantities of water escape from subglacial reservoirs (Neall, 1996). These glacial outbursts, named jökulhlaups in Icelandic, have been responsible for creating the extensive sandur of the south Icelandic coastline. Their immense size is illustrated by the 1918 jökulhlaup generated by the Katla subglacial volcano and by the November 1996 jökulhlaup resulting from Gjálp, the new volcano under the Vatnajökull ice cap in Iceland. Given the intense scientific and media activity before the Gjálp event, no casualties were incurred but that was not the case in many historical subglacial eruptions. For further information on jökulhlaups see Björnsson (this volume).

3.2.3.2 Syneruptive sediment-water flows

Emplacement of hot pyroclastic rock debris on snow packs on volcanoes can trigger hazardous rapid flows of sediment and water which can extend far beyond the flanks of the volcano, producing catastrophic consequences over 100 km downstream.

Sediment-water flows encompass three categories (Pierson and Costa, 1987):

1 Unsaturated, predominantly granular flows of snow and admixed rock debris, termed 'volcanic mixed avalanches', can occur as a consequence of explosive events, such as the 13 November

1985 eruption of Nevado del Ruiz, Colombia (Pierson and Janda, 1994; Thouret *et al.*, 1995). Unusual 'ice diamicts', comprising clasts of glacier ice and subordinate rock debris in a matrix of ice, snow and coarse ash, were emplaced during the 15 December 1989 eruption of Redoubt volcano in Alaska (Waitt *et al.*, 1994). Transient, mixed avalanches transformed to initial 'snow slurry' lahars, then either to dilute or concentrated lahars on Ruapehu, New Zealand, in 1995 (Cronin *et al.*, 1996). These hybrid wet flows are hazardous because they extend beyond the reach of dry pyroclastic currents.

2 Debris flows, involving saturated slurries of rock debris and water.

3 Hyperconcentrated flows occur as the more dilute downstream run-out flows of some large debris flows.

Historical eruptions at five snow-clad volcanoes (Tokachidake, Japan, in 1926; Nevado del Ruiz, Colombia, in 1985; Cotopaxi, Ecuador, in 1877; Mt St Helens, USA, in 1982–84; Ruapehu, New Zealand, in 1995–96) have demonstrated that snowmelt-generated debris flows can have peak discharges as large as 10^5 m^3 s^{-1}, attain velocities as high as 20–40 m s^{-1}, mobilize as much as 10^8 m^3 of debris, and travel more than 100 km in valleys draining the edifices (Pierson, 1999).

The risk to human life from such events was tragically demonstrated on 13 November 1985, at Nevado del Ruiz, where snowmelt-triggered lahars rushed down adjoining canyons to nearby villages and inundated the town of Armero and 23 000 of its inhabitants (Figure 11.5). This was the second worst volcanic disaster of the twentieth century and the fourth most disastrous eruption in recorded history. The total financial loss was over US\$ 1 billion (Voight, 1990). The key findings from this disaster are as follows (Pierson *et al.*, 1990; Thouret, 1990).

1 A small eruption ($<5 \times 10^6$ m^3 of magma ejected) was able to generate as much as 9×10^7 m^3 of lahar from about 2×10^7 m^3 of meltwater.

2 Pyroclastic flows were mainly responsible for interacting vigorously with the snow and ice.

3 The hot eruptive products scoured about 10 km^2 of snow and ice. About 16% of the surface ice and snow area was lost, amounting to 9% of the total volume.

4 A volume of meltwater as large as 38–44 $\times 10^6$ m^3 released in 20–90 minutes implies a vigorous heat transfer from hot eruptive products to snow and ice.

5 Mixed avalanches of varying proportions of snow, ice, liquid water, rock and colluvium were also initiated, contributing slush, meltwater and rocky debris into the headwaters of lahar channels.

6 The lahars were initiated within minutes of the onset of the eruptions. The lahars incorporated unconsolidated materials, increasing the total flow and peak flow rate, a process termed 'bulking'.

7 Over the 104 km distance travelled by lahars, net flow-volume increases occurred by factors of two to four. This finding has a major social implication because lahars may far exceed their initial volume during their passage and thus inundate inhabited regions over 100 km from the mountain.

8 Peak-flow mean velocities ranged from 5 to 15 m s^{-1}. The Ruiz lahars moved slowly close to the volcano, but travelled relatively rapidly in distal channels.

9 Maximum computed peak discharges for all lahars occurred at locations within 10–20 km of the crater. The maximum discharge is estimated at 48 000 m^3 s^{-1} at a point 9.6 km from source (Pierson *et al.*, 1990).

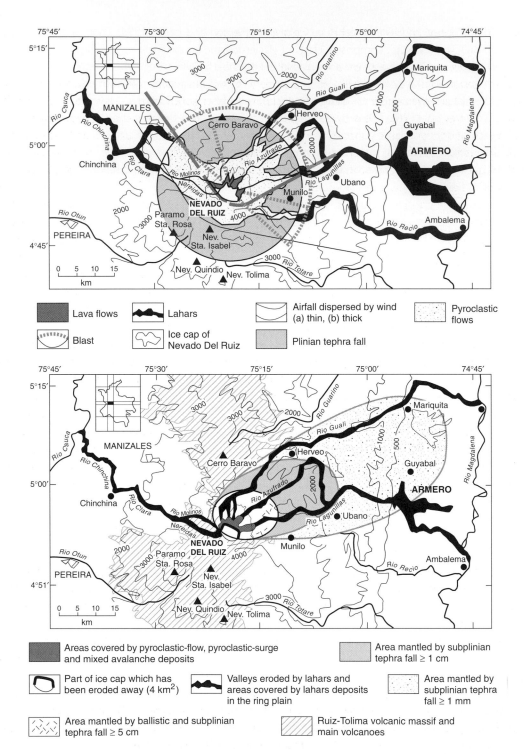

Figure 11.5 *Volcanological maps preceding and following the 1985 eruption of Nevado del Ruiz, Colombia, with emphasis on lahars. (A) Map showing hazard zones potentially affected by an eruption of Nevado del Ruiz, released on 7 October 1985. (B) Map of actual areas affected by tephra fall and lahars on 13 November 1985*

Source: *reprinted from Thouret, J.-C., Méthodes de zonage des menaces et des risques volcaniques. In Bourdier, J.-L. (dir.), Le volcanisme. Manuels et Méthodes, vol. 25, pp. 267–83. © 1994, with permission from Editions B.R.G.M., Orléans.*

3.2.4 Landslides and debris avalanches on volcanic mountains

3.2.4.1 Catastrophic landslides on oceanic shield volcanoes

Oceanic shield volcanoes grow relatively rapidly to great heights, several kilometres above the ocean floor. Mauna Loa rises up to 9 km in Big Island, Hawaii, and has grown at an average rate of about 0.02 $km^2 a^{-1}$ for the past 600 ka (Moore and Clague, 1992). The air of permanence of large, low-angle shield volcanoes belies their inherent instability. Instability and collapse are a major process and hazard in the evolution of the basaltic oceanic islands. Particularly noteworthy are 70 landslides on the Hawaiian ridge, which have removed volcano-flank sectors that exceed 1000 km^3 in volume (Moore et al., 1994).

The Canarian and Hawaiian volcanoes share common constructional and structural features, such as rift zones, progressive volcano instability and multiple gravitational collapses (Carracedo, 1999). Gravitational stresses and dyke injections progressively increase the mechanical instability of these edifices, especially in the most active shield-stage phases of growth. Lower eruptive rates in the Canaries than in the Hawaiian islands result in higher aspect ratio volcanoes. Slopes are generally steeper in the younger shields. Catastrophic flank failures in the Canaries have shallower slide planes and collapses have lower mass volumes than those of the Hawaiian shields. The Canaries will end their emergent phase after a very long time compared with the Hawaiian shields, because of catastrophic mass-wasting processes in the early stages of development and erosion later (Carracedo, 1999).

What processes trigger slope instability on low-angle shield volcanoes? Keating and McGuire (2000) have identified 23 endogenetic and exogenetic processes that contribute to edifice collapse on volcanic ocean islands (Figure 11.6). Endogenetic causes of failure dominate during periods of active volcanism; exogenetic sources may cause failure at any time. Endogenetic instability includes unstable foundations, detachment surfaces linked to volcanic intrusions, thermal alteration and faulting (Figure 11.6(A)). Other endogenetic factors are: increased interstitial pore pressure related to intrusives; thermal alteration making rocks susceptible to failure, either catastrophically or by creep of the flanks of the volcano. Figure 11.6(B) illustrates exogenetic sources of instability and failure. The heterogeneous combination of rock types having poor cohesion and low strength all facilitate failure. The steepness of an edifice appears to determine how quickly it becomes unstable and susceptible to lateral failure.

3.2.4.2 Flank failure on stratovolcanoes

Because of rapid growth, many stratovolcanoes are liable to massive flank failure. More than 20 major slope failures have occurred globally during the past 500 years, a rate exceeding that of caldera collapse (Siebert, 1996). Flank failures produce extremely mobile debris avalanches that can travel long distances beyond the flanks of cones at high velocities (McGuire et al., 1996).

Massive landslides create specific morphology and deposits: horseshoe-shaped re-entrants into the edifice, with a high, steep break-away scarp. The debris-avalanche deposits form a large talus apron (e.g., 50 km^3 at Mt Shasta, California) of hummocky terrain with water-filled depressions, steep flow margins, and thick deposits of unsorted, unstratified angular-to-subangular debris. A relationship exists between the distance travelled by an avalanche and the failure volume. Collapsed portions exceptionally exceed 10% of the volume of the pre-failure edifice. The ratio of vertical drop H to travel length L range from 0.09 to 0.18 for Quaternary volcanic avalanches <1 km^3 in volume and from 0.04 to 0.13 for avalanches >1 km^3 (Siebert, 1996). The ratio of H to L for volcanic avalanches is much lower than the ratio for nonvolcanic deposits of similar volume, suggesting that low-rigidity, partially fluidized avalanches are capable of travelling great distances, e.g., in excess of 100 km from Nevado del Colima in Mexico (Stoopes and Sheridan, 1992).

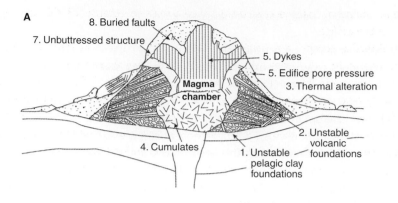

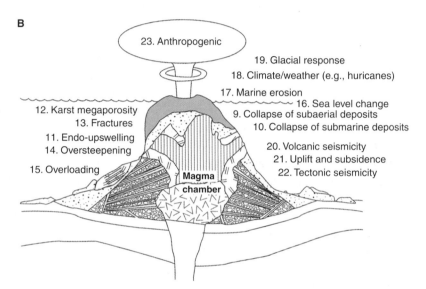

Figure 11.6 *Sources of failures on oceanic volcanoes. (A) Schematic profile of an oceanic volcano, illustrating the distribution of endogenetic sources of failure. Symbols used in the illustration: vertical lines indicate dykes, the double dash pattern indicates thermally altered zones, the v-pattern indicates a magma chamber, dotted patterns indicate subaerial and submarine debris, and the patterns of lines and ovals illustrate lava flows and pillow basalts. The eight endogenetic sources of failure are numbered. (B) Schematic profile of an oceanic island indicating the distribution of exogenetic sources of failure (symbols used in the illustration are similar to those used in (A). The 15 exogenetic sources of failure are numbered* Source: *from Keating and McGuire (2000); reproduced with the permission of Birkhäuser Verlag, Basel.*

Widespread slope failure in a variety of tectonic settings suggests that it may be the dominant catastrophic process modifying volcanoes. Not only are stratovolcanoes susceptible to slope failure, but they are also less voluminous and steep-sided lava-dome complexes. The summit of Mt Augustine, Alaska, has repeatedly collapsed and regenerated, averaging 150–200 years per cycle, during the past 2000 years (Beget and Kienle, 1992). The unprecedented frequency of summit failure was made possible by sustained lava effusion rates over ten times greater than is typical of plate-margin volcanoes. Regardless of how extensively a volcano is eviscerated, the conical shape is re-established very rapidly.

At Parinacota in northern Chile, reconstruction since a massive collapse 13000 years ago has virtually rebuilt the entire edifice.

The causes of edifice collapse are uncertain. Structural factors such as steep dip slopes, zones of weakness and local extension promoted by parallel dyke swarms can contribute to flank failure. Siebert (1996) emphasized differences in eruption style associated with flank failure: magmatic eruption of Bezymianny type, nonmagmatic explosions of Bandai type and cold avalanches of Ontake type (Table 11.1). But van Wyk de Vries and Francis (1997) argue that the volcanic edifice itself contributes to the weakness of its bedrock. In contrast to radial spreading, preferential spreading in one direction is critical to collapse development. Radial spreading induces inward-dipping faults which inhibit collapse, whereas sector spreading generates failure-prone outward-dipping structures. Spreading in a preferential direction may be caused by buttressing, the regional slope of basement beds, regional stress, weak basement or by high fluid pressures under one side.

Socompa volcano in northern Chile has a wedge-shaped collapse scar 12 km wide and 300 m deep and an avalanche deposit covering almost 500 km^2 with a volume of 36 km^3. van Wyk de Vries et al. (2001) attribute the giant sector collapse and debris-avalanche deposit to spreading and liquefaction of the substrata. The authors claim that the collapse was triggered by failure of active thrust-anticlines in sediments and ignimbrites underlying the volcano. The thrust-anticlines were a result of gravitational spreading of substrata under the volcano load. About 80% of the resulting avalanche deposit is composed of substrata formerly residing under the volcano and in the anticlines. Structural relations indicate that immediately prior to collapse the substrata disintegrated, liquefied and were ejected from beneath the edifice.

The relationship of spreading to collapse has been described at just one other volcano so far: Mombacho, Nicaragua (van Wyk de Vries and Francis, 1997). There are many sector collapses, however, with similar wedge-shaped scars, which could have involved such processes. Among many edifices standing on sedimentary and pyroclastic substrata, some are known to be spreading, such as Momotombo and many Javan volcanoes. Thus not only have spreading and substrata liquefaction probably generated previous collapses, but they may occur at many intact volcanoes.

3.3 Hazardous geomorphic responses to eruptions

Rain-generated lahars and floods are a serious hazard at any time during the rainy seasons following an eruption. On volcanic mountains devastating geomorphic impacts are complex, with an initial stage of accelerated erosion and associated sedimentation, followed by an exponential decrease over a few years.

3.3.1 Hazardous geomorphic response to small eruptions

Geomorphic hazards following small to moderate eruptions are severe but restricted in time, usually only one year, whereas geomorphic hazards after large-scale eruptions are worse and protracted. Many studies of recent tephra erosion, such as Paricutin in Mexico 1943–52 (Segerstrom, 1950), Irazu in Costa Rica 1963–65 (Waldron, 1967), and Usu in Japan 1977–82 (Chinen and Kadomura, 1986), suggest that the erosion rate peaked soon after the tephra erupted and then declined rapidly, from 25–100 mm a^{-1} in the first two years to 1–5 mm a^{-1} within five years of the eruption.

Sediment yields calculated at several stratocones following moderate eruptions range between 1.1 and 2.7×10^5 m^3 km^{-2} (e.g., Galunggung in 1982–83, Merapi in 1994–95 and Semeru in 2000, in Java; Unzen, Japan, in 1991–93). The range of sediment yields compares well with data compiled by Major et al. (2000) for catchments affected by eruptions on composite volcanoes in a wide range of humid climatic settings, further demonstrated in Table 11.4. Persistently active stratovolcanoes characterized

Table 11.4 – Comparison of sediment discharge and hydro-geomorphological characteristics on three volcanic mountains

Geomorphic and hydrologic parameters	Yakedake Kamikamihori	Merapi Bebeng river	Unzen Mizunashi river
Summit elevation (m asl)	2445	2911	1473
Relative height between summit and observation site (m)	865	2100	1392
Full length of the stream (m)	2500	14000	7500
Mean stream gradient (°)	20.7	9.9	10.9
Stream gradient at the observation site (°)	7	2.8	2.7
Catchment area for the observation site (km^2)	0.83	5	12
Mean annual precipitation (mm)	2600	4500	3100
Time after last effective eruption (eruption year)	33 (1962)	11 (1984)	0 (1991–1994)
Permeability at the surface (cm s^{-1})	10^{-2}–10^{-3}	10^{-2}–10^{-3}	10^{-3}–10^{-5}
Frequency of debris flow (per year)	0.7	5	23
Total bulk volume of sediment transported (10^4 m^3 a^{-1})	0.5	20	210
Specific sediment yield ($\times 10^4$ m^3 km^{-2} a^{-1})	0.6	4	17.5
Annual depth of sediment yield (mm a^{-1})	6	40	175

Notes: Mt Yakedake (Japan) is dormant, Mt Merapi (Java) and Mt Unzen (Japan) are active volcanoes.
Source: from Suwa and Sumaryono (1996).

by small but frequent eruptions (e.g., Sakurajima and Semeru) deliver a very large amount of sediment to rivers worldwide.

After deposition of tephra, geomorphic hazards tend to concentrate in valley channels but they decline rapidly on hillslopes and divides (Collins and Dunne, 1986). At Usu volcano, Chinen and Kadomura (1986) reported a rate of erosion immediately after the 1977 eruption of 136 mm a^{-1} owing to the upheaval of the Ogari-yama dome (Figure 11.7). But sheet wash and rill erosion rapidly declined after the first rainy season. This decline was not caused by a recovery of vegetation, but by increased infiltration and decreased erodibility of the tephra layer, by exposure of more permeable and less erodible substrates, and by the development of a stable rill network. If the 20-year perspective from Mt St Helens (Major et al., 2000) can serve as a guide, yields from basins affected solely by hillslope disturbance will diminish rapidly, probably within one or two years, whereas yields from basins that experience dominantly channel disturbance will likely remain elevated for as much as several decades. Erosion rates and sediment yields decline as drainage networks extend and integrate. In contrast, erosion of tephra-mantled hillslopes typically subsides within several months.

3.3.2 Severe geomorphic hazards in catchments disturbed by large eruptions

Some of the largest sediment yields worldwide (commonly from 10^3 to 10^6 t km^{-2} a^{-1}) are delivered in mountainous catchments affected by debris flows in the aftermath of explosive eruptions of large magnitude. Post-eruption sediment yields can exceed pre-eruption yields by several orders of magnitude. Annual suspended sediment yields following the 1980 Mt St Helens eruption were as much as 500 times greater than typical background levels (Major et al., 2000). After 20 years, the average annual

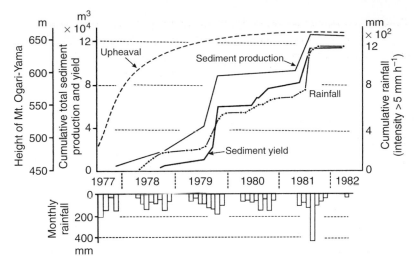

Figure 11.7 *Sediment production, sediment yield, rainfall, and upheaval of Mt Ogari-yama, August 1977–May 1982, Usu volcano, Hokkaido, Japan. Rainfall data from Nishiyama-gawa watershed gauge, 1 km distant from the summit atrio*
Source: *from Chinen and Kadomura (1986); reproduced with the permission of Gebrüder Bornatraeger Verlagsbuchhandlung, Stuttgart.*

suspended sediment yield from the 1980 debris-avalanche deposit remains 100 times (10^4 t km^{-2}) above typical background levels (10^2 t km^{-2}). Within five years of the eruption, annual yields from valleys coated by lahar deposits remained roughly steady, and average yields remain approximately ten times above background levels.

The long-term instability of eruption-generated debris means that effective mitigation measures must remain functional for decades. Prolonged excessive sediment transport after an eruption can cause environmental and socio-economic harm exceeding that caused directly by the eruption, as shown by the protracted lahars around Pinatubo, described below.

3.3.2.1 The geomorphic impact around Pinatubo, Philippines

The geomorphic response to large eruptions (>10 km^3 of tephra) had not been documented prior to the study of the Mt Pinatubo lahars. The explosive eruption of Mt Pinatubo on 15 June 1991, which erupted a total bulk volume of 8.4–10.4 km^3, deposited 5–6 km^3 of loose pyroclastic-flow deposits in the heads of valleys draining the volcano and about 0.2 km^3 of tephra on the volcano's flanks that would later be the primary source of sediment for lahars (Figure 11.8). Numerous debris flows and hyperconcentrated flows were triggered during and following 1991 and affected eight major drainages of Mt Pinatubo. Lahars have been flowing into densely populated areas of central Luzon over the past 10 years, causing deaths, leaving more than 50000 people homeless, affecting more than 1 350 000 people and causing enormous property losses and social disruption (Janda *et al.*, 1997).

The sediment budget at Mt Pinatubo was evaluated by Pierson *et al.* (1992). Sediment yields in 1991 were of the order of 10^6 m^3 km^{-2} a^{-1}, nearly an order of magnitude greater than the maximum sediment yield computed following the 18 May 1980 eruption of Mt St Helens. To quantify the annual erosion in the Sacobia catchment on the eastern slope of Mt Pinatubo, GIS-based geomorphological maps superimposed on digital elevation models (DEMs) were constructed by Daag and van Westen (1996) before and during the eruption and for three consecutive years afterward. In the catchment of

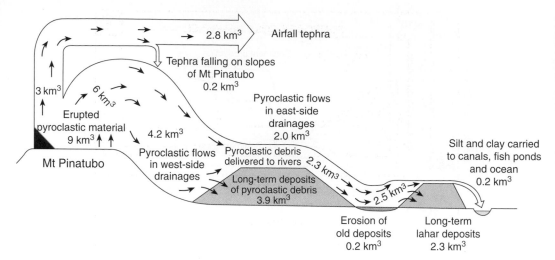

Figure 11.8 *Schematic portrayal of transportation and distribution of pyroclastic material produced by the June 1991 eruptions of Mt Pinatubo, Philippines. Thickness of the flux arrows and deposits is proportional to the estimated material volume*
Source: *Pierson et al. (1992); reproduced with permission of the USGS.*

24 km², erosion rates were calculated to be in the range 1.3–2.2 × 10⁷ m³ a⁻¹ and the sediment yield was approximately 5.6–9.1 × 10⁶ m³ km⁻² a⁻¹.

The prodigious sediment yield from Pinatubo had geomorphic after-effects such as channel piracy and avulsion, and blockage of tributaries. Blockage of tributaries at their confluence with the main channel formed temporary lakes and impoundments. Floods triggered by breaching of these lakes provided an additional hazard. Unlike rain-induced lahars, these floods can occur in the absence of rainfall, and thus limit the capability to warn threatened areas. Although the areas affected by lahars have expanded, the frequency of events has decreased and the number of impacted river systems had dwindled to four in 1995, as source materials were depleted.

The Pinatubo case study raises at least three important points (Major *et al.*, 1997): (1) heavy rainfall alone was not responsible for generating the lahars in 1991; (2) geomorphic processes affecting catchments and channels play a significant role in the erosion and redistribution of sediments. These have fostered more lahars for the ensuing years and more flooding beyond the alluvial fans onto the densely populated plains; and (3) after the geomorphic impact of the devastating 1991 lahars, the subsequent lahars triggered by the seasonal monsoon rains and other geomorphic processes had far greater social and economic impacts.

3.4 Delayed geomorphic hazards

3.4.1 Delayed geomorphic hazards in basaltic and caldera contexts
In basaltic regions tephra is commonly coarse and post-eruptive sedimentary response is muted. The 1945 eruption of Paricutin in Mexico accumulated coarse tephra in many areas. Deep gullying and downstream aggradation were strongly retarded on coarse, highly permeable tephra (Segerstrom, 1950). Response thresholds are more sensitive to grain size and sorting than to the volume of tephra (Collins and Dunne, 1986). Coarse, well-sorted, highly permeable tephra can form very stable deposits. Such situations require some outside influence to change stability, such as lowering local base levels or altering the drainage network.

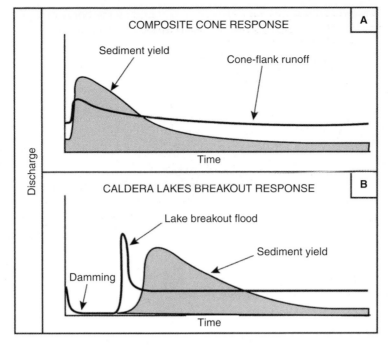

Figure 11.9 *Contrast between hydrolic and sedimentary response to composite cone eruptions versus intracaldera eruptions as exemplified by Tarawera, New Zealand. (A) On a volcanic cone, the hydrograph curve shows an initial increase in runoff owing to soil and vegetation damage followed by a gradual return to background levels. (B) Following caldera eruption, a caldera lake is initially dammed for years to decades, a breakout flood follows, then there is a return to pre-eruption flow* Source: *from White* et al. *(1997); reproduced with the permission of the Geological Society of America.*

The contrast between hydrolic and sedimentary response to intracaldera eruptions with respect to composite cones is exemplified by the Tarawera (Figure 11.9; White *et al.*, 1997) and Taupo calderas (Manville *et al.*, 1999) in New Zealand. Temporary storage of water, which is later released in catastrophic floods, commonly takes place in lakes of caldera and dome complexes in temperate to humid environments. The delayed channel disturbance following the 1886 Tarawera event may typify response to Plinian eruptions in humid-environment caldera complexes, which have modest relief, weakly integrated drainage systems and numerous lakes. Sediment yield is minimal prior to a breakout flood, then increases to a maximum only after the breakout flood modifies the trunk stream valley. Sediment yield rises rapidly as gullies sap headwards from the flood-scoured valley, and the increase is reflected by rapid downstream aggradation of stream deposits. Sediment yield declines gradually as streams restabilize but the gradual decline in sediment yield is distinct at calderas in having an initial period of stability.

3.4.2 *Geomorphic hazards on inactive volcanic mountains*
Geomorphic processes acting on dormant (i.e., could become active again) and extinct (i.e., not expected to erupt again) volcanic mountains are related to debris flows and mass movements, common on all steep mountains but exacerbated by unconsolidated material. These hazards are unrelated to eruptions and stem from the remobilization of old tephra. On dormant volcanoes, catastrophic erosion processes are uncommon but can reach dramatic proportions. Examples of landslide disasters on dissected volcanoes are Mt Ontake (Japan), triggered by an earthquake in 1984 (Voight

and Sousa, 1994), and at Casita volcano (Nicaragua), induced by the heavy rains of Hurricane Mitch in 1998 (Scott, 2000). Characteristically, the volume transported changes during movement owing to entrainment and deposition along the flow path, and the sequence of movement commonly shows stop–start behaviour as a result of landslide dam formation and breaching.

In 1984 a mass of Quaternary pyroclastic rock slid from the western flank of Mt Cayley volcano in southwest British Columbia (Evans et al., 2001). The disintegrating rock mass entrained more material and formed a rock avalanche that travelled at 42 m s^{-1} a horizontal distance of 3.45 km from its source over a vertical elevation difference of 1.2 km. The rock avalanche was partially transformed into a distal debris flow that travelled a further 2.6 km down a narrow path to the main river, temporarily blocking it. The landslide is one of seven high-velocity rock avalanches that have occurred in the Garibaldi volcanic belt of southwest British Columbia since 1855. The 1984 event illustrates movement mechanisms typical of rapid landslides on dissected volcanoes, i.e., significant entrainment of debris following initial failure, high velocity and a reach extended by a distal debris flow.

The 1998 Casita disaster further emphasizes the hazards of highly mobile debris flows beginning as landslides, their potential to transform, and their ability to amplify during flow. A single wave of debris flow, triggered by hurricane precipitation, killed more than 2000 residents of two towns near Casita volcano on 30 October 1998 (Scott, 2000). The flow wave began as a small flank failure near the summit of the inactive edifice. Peak precipitation from Hurricane Mitch was the flow trigger: almost 1 m of rain had fallen between 26 and 30 October. Downstream, the flow volume increased by a factor of at least nine, transforming to a catastrophic debris flow >1 km wide and 3–6 m deep. Only land-use planning based on debris-flow hazard assessment could have prevented the Casita disaster. Afterwards the towns – established in the 1980s on apparently safe terrain 4 km from the volcano – were found to have been located in a pathway of prehistoric debris flows.

4 Assessment of hazards on populated volcanic mountains

4.1 People and risk around hazardous volcanoes

In the year 2002, at least 500 million people were living in the shadow of a volcano. Twice during the twentieth century large towns were destroyed by eruptions (St Pierre, Martinique, in 1902 and Armero, Colombia, in 1985). Major population centres lie just tens of kilometres from several large volcanoes with a likelihood of eruption during this century, e.g., Naples near Vesuvius (Chester et al., 2001; Figure 11.10).

Blong (2000) points out three methods of adjustment that can reduce the risk associated with the occurrence of hazards. (1) Modify the hazard itself, though this is rarely possible with volcanic hazards. The most publicized diversion of a lava flow occurred at Etna in May 1992. (2) Modify vulnerability to the impact of the hazard. Examples include building roofs that withstand ash loads, protecting crops under plastic sheeting, diverting lahars, land-use planning, and broadcasting volcano forecasts which allow inhabitants to plan daily activities. (3) 'Distribute' the loss associated with eruption damage to a wider community through disaster relief and insurance.

Vulnerability varies significantly according to the type of volcanic hazard. People can be evacuated from the path of an advancing lava flow but may not escape a rapidly flowing lahar, even in an apparently safe place several tens of kilometres from the summit. Nearly everything is vulnerable to pyroclastic flows. Factors contributing to vulnerability include locations in marginal and potentially hazardous areas, such as valleys where lahars, pyroclastic flows, lava flows and CO_2 inundation may occur, and densely populated areas with old residential buildings that are poorly maintained, illegally altered or extended, and built to standards now considered inadequate.

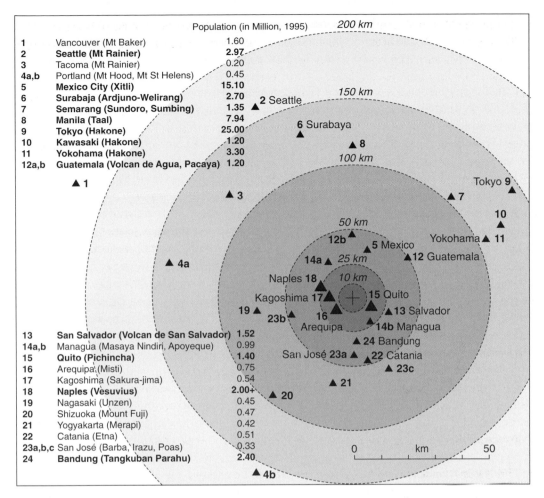

Figure 11.10 *A selection of the world's most highly exposed conurbations, plotted according to relative distance and direction from the nearest volcano*
Source: *reprinted from Global Environmental Change, vol. 2, Chester, D.K., Degg, M., Duncan, A.M. and Guest, J.E., The increasing exposure of cities to the effects of volcanic eruptions: a global survey, pp. 89–103. © 2001, with permission from Elsevier.*

4.2 The role of geomorphic surveys in hazard and risk assessment

Geomorphology can contribute to risk assessment through two zonation approaches: geomorphic hazard zonation and composite risk zonation. Geomorphic hazard domains are established according to the capacity of each hazard to affect geomorphic stability, the perceived vulnerability of people, and the priority of elements at stake, such as urban settlements and land-use types. A composite risk zonation, incorporating geomorphic mapping and analysis of risk and of vulnerability, can therefore be achieved.

Geomorphic surveys with the aid of satellite imagery form a logical starting point for natural hazard zoning. Geomorphic hazard zonation aims at recognizing old deposits, mapping flow paths and delineating hazard zones, which are primary inputs in elaborating eruption scenarios. Potential inundation areas can be delineated based on the palaeohydrologic records of flows in specific areas, on relationships between failure volumes and corresponding inundation areas, and with the help of slope stability analysis using a DEM. Additional risk assessment requires the development of a series of scenarios in which eruption magnitudes, hazard types, composite risk zonation and the vulnerability of people and infrastructure are adequately considered. Eruption scenarios are useful for the preparation of emergency evacuation plans and long-term land-use planning.

4.2.1 A case study: eruption scenarios and risk assessment at Merapi volcano, Java

Of 1.1 million people living on the flanks of the active Merapi volcano, 440000 are at relatively high risk in areas prone to pyroclastic flows, surges and lahars, according to four probable eruption scenarios (Thouret et al., 2000). Figure 11.11 displays the two most common eruption scenarios. The first scenario, termed 'Merapi' type, considers relatively small eruptions (volume of tephra $<4 \times 10^6$ m^3) and block-and-ash flows at 2- to 8-year intervals. Relatively moderate eruptions (4–10×10^6 m^3) at 20- to 50-year intervals, such as the 1953–54 and 1969 events, including large pyroclastic flows and surges and lahars, are also considered in the first scenario. Figure 11.11 also shows the zones likely to be affected in the event of the second eruption scenario, which encompasses relatively large eruptions ($\leq 10^7$ m^3) such as the 'mixed' effusive and pelean 1930–31 events. Additional effects of large eruptions such as the sub-Plinian 1872 event, which can be expected once or twice per century, are not shown in Figure 11.11, nor those of catastrophic eruptions ($>10^7$ m^3), which are unknown throughout the historical record but probably occurred in prehistoric time. Although areas likely to be devastated cannot be delineated on a reliable basis in the event of a catastrophic scenario, the eruption might encompass a failure of the southwest flank and of the entire summit dome, with subsequent voluminous debris avalanches and debris flows, and a laterally directed explosion with powerful pyroclastic surges.

Hazard microzonation (scale 1:2000) has been carried out in three currently lahar-prone drainages, the Bebeng/Krasak and the Boyong valleys and the Code valley across the suburbs of Yogyakarta. Microzonation was based on lahar scenarios in which four discharge categories correspond to four lahar-prone areas. To assess the lahar threat at Merapi, Lavigne et al. (2000) have combined: (1) the distribution and frequency of recent lahar deposits, and reported damages since the mid-1600s; (2) the DEM-based topography of valley channels likely to be affected, the location of sites of overflow and avulsion of the drainage system; and (3) four rainfall parameters: intensity, duration, total amount and the most intense 10-min rainfall.

The analysis of vulnerability completes the hazard zonation in order to achieve a risk assessment in densely populated lahar-prone areas. Three groups of factors render people and property vulnerable: (1) population density and growth, derived from a database collected in 38 communes; (2) technical factors such as the location and construction quality of houses and other buildings, as well as additional information on local disaster relief organizations and civil defence countermeasures at the commune scale; and (3) economic values at stake such as the infrastructure, facilities and businesses. The total elements at risk in the Boyong Code valley are estimated to be as much as US$100 million (Lavigne et al., 2000).

4.3 Geomorphic methods for delineating hazard zones

An alternative estimate of hazard zones can be obtained with the aid of mathematical models that simulate the evolution of volcanic phenomena and compute the effects at ground level, allowing estimation of the area affected by an event according to a certain scenario. Geomorphic and hydrologic parameters are critical input requirements for the use of DEMs and GIS in long-term planning. The use of DEMs and of simulation models such as LAHARZ and FLOW3D have enabled Iverson et al. (1998), Pareschi et al. (2000) and Sheridan et al. (2001) to gauge volcanic flow hazards in densely populated areas around Mt Rainier, Vesuvius and Popocatepetl volcanoes, respectively.

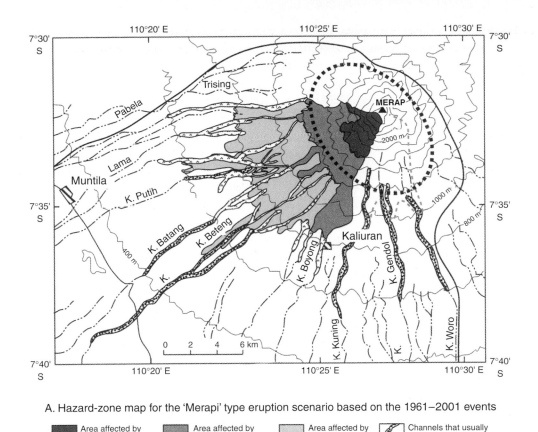

A. Hazard-zone map for the 'Merapi' type eruption scenario based on the 1961–2001 events

| | Area affected by lava dome rocksliding and non-explosive avalanching | | Area affected by frequent small- to moderate-scale eruptive dome avalanching, block-and-ash flows, and companion ash-cloud surges | | Area affected by moderately large, less common, and more common pyroclastic flows and surges | | Channels that usually convey small- to moderate-scale, mostly rain-triggered lahar events |

B. Approximate hazard zones in the event of the 'mixed', effusive and pelean, eruption based on the 1930–31 events

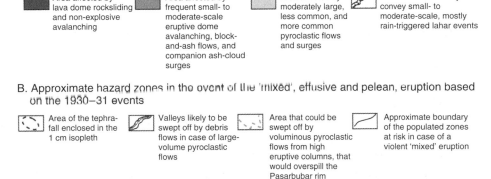

| | Area of the tephra-fall enclosed in the 1 cm isopleth | | Valleys likely to be swept off by debris flows in case of large-volume pyroclastic flows | | Area that could be swept off by voluminous pyroclastic flows from high eruptive columns, that would overspill the Pasarbubar rim | | Approximate boundary of the populated zones at risk in case of a violent 'mixed' eruption |

Figure 11.11 *Hazard-zone map showing the extent of areas likely to be affected in the event of future eruption scenarios at Merapi. (A) Hazard-zone map for the 'Merapi' type eruption scenario based on the 1961–2001 events; (B) approximate hazard zones in the event of the 'mixed', effusive and pelean, eruption scenario based on the 1930–31 events*
Source: *reprinted from Journal of Volcanology and Geothermal Research, vol. 100, Thouret, J.-C., Lavigne, F., Kelfoun, K. and Bronto, S., Toward a revised assessment of volcanic hazards at Merapi, Central Java, pp. 479–502. © 2000, with permission from Elsevier.*

Iverson *et al.* (1998) proposed a method of delineating lahar hazard zones in valleys that head on volcano flanks, which provides a rapid and reproducible alternative to traditional methods. The rationale derives from scaling analyses of generic lahar paths and statistical analyses of 27 lahar paths documented at nine edifices. These analyses yield semi-empirical equations that predict inundated valley cross-sectional areas (A) and planimetric areas (B) as functions of lahar volume (V). The predictive equations ($A = 0.05V^{2/3}$ and $B = 200V^{2/3}$) provide all the information necessary to calculate and plot inundation limits on topographic maps. By using a range of prospective lahar volumes to evaluate *A* and *B*, a range of inundation limits can be plotted for lahars of increasing volume and decreasing probability.

Iverson *et al.* (1998) automated hazard zone delineation by embedding the predictive equations in a GIS computer program that uses a DEM of topography. The simulation model LAHARZ provides a rapid, automated means of applying predictive equations to regions around edifices and compare the results with the hazard-zone boundaries established in the field by mapping lahar deposits. Lahar hazard zones computed for Mt Rainier, USA, mimic those constructed on the basis of intense field investigations. The computed hazard zones illustrate the potentially widespread impact of large lahars, which on average inundate planimetric areas 20 times larger than those inundated by rock avalanches of comparable volume.

4.3.1 Computer simulations of hazardous areas

Pareschi *et al.* (2000) used a computer simulation approach to deal with ongoing volcanic hazards controlled by topography, such as lava flows. A maximum slope-statistical approach allows the authors to assign the lava vent and to estimate the zonation of hazards using a map superimposed on a georeferenced image of Mt Etna and other GIS layers. The authors also present a GIS hazard map for Vesuvius, using the 1631 event as the maximum expected eruption. Three main regions are outlined: (1) the area potentially exposed to proximal destructive flows; (2) the zone potentially exposed to roof collapse due to ash fallout of 400, 300 and 200 kg m^{-2}; and (3) areas of flooding.

Wadge *et al.* (1998) used a physically based simulation of avalanche based on a three-fold parameterization of flow acceleration for which they chose values using an inverse method. Multiple simulations based on uncertainty of the starting conditions and parameters, specifically location and mass flux, have been used to map hazard zones and successfully model the pyroclastic flows generated on 12 May 1996 by the Soufrière Hills volcano, Montserrat. Hooper and Mattioli (2001) used a kinematic modelling of 'dense' flows initiated by nonexplosive, gravitational collapse pyroclastic flows and employed a FLOW3D model to simulate this type of volcanic flow. The program constructs a digital terrain model based upon a 3D network of triplets, while a synthetic dome was added to the topographic model to improve the accuracy of the simulations. Simulated flow pathways, run-out distances and velocities closely approximated observed block-and-ash flows on Soufrière Hills. While the simulations presented do not elucidate the physics of pyroclastic flows, this type of kinematic modelling can be completed easily and without extensive knowledge of specific parameters other than topography. It thus serves as a rapid and inexpensive first approach for initial hazards assessment. However, such computer-based simulations require accurate digital topographic information, because initial topographic boundary conditions critically influence the extent and velocity of pyroclastic flows.

DEM- and GIS-based computer models for simulating lahars and pyroclastic flows have been used by Sheridan *et al.* (2001) to gauge volcanic hazards at Popocatepetl in Mexico, in addition to a detailed survey of the past eruptive history and a close monitoring of the present activity. Popocatepetl stands 60 km southeast of Mexico City and 40 km west of the city of Puebla. The combined population of these two metropolitan areas exceeds 30 million, so the 2000–2001 activity has received intense public scrutiny. The present eruptive activity, which started in December 1994, has already forced the government to evacuate 20000–50000 people at a time from towns in the state of Puebla. Several domes have formed in the crater and domes will probably continue to grow until the crater is entirely filled by lava, a process which may take a few more years at the current rate of dome emplacement.

At Popocatepetl, the major hazard associated with the current activity must be considered to be lahars on surrounding villages. Rock avalanches and pyroclastic flows could produce materials for lahars. Assuming likely source areas, inundation zones for lahar volumes of 10^7 m^3 and 10^8 m^3 were simulated with ARCINFO using the LAHARZ model developed by Iverson et al. (1998). The smaller-volume model lahar represents the water available from glacier ablation by pyroclastic flows and the larger volume represents the water from complete melting of the current glacier. The maximum possible lahar, based on the size of the Ventorillo glacier, is about 10^8 m^3.

Simulations of pyroclastic flows and rock avalanches used the FLOW3D model, which provides velocity histories of particle streams along flow paths in three dimensions (Sheridan et al., 2001). Simulations were used to create the hazards map at Popocatepetl. Bit-mapped and colour-coded overlays of multiple themes, including the flow paths and velocities, were used to produce a realistic image useful for nonprofessional observers. The interactive platform of FLOW3D allows the observer to adjust the perspective and distance for the desired view.

5 Conclusion and perspective

The geomorphic hazards of volcanic mountains are those specific to volcanic eruptions in addition to those which are common to all mountains. Geomorphic hazards of volcanic origin are direct, related to explosive activity, and indirect, resulting from secondary processes such as lahars. Geomorphic hazards often recur many years after an eruption. The geomorphic consequences of large volcanic eruptions are severe, long-lasting, and disturb slopes and drainage processes on and around the volcanic mountain for decades. Rates of growth and erosion are very rapid at active volcanic mountains, especially at composite volcanoes, thus increasing noneruptive hazards on slopes, in channels and in ring plains. Furthermore, geomorphic hazards can also be severe on those volcanic mountains without eruptive activity.

Recent case studies, such as the protracted post-eruption crisis around Mt Pinatubo, indicate that, despite some success in forecasting events and in mitigating hazardous volcanic processes at individual volcanoes, the losses and damage still remain too high. Even in developed countries, damage increases with respect to the national GNP because of the increasing value of the resources at stake. Over the past 50 years, the risk to society from volcanic eruptions has increased sharply because of population growth, more developed and diversified economies, and a more technologically advanced infrastructure. Correct land-use planning is fundamental in minimizing both loss of life and damage to property. Remote-sensing technology, numerical models warning systems and GIS have emerged as the most promising tools to support the decision-making process.

People and decision-makers are increasingly facing problems of mitigating volcanic hazards, especially in large cities located near potentially destructive volcanic mountains, such as Naples and Mt Vesuvius, Puebla and Mt Popocatepetl, Arequipa and El Misti, Tacoma-Seattle and Mt Rainier (Chester et al., 2001). With the continuing rise of global urbanization, the management of volcanic hazards in urban and densely populated areas will present major challenges for scientists and disaster workers in this millennium. Policy decisions will require improvements in awareness of potential threats as well as timely and appropriate response.

Acknowledgements

I would like to thank Professor C.D. Ollier, Dr P.N. Owens and Professor O. Slaymaker for improving an early draft of this manuscript. Two referees, Drs D.K. Chester and T.C. Pierson, are also thanked for their constructive remarks.

References

Beget, J.E. and Kienle, J., 1992. Cyclic formation of debris avalanches at Mount St Augustine volcano. *Nature*, **356**: 701–704.

Blong, R., 1984. *Volcanic hazards – a sourcebook on the effects of eruptions*. Sydney: Academic Press.

Blong, R., 2000. Volcanic hazards and risk management. In Sigurdsson, H., Houghton, B., McNutt, S.R., Rymer, H. and Stix, J. (eds), *Encyclopedia of volcanoes*. San Diego, CA: Academic Press, 1215–27.

Carracedo, J.C., 1999. Growth, structure, instability and collapse of Canarian volcanoes and comparison with Hawaiian volcanoes. *Journal of Volcanology and Geothermal Research*, **94**: 1–19.

Cas, R.A.F. and Wright, J.V., 1987. *Volcanic successions, modern and ancient*. London: Unwin Hyman.

Chester, D.K., Degg, M., Duncan, A.M. and Guest, J.E., 2001. The increasing exposure of cities to the effects of volcanic eruptions: a global survey. *Global Environmental Change*, **2**: 89–103.

Chinen, T. and Kadomura, H., 1986. Post-eruption sediment budget of a small catchment on Mt. Usu, Hokkaido. *Zeitschrift für Geomorphologie, Supplementband*, **60**: 217–32.

Collins, B.D. and Dunne, T., 1986. Erosion of tephra from the 1980 eruption of Mount St. Helens. *Geological Society of America Bulletin*, **97**: 896–905.

Cronin, S.J., Neall, V.E., Lecointre, J.A. and Palmer, A.S., 1996. Unusual 'snow-slurry' lahars from Ruapehu volcano, New Zealand, September 1995. *Geology*, **24**: 1107–10.

Daag, A. and van Westen, C.J., 1996. Cartographic modelling of erosion in pyroclastic-flow deposits of Mount Pinatubo, Philippines. *ITC Journal*, **2**: 110–24.

Davidson, J. and de Silva, S., 2000. Composite volcanoes. In Sigurdsson, H., Houghton, B., McNutt, S.R., Rymer, H. and Stix, J. (eds), *Encyclopedia of volcanoes*. San Diego, CA: Academic Press, 663–81.

Evans, S.G., Hungr, O. and Clague, J.J., 2001. Dynamics of the 1984 rock avalanche and associated distal debris flow on Mount Cayley, British Columbia, Canada; implications for landslide hazard assessment on dissected volcanoes. *Engineering Geology*, **61**: 29–51.

Fournier d'Albe, E.M., 1979. Objectives of volcanic monitoring and prediction. *Journal of the Geological Society of London*, **136**: 321–26.

Francis, P., 1993. *Volcanoes – a planetary perspective*. Oxford: Oxford University Press.

Freundt, A., Wilson, C.J.N. and Carey, S.N., 2000. Ignimbrites and block-and-ash deposits. In Sigurdsson, H., Houghton, B., McNutt, S.R., Rymer, H. and Stix, J. (eds), *Encyclopedia of volcanoes*. San Diego CA: Academic Press, 582–99.

Hildreth, W. and Lanphere, M.A., 1994. Potassium-argon geochronology of a basalt–andesite–dacite system: the Mount Adams volcanic field, Cascade Range of southern Washington. *Geological Society of America Bulletin*, **106**: 1413–29.

Hooper, D.M. and Mattioli, G.S., 2001. Kinematic modeling of pyroclastic flows produced by gravitational dome collapse at Soufrière Hills volcano, Monserrat. *Natural Hazards*, **23**: 65–86.

Iverson, R.M., Schilling, S.P. and Vallance, J.W., 1998. Objective delineation of lahar-inundation hazard zones. *Geological Society of America Bulletin*, **110**: 972–84.

James, P.A., Chester, D.K. and Duncan, A.M., 2000. Soils of Quaternary volcanic lavas: some archaeological considerations. In McGuire, B., Griffiths, D. and Stewart, I. (eds), *The archaeology of geological catastrophes*. Geological Society Special Publication, 171. London: Geographical Society of London, 317–38.

Janda, R.J., Daag, A.S., Delos Reyes, P.J., Newhall, C.G., Pierson, T.C., Punongbayan, R.S., Rodolfo, K.S., Solidum, R.U. and Umbal, J.V., 1997. Assessment and response to lahar hazard around Mount Pinatubo, 1991 to 1993. In Newhall, C.G. and Punongbayan, R.S. (eds), *Fire and mud: eruptions and lahars of Mt Pinatubo, Philippines*. Washington, DC: University of Washington Press, 107–39.

Keating, B.H. and McGuire, W.J., 2000. Island edifice failures and associated tsunami hazards. *Pure and Applied Geophysics*, **157**: 899–955.

Lavigne, F., Thouret, J.-C. and Suwa, H., 2000. Lahars at Merapi volcano, Central Java: an overview. *Journal of Volcanology and Geothermal Research*, **100**: 423–56.

Lipman, P.W., 2000. Calderas. In Sigurdsson, H., Houghton, B., McNutt, S.R., Rymer, H. and Stix, J. (eds), *Encyclopedia of volcanoes*. San Diego, CA: Academic Press, 643–62.

Major, J.J., Janda, R.J. and Daag, A.S., 1997. Watershed disturbance and lahars on the east side of Mount Pinatubo during the mid-June 1991 eruptions. In Newhall, C.G. and Punongbayan, R.S. (eds), *Fire and mud: eruptions and lahars of Mt Pinatubo, Philippines*. Seattle, WA: University of Washington Press, 895–918.

Major, J.J., Pierson, T.C., Dinehart, R.L. and Costa, J.E., 2000. Sediment yield following severe disturbance – a two decade perspective from Mount St. Helens. *Geology*, **28**: 819–22.

Manville, V., White, J.D.L., Houghton, B.F. and Wilson, C.J.N., 1999. Paleohydrology and sedimentology of a post-1.8 ka breakout flood from intracaldera Lake Taupo, North Island, New Zealand. *Geological Society of America Bulletin*, **111**: 1435–47.

McGuire, W.J., Jones, A.P. and Neuberg, J. (eds), 1996. *Volcano instability on the Earth and other planets*. Geological Society Special Publication 110. London: Geographical Society of London.

Moore, J.G. and Clague, D.A., 1992. Volcano growth and evolution of the island of Hawaii. *Geological Society of America Bulletin*, **104**: 1471–84.

Moore, J.G., Normak, W.R. and Holcomb, R.T., 1994. Giant Hawaiian underwater landslides. *Science*, **264**: 46–47.

Myers, B., Brantley, S.R., Stauffer, P. and Hendley, J.W., 1998. What are volcano hazards? *U.S. Geological Survey – Reducing the risk from volcano hazards*. USGS Fact Sheet 002-97. Washington, DC: US Geological Survey.

Nakada, S., 2000. Hazards from pyroclastic flows and surges. In Sigurdsson, H., Houghton, B., McNutt, S.R., Rymer, H. and Stix, J. (eds), *Encyclopedia of volcanoes*. San Diego, CA: Academic Press, 945–55.

Neall, V.E., 1996. Hydrological disasters associated with volcanoes. In Singh, V.P. (ed.), *Hydrology of disasters*. Dordrecht: Kluwer Academic Publishers, 395–425.

Ollier, C.D., 1988. *Volcanoes*. London: Blackwell.

Pareschi, M.T., Cavarra, L., Favalli, M., Giannini, F. and Meriggi, A., 2000. GIS and volcanic risk management. *Natural Hazards*, **21**: 361–79.

Pierson, T.C. (ed.), 1999. *Hydrologic consequences of hot-rock/snowpack interactions at Mount St. Helens volcano, Washington 1982–84*. US Geological Survey Professional Paper 1586. Washington, DC: Government Printing Office.

Pierson, T.C. and Costa, J.E., 1987. A rheologic classification of subaerial sediment–water flows. In Costa, J.E. and Wieczorek, G.E. (eds), Debris flows/avalanches: process, recognition, and mitigation. *Geological Society of America, Review of Engineering Geology*, **7**: 1–12.

Pierson, T.C. and Janda, R.J., 1994. Volcanic mixed avalanches: a disaster eruption-triggered mass-flow process at snow-clad volcanoes. *Geological Society of America Bulletin*, **106**: 1351–58.

Pierson, T.C., Janda, R.J., Thouret, J.-C. and Borrero, C.A., 1990. Perturbation and melting of snow and ice by the 13 November 1985 eruption of Nevado del Ruiz, Colombia, and consequent mobilization, flow, and deposition of lahars. *Journal of Volcanology and Geothermal Research*, **41**: 17–66.

Pierson, T.C., Janda, R.J., Umbal, J.V. and Daag, A.S., 1992. *Immediate and long-term hazards from lahars and excess sedimentation in rivers draining Mount Pinatubo, Philippines*. US Geological Survey Water-Resources Investigations Report 92-4039. Washington, DC: Government Printing Office.

Scott, K.M., 2000. Precipitation-triggered debris-flow at Casita volcano, Nicaragua: implications for mitigation strategies in volcanic and tectonically active steeplands. In Wieczorek, G.F. and Naeser, N.D. (eds), *Debris-flow hazard mitigation: mechanics, prediction, and assessment*. Rotterdam: Balkema, 3–13.

Segerstrom, K., 1950. *Erosion studies at Paricutin, state of Michoacan, Mexico*. US Geological Survey Bulletin, 965-A.

Sheridan, M., Hubbard, B., Bursik, M.I., Abrams, M., Siebe, C., Macias, J.L. and Delgado, H., 2001. Gauging short-term volcanic hazards at Popocatepetl. *EOS*, **185**: 187–88.

Siebert, L., 1996. Hazards of large volcanic debris avalanches and associated eruptive phenomena. In Scarpa, R. and Tilling, R.I. (eds), *Monitoring and mitigation of volcano hazards*. Berlin: Springer-Verlag, 541–658.

Slaymaker, O. (ed.), 1996. *Geomorphic hazards*. Chichester: Wiley.

Stoopes, G.R. and Sheridan, M.F., 1992. Giant debris avalanches from the Colima volcanic complex, Mexico: implication for long-runout landslides (>100 km) and hazard assessment. *Geology*, **20**: 299–302.

Suwa, H. and Sumaryono, A., 1996. Sediment discharge by storm runoff from a creek on Merapi volcano. *Journal of the Japan Society of Erosion Control Engineering*, **48**: 117–28.

Suzuki, T., 1977. Volcano types and their global population percentages. *Bulletin of the Volcanological Society of Japan*, **22**: 27–40.

Tanguy, J.-C., Ribière, Ch., Scarth, A. and Tjetjep, W.S., 1998. Victims from volcanic eruptions: a revised database. *Bulletin of Volcanology*, **60**: 137–44.

Thouret, J.-C., 1990. Effects of the 13 November 1985 eruption on the ice cap and snow pack of Nevado del Ruiz, Colombia. *Journal of Volcanology and Geothermal Research*, **41**: 177–201.

Thouret, J.-C., 1994. Méthodes de zonage des menaces et des risques volcaniques. In Bourdier, J.-L. (dir), *Le volcanisme*, Orléans: éditions B.R.G.M. Manuels et Méthodes, vol. 25, 267–83.

Thouret, J.-C., 1999. Volcanic geomorphology – an overview. *Earth-Science Reviews*, **47**: 95–131.

Thouret, J.-C., Lavigne, F., Kelfoun, K. and Bronto, S., 2000. Toward a revised assessment of volcanic hazards at Merapi, Central Java. *Journal of Volcanology and Geothermal Research*, **100**: 479–502.

Thouret, J.-C., Vandemeulebrouck, J., Komorowski, J.C. and Valla, F., 1995. Volcano–glacier interactions: field survey, remote sensing and modelling – a case study (Nevado del Ruiz, Colombia). In Slaymaker, O. (ed.), *Steepland geomorphology*, Chichester: Wiley, 63–88.

Tilling, R.I., 1989. Volcanic hazards and their mitigation: progress and problems. *Reviews of Geophysics*, **27**: 237–67.

Vallance, J.W., 2000. Lahars. In Sigurdsson, H., Houghton, B., McNutt, S.R., Rymer, H. and Stix, J. (eds), *Encyclopedia of volcanoes*. San Diego CA: Academic Press, 601–16.

Vallance, J.W. and Scott, K.M., 1997. The Osceola mudflow from Mount Rainier: sedimentology and hazard implications of a huge clay-rich debris flow. *Geological Society of America Bulletin*, **109**: 143–63.

van Wyk de Vries, B. and Francis, P.W., 1997. Catastrophic collapse at stratovolcanoes induced by gradual volcano spreading. *Nature*, **387**: 387–90.

van Wyk de Vries, B., Self, S., Francis, P.W. and Keszthelyi, K., 2001. A gravitational spreading origin for the Socompa debris avalanche. *Journal of Volcanology and Geothermal Research*, **105**: 225–47.

Voight, B., 1990. The 1985 Nevado del Ruiz volcano catastrophe: anatomy and retrospection. *Journal of Volcanology and Geothermal Research*, **42**: 151–88.

Voight, B. and Sousa, J., 1994. Lessons from Ontake-san: a comparative analysis of debris avalanche dynamics. *Engineering Geology*, **38**: 261–97.

Wadge, G., Jackson, P., Bower, S.M., Woods, A.W. and Calder, E., 1998. Computer simulation of pyroclastic flows from dome collapse. *Geophysical Research Letters*, **25**: 3677–80.

Waitt, R.B., Gardner, C.A., Pierson, T.C., Major, J.J. and Neal, C.A., 1994. Unusual ice diamicts emplaced during the December 15, 1989 eruption of Redoubt volcano, Alaska. *Journal of Volcanology and Geothermal Research*, **62**: 409–28.

Waldron, H.H., 1967. *Debris flow and erosion control problems caused by the ash eruptions of Irazu volcano, Costa Rica.* US Geological Survey Bulletin, 1241-I.

White, J.D.L., Houghton, B.F., Hodgson, K.A. and Wilson, C.J.N., 1997. Delayed sedimentary response to the AD 1886 eruption of Tarawera, New Zealand. *Geology*, **25**: 459–62.

PART 5 Mountain Geomorphology and Global Environmental Change

Guilin, Guangxi Province, China
Photo: P.N. Owens

12
Mountain geomorphology and global environmental change

Olav Slaymaker and Philip N. Owens

1 Introduction

The purpose of this chapter is (a) to situate mountain geomorphology in the context of global environmental change, and to show (b) that mountains are drivers of global environmental change over longer geological timescales, (c) that mountains regulate the rhythm of geomorphic work over a Quaternary timescale, (d) that mountain environments are sensitive in their response to climatic and anthropogenic changes over Late Glacial, Holocene and contemporary timescales, (e) that the reflexive relation between human activity and geomorphic processes in mountains raises the question of how to deal with complexity, and (f) that monitoring of socio-economic and geomorphic processes in mountains will be needed to provide an early warning of future environmental change. Implicit in this requirement is a need for flexible institutional mechanisms that transcend political boundaries. Global environmental change is a constant feature of Earth history. It continues today and will be even more complex in the future as the twin driving forces of tectonics and climate are reinforced by the increasingly pervasive role of human activity (Slaymaker, 1991; Slaymaker and Spencer, 1998).

2 Global environmental change, mountains and the question of scale

2.1 A definition of global environmental change

Global environmental change is commonly, but not universally, distinguished from 'global change' and 'global climate change'. The concept is narrower than global change in the sense that much of the global change literature emphasizes the cultural, economic and geopolitical global processes as the drivers of change and considers the biogeochemical processes to be passive participants (Slaymaker, 2000). The concept is broader than global climate change, which excludes most of the biogeochemical processes and their modification by human agency. Specifically, global environmental change as used consistently in this chapter emphasizes the biogeochemical global processes over longer timescales (i.e., including the pre-Quaternary and Quaternary) and the interactions of those processes with human activity over more recent timescales. The adaptive cycle (Holling, 1986) is a model drawn from ecology that attempts to integrate socio-economic and ecologic change, and will be described below in terms of its relevance to global environmental change. The concepts of reflexivity and feedback mechanisms are also central to global environmental change over all temporal and spatial scales.

2.2 Spatial scale

Turner et al. (1990) have made the helpful distinction between 'systemic' and 'cumulative' ways of studying global problems. They define the systemic approach as one which interrogates the whole

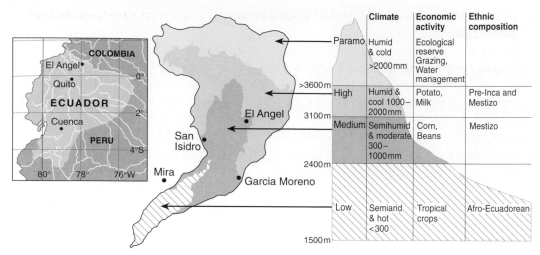

Figure 12.1 *Distinctive altitudinal zones in the El Angel watershed, Ecuador, and the relation of land degradation to global environmental change*
Source: *modified from Trottier (2001).*

global system, whether anthropospheric, atmospheric, biospheric, hydrospheric or lithospheric. Included in this is the global process of globalization, which views all global change as taking place within a global system of political economy and in which the environment plays a passive role. The mountain geomorphologist is unlikely to subscribe to this view, but cannot afford to ignore well-documented ways in which political economy strongly influences the environment (Rajaee, 2000). The systemic approach is characterized by the use of remote-sensing technology, global modelling techniques and global networks of information flow. The cumulative approach identifies local and regional processes whose cumulative impacts have become globally pervasive. These processes include land-cover and land-use changes that are sufficiently widespread to affect global energy and mass balances (Land Use and Cover Change (LUCC) Scientific Steering Committee, 1999). One example of such a process is that of land degradation in the Ecuadorean Andes (Figure 12.1). Deforestation, erosion, water use and management, loss of biodiversity, agricultural expansion towards the paramo, agricultural intensification, and impacts of pesticides on human health are seven of the nine priority issues identified at the local scale; in the generic form of land degradation, these constitute a global issue because of their global pervasiveness and their interruption of the global mass balance.

2.3 Temporal scale

For temporal scales, it is convenient to think in terms of pre-Quaternary, Quaternary, Holocene and contemporary (defined here to include the last few hundred years) timescales. The status of explanatory variables changes from one timescale to another (Schumm and Lichty, 1965). At pre-Quaternary timescales, mountains and mountain building drive global environmental change (Summerfield, 2000); at Quaternary timescales, mountains control the rhythm of geomorphic work, especially in so far as they are ice-sheet covered, glacierized or unglaciated (e.g., Hallet *et al.*, 1996); at Holocene timescales, mountain environments provide sensitive records of global climate change (e.g., Leonard, 1997); and at contemporary timescales, mountains provide some of the most detailed accounts of land-use and land-cover changes superimposed on climate change (e.g., Chang and Slaymaker, 2002).

3 Mountains as drivers of global environmental change at pre-Quaternary timescales

There are two somewhat distinct approaches at the pre-Quaternary timescale, both of which are represented in this volume. There are those who take a global systemic view, and demonstrate a direct relation between plate tectonic movements and mountains in active plate margins (Summerfield, 2000; Owen, this volume; Williams, this volume). There are also those who begin from regional morphological and stratigraphic evidence, reason from geomorphic theory and define cumulative impacts which are globally consistent (Ollier and Pain, 2000; Ollier, this volume). The latter approach does not, of course, neglect global plate movements, but it attaches greater significance to regional morphology as observed on the ground than to plate tectonic theory in those places where contradictions appear. Pre-Quaternary global environmental change is thought to be a result of some combination of magmatic processes, continental drift and tectonic uplift, in addition to astrophysical forces, which are not considered here. The first three sets of processes are all a function of global plate tectonics but can be considered in the following ways.

3.1 Magmatic processes

A decrease in the rate of ocean ridge formation, for example, can favour ice accumulation at high latitudes. This results from a fall in global sea level and increasing seasonality of climates. A global reduction in volcanism can produce global cooling by lowering the carbon dioxide concentration of the atmosphere. An increase in volcanism in northern latitudes can be a triggering mechanism to tip the climate across a glacial threshold (Bray, 1977).

3.2 Continental drift

The high latitudes are especially sensitive to the astrophysical forces summarized by the Milankovitch cycles (Milankovitch, 1920). If poleward heat transport by ocean currents is disrupted by changing latitudinal distribution of continents, massive environmental change can result (Ruddiman and Raymo, 1988).

3.3 Tectonic uplift

The rise and growth of the Tibetan Plateau, for example, can change global circulation patterns, increase weathering rates and thereby reduce carbon dioxide levels in the atmosphere (Molnar and England, 1990). Uplifted coastal areas around the North Pacific and the North Atlantic can produce a glacial triggering mechanism. In all the above examples, the changing distribution of land masses and mountains is a driver of global environmental change at pre-Quaternary timescales; climate is a dependent variable (Figure 12.2).

4 Global environmental change at the Quaternary timescale

4.1 Quaternary glacial cycles

Quaternary glacial/interglacial oscillations are thought to be a response to the solid geometry of the Earth, its rotation on its axis and its revolution around the Sun (Milankovitch, 1920). Variations in the eccentricity of the Earth's orbit (100 ka periodicity), the obliquity of its rotational axis (41 ka) and precessional changes (23 and 19 ka) generate regular periodicities of change. Typically, the transition from glacial to interglacial stage is abrupt (exactly how abrupt is not known) and the transition from interglacial to glacial stage seems more gradual, via a series of stades with progressively colder conditions (Petit et al., 1990).

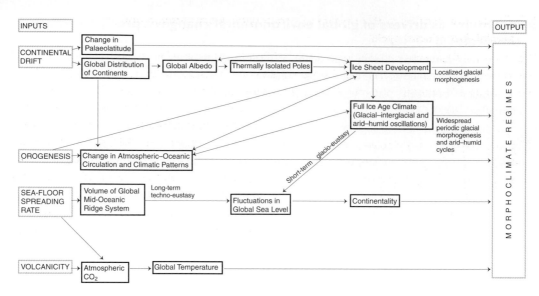

Figure 12.2 *Tectonic processes and morphoclimatic regimes in mountains*
Source: *modified from Summerfield (1991).*

Unfortunately there is some disagreement in evidence generated by proxy data (CLIMAP Project Members, 1976, 1981; COHMAP Members, 1988) and the geomorphic evidence of former glacier extent and downward displacement of vegetation zones in tropical mountain areas (e.g., Bowler *et al.*, 1976; Porter, 1979). The terrestrial morphological evidence suggests temperature lowering of 5–6°C whereas the oceanic proxy data (e.g., planktonic foraminifera) suggest only a modest cooling no greater than 2°C at the glacial maximum. The modelling and proxy data are being continuously re-evaluated and, in some locations, have been shown to have a warm bias (Rind, 1988). The morphological evidence of glacial processes at more than 1 km below present glaciers remains unambiguous evidence of global environmental change, but the precise mix of temperature decrease and precipitation increase which caused this change is the object of continuing research.

4.2 Distinctive alpine glacial processes and landforms

Mountain geomorphology in the Quaternary period is dominated by the unique assemblage of alpine glacial processes, the intensity of which has alternated with glacial/interglacial stages and depends in detail on the glaciation limit at a particular time and place. The glaciation limit is itself a function of latitude, the presence of other mountains and the proximity of oceans (Ostrem, 1966, 1972). High northern latitudes, particularly those with mountainous terrain, are the sites that have triggered the development of ice sheets during the Quaternary period (Syvitski *et al.*, 1987). The spatial distribution of alpine glacial processes is dependent on the presence of mountains and, in turn, the intensity of geomorphic work is dependent on the location of the glaciation limit.

4.3 The most recent glacial–interglacial cycle

The most recent glacial–interglacial cycle occupies the last 125 000 years or so, and it is the cycle about which most detailed information is available (Figure 12.3(A)). In Europe, the last glacial maximum occurred around 21 000 years BP; elsewhere a figure of 18 000 years BP is commonly quoted (Ruddiman, 1984).

4.4 Conceptual models of the rhythm of geomorphic work through one glacial–interglacial cycle

The terrestrial Quaternary stratigraphic record of British Columbia, Canada, is largely a product of brief depositional events separated by long periods of nondeposition (Clague, 1986; Figure 12.3(B)). The paraglacial hypothesis of Ryder (1971) (elaborated by Church and Ryder, 1972; Church and Slaymaker, 1989) derives from nonglacial evidence immediately following the last glaciation in British Columbia. Although the climate was constantly changing throughout the Quaternary, mountain geomorphic processes have been conceptualized as most closely approximating a punctuated equilibrium/disequilibrium model (Clague, 1986). Long periods of comparative quiescence have alternated with shorter periods of intense geomorphic activity. This is not dissimilar to the geologic norm for all erosion and sedimentation (Gage, 1970; Ager, 1981). Schumm's complex response model emphasizes erosion following threshold exceedances at specific locations, alongside slopes and basins that are inactive (Schumm, 1973). Starkel (1987) summarized changes in geomorphological and soil systems through one glacial–interglacial cycle (Figure 12.3(C)).

At the height of each glacial stage, mountain environments were ice-sheet covered, glacierized or unglaciated. Assumptions of the present model (Figure 12.3(D)) are that at the height of each glacial stage, when continent-wide ice sheets prevailed, alpine glaciation was minimized because glaciers were frozen to their beds. Likewise, towards the end of each interglacial stage, alpine glaciers were at their most restricted and paraglacial effects were minimized. However, during glacier growth (deposition of advance outwash) and glacier recession (deposition of recessional outwash) and at early stages of the interglacials, when paraglacial environments were most active (Church and Slaymaker, 1989), it is hypothesized that geomorphic activity was at its most intense.

There are several contested assumptions in this conceptual model and it cannot be assumed that the model is universally applicable. It is only within the last decade that time-dependent

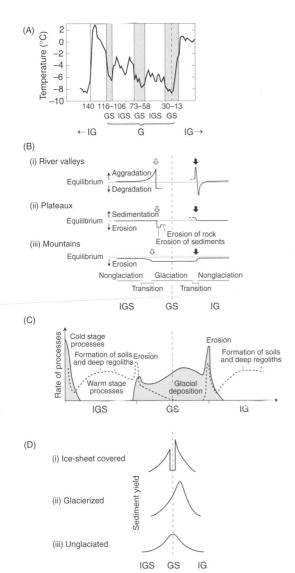

Figure 12.3 Rhythm of glacial–interglacial cycles and geomorphic work. (A) The most recent glacial–interglacial cycle; (B) summary of patterns of sedimentation and erosion in British Columbia during the Quaternary period (after Clague, 1986); (C) changes in geomorphological and soil systems during a glacial–interglacial cycle (after Starkel, 1987); and (D) conceptual model of the rhythm of geomorphic work during one glacial–interglacial cycle in (i) ice-sheet covered mountains; (ii) glacierized mountains and (iii) unglaciated mountains

fluxes of meltwater, surface runoff and icebergs have been incorporated into realistic models of the ice-sheet decay process (e.g., Marshall and Clarke, 1999) and only semi-quantitative data on sediment flux are available. Binge-purge models depend critically on alternations of cold-based and warm-based ice (MacAyeal, 1993), a sequence of events most likely associated with general ice-sheet retreat conditions. A similar set of assumptions has to do with the coincidence in timing of glacial episodes in high latitudes and pluvial (or wet) conditions in lower latitudes. Although this has been shown to be the case for the American Southwest (Smith and Street-Perrott, 1983), in most documented cases in the tropics the last wet period occurred during terminal Pleistocene and early Holocene times. The conceptual model then should be understood with these assumptions clearly in mind.

If this general model is confirmed, then it follows that patterns of Quaternary geomorphic activity will differ depending on whether mountains were ice-sheet covered (geomorphic work almost ceasing during ice-sheet cover), whether they were simply glacierized (remaining geomorphically active throughout the glacial stage), or whether they were unglaciated throughout the Quaternary (in which case they responded primarily to increased precipitation intensity during the pluvial stages) (Figure 12.3(D)).

5 Mountains as sensitive indicators of global environmental change at Late Glacial, Holocene and contemporary timescales

5.1 Definitions of landscape resistance and vulnerability

The landscape resistance concept of Brunsden (Brunsden and Thornes, 1979; Brunsden, 1980, 1993) invokes five components that resist the driving forces of geomorphic change: strength resistance (bedrock and surficial materials); morphological resistance (first, second and third derivatives of altitude with respect to horizontal distance); system state resistance (patterns of historical disturbance); structural resistance (slope-channel coupling, connectivity of channels); and filter resistance (storage elements such as floodplains, terraces and fans). From the ecological literature, we have the definition of vulnerability as a function of both sensitivity and resilience, where sensitivity is the rate and magnitude of response to system disturbance, and resilience is the rate at which an ecosystem recovers from perturbation (Westman, 1978; Barrow, 1995). Bringing these two concepts of landscape resistance and vulnerability together in the mountain context has still to be achieved in detail, but the general point to be made here is that mountain environments are the most variable environments at the Earth's surface in both resistance and vulnerability to global environmental change. This extreme spatial variability offers the opportunity to identify driving forces that are both systemic and cumulative (Slaymaker, 1996). The following kinds of evidence, drawn specifically from mountain environments, illustrate the application of these concepts.

5.2 Varieties of evidence of global environmental change in mountains

The magnitude of past changes in the Earth's natural environments was first appreciated from the scale of apparent oscillation of European glaciers, as evidenced by the presence of glacial erratics at considerable distances from present ice sheets and glaciers (Grove, 1988). Rapidly thereafter, stratigraphic, lithological and morphological studies allowed the determination of sequences of regional environmental changes. It was a long time, however, before this information could be placed in the context of global change because of the lack of suitable dating devices and the spatial discontinuities in the terrestrial evidence (Bradley and Jones, 1992). Today, the Past Global Changes (PAGES) Project of the International Geosphere–Biosphere Programme (IGBP) is engaged in reconstructing (a) a history of climatic and environmental change through a full glacial cycle to improve our understanding

Table 12.1 – Varieties of evidence of global environmental change in mountain regions

Temporal scale (years)	Evidence	Inference
>10^6	Glacial deposits and landforms	Temperature
	Periglacial deposits and landforms	Temperature
	Relict soils and palaeosols	Precipitation/temperature
	Speleothems	Temperature/groundwater
10^4–10^6	Alluvial stratigraphy	Sediment transport
	Alluvial terraces and fans	Base level/climate/sediment
	Lake sediment	Sediment supply/water chemistry
	Colluvial fans and cones	Sediment supply/surface erosion
	Evaporites and tufas	Temperature/moisture
	Tephra	Dated events
	Pollen	Temperature/ecology
	Tree rings	Aridity, temperature
10^2–10^4	Mountain glaciers	Mass balance
	Layered ice cores	Temperature
	Permafrost	Temperature
	Lake sediment	Sediment supply, water chemistry
	Closed basin lakes	Water balance, E–T
	River planform	Discharge, sediment supply
	Water table	Water balance
	Peats	Water balance
	Historical records	Moisture, temperature, habitability, technology
	Monitored and remotely sensed data	Snow, ice, temperature, soil moisture

of global climate change, and (b) a detailed history of climatic and environmental change for the entire globe for the period since 2000 years BP, with a temporal resolution that is at least decadal and, ideally, annual or seasonal (IGBP, 1990).

It is convenient to identify: Late Glacial evidence, derived primarily from landforms and glacial sediments; Holocene evidence, derived largely but not exclusively from layered sediments, tree rings, tephra, pollen and historical records; and contemporary evidence, derived from the behaviour of glaciers, permafrost, and treelines, and from monitored and remotely sensed data (Table 12.1). The evidence associated with each of these three periods is discussed in more detail in Sections 5.3–5.5.

5.3 Late Glacial evidence of global environmental change

The presence of glacial landforms where glaciers no longer exist is *prima facie* evidence of global environmental change. The presence of glacial cirques is commonly used to infer the position of former snowlines, which are themselves climatically controlled. The limits of past glaciations and the extent of temperature depression at specific sites are established in this way. Yet spatial discontinuity remains a problem. Ice-sheet dynamics can be inferred from striations, flow-lines, till and erratic distribution, but how many times the evidence has been erased is normally impossible to resolve (Benn and Evans, 1998). Fossil pingos, palsas, ice wedge casts, and sorted and unsorted polygons provide evidence of significantly reduced mean annual temperatures in the past (French, 1996). Carbonate-rich sediments, which are precipitated regularly as a function of changing temperature and pressure conditions (speleothems), preserve ancient environmental information. Ford (1987) has interpreted the history of environmental change in the southern Rocky Mountains of British Columbia and Alberta via speleothem analysis. Similarly, soils formed under past environmental conditions retain features such as gleying, carbonate accumulation, varying degrees of leaching and frost structures which can be used to measure environmental change.

5.4 Holocene evidence of global environmental change

Lake basins are natural sediment traps and frequently contain a history of deposition spanning thousands of years (Oldfield, 1977). In particular, the nature of the sediments themselves provides substantial palaeoenvironmental information (Souch and Slaymaker, 1986; Owens and Slaymaker, 1994; Desloges and Gilbert, 1998). Paraglacial sediments (Ryder, 1971) have been increasingly recognized as important archives of information about a wide variety of past environmental changes (Ballantyne, 2002), especially with respect to interpreting the duration of system recovery time after glacial recession (Church and Slaymaker, 1989). Floristic assemblages during interglacials and interstadials are enriched. Hence the pollen archive is more useful for interpreting warmer environments than glacial episodes. The underlying principle of interpretation is that some sediments contain pollen grains and spores that mostly come from the air by fall-out and are representative of both regional and local vegetation. In order to achieve palaeoclimatic reconstruction, one or more of the following approaches is used: (a) identification of climatic indicator species; (b) reconstruction of ecological communities, whose present distribution is clearly related to climatic parameters; (c) measurement of climatically linked growth functions; and (d) development of numerical transfer functions linked to climatic parameters controlling its present-day distribution. Land-cover changes are often interpreted through pollen analysis, fossil plant assemblages and palaeoecology. An instructive mountain environment example is that of Japan where the island of Hokkaido was tundra and the island of Honshu was covered with conifer forest at Late Glacial times. Today Hokkaido has boreal conifers and Honshu is dominated by deciduous and evergreen broadleaf forest (Dickinson, 1995). Sensitive tree-ring analysis (as opposed to complacent tree rings) has been exploited to great effect by Luckman (1996) in the Canadian Cordillera. His work on spruce trees at treeline localities records four centuries of temperature change (Figure 12.4).

Changing land use in response to global environmental change has been well explored in the archaeological literature. The main point to note is that people were active in environmental change on most landmasses from the very start of the Holocene. By 11000 years BP, wild grains had reached the Anatolian uplands and Clovis immigrants to the Americas had spread from Alaska to Patagonia (Dickinson, 1995). Deliberate cultivation of wheat and barley had begun in the Zagros Mountains by 10000 years BP and people were tending gardens in the New Guinea Highlands by 9000 years BP. For historical data relevant to the late Holocene, the reader is recommended to consult Grove (1988), Bradley and Jones (1992) and Lamb (1995).

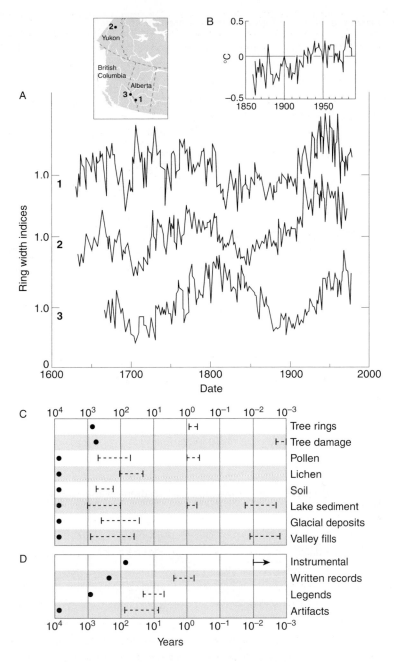

Figure 12.4 *Tree-ring chronologies for spruce trees at treeline. (A) Tree-ring chronolgies for spruce trees at tree-line localities in western Canada. (B) Standardized mean annual Nothern Hemisphere temperatures 1861–1984. (C) Biological and geographical attributes and time scales used for environmental reconstruction. (D) Historical and archaeological sources used for reconstruction.*
Source: *from Luckman (1990) and reprinted from Slaymaker, O. and French, H.M., 1993, Cold environments and global change. In French, H.M. and Slaymaker, O. (eds), Canada's cold environments.* Reproduced with permission of McGill-Queens University Press, Montreal.

5.5 Contemporary evidence of global environmental change

Many components of cryospheric systems (including snow, ice, glaciers and mountain permafrost) are particularly sensitive to changing environmental conditions because of their thermal proximity to melting conditions. The advance and retreat of mountain glaciers respond to changes in their mass balance, which in turn responds to climatic fluctuations. Historical changes are well documented, especially those associated with the Little Ice Age (Luckman and Kavanagh, 2000). In the Coast Mountains of British Columbia, for example, 100 m high Neoglacial moraines, left stranded by ice retreat of more than 1 km in the last century, are not uncommon (Figure 12.5).

A relatively unexploited archive of contemporary environmental information is that of the temperature gradient within permafrost (Figure 12.6). Near-surface temperature warming of 1.5°C (±0.5°C) has occurred during the twentieth century. Information on maximum and minimum surface temperatures from several hundred years past is contained within 8–10 m of the surface (Deming, 1995). Discontinuous mountain permafrost exists at ground temperature just below 0°C (Haeberli and Beniston, 1998). A slight increase in the mean annual temperature leads to ice melting and this may destabilize the slope. Active research in the Swiss and Austrian Alps, in Mexico and in the Bolivian Andes is attempting to detect evidence of slight temperature shifts using this kind of evidence (Barsch, 1996; see also Section 7.1). Monitored and remotely sensed data from orbiting satellites are becoming increasingly available and, where the same site is sensed repeatedly over a relatively short period, data on seasonal changes in snow and ice cover, biomass and hydrologic conditions are readily available. More elusive is the establishing of trends over decades because the variability of the monitored parameters is high. There are also technical difficulties with determining snow and ice volume changes, as opposed to areal change, but significant developments are being made (Fitzharris, 1996).

Anticipated climate change following a doubling of CO_2 concentration in the atmosphere indicates even greater complexity of mountain environment response to change. Beniston (2000) quotes the results of simulating a 1.5°C mean temperature increase and shows the global spatial patterns of temperature and precipitation changes that result (Figures 12.7 and 12.8). Although the reliability of these simulations is quite speculative in detail, they serve to illustrate the highly variable spatial response to global warming. This spatial variation shows no clear relation to the spatial pattern of mountains.

Figure 12.5 *Neoglacial moraines, British Columbia, Canada*
Photo: P.N. Owens.

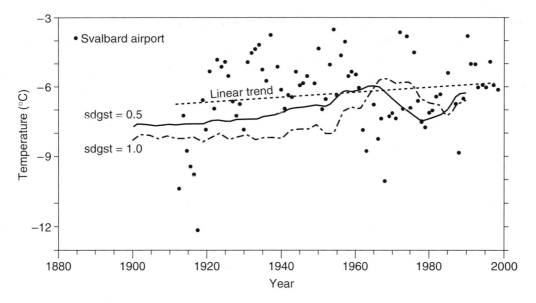

Figure 12.6 *Svalbard deep borehole data. The solid line and the dot-dash broken line are estimates of ground surface temperature based on two different cases: sdgst = 1, which assumes a normal variation of the standard deviation of the T-2 data and a thermal diffusivity of 0.9 × 10⁻⁶ m² s⁻¹ and sdgst = 0.5, assuming a narrow variation of the T-2 data and a thermal diffusivity of 0.5 × 10⁻⁶ m²s⁻¹ (where T-2 data refers to variation of temperature with depth)*
Source: *reprinted from Isaksen, K., Holmlund, P., Sollid, J.L. and Harris, C., Three deep alpine permafrost boreholes in Svalbard and Scandinavia.* Permafrost and Periglacial Processes, *vol. 12, pp. 13–25. © 2001. John Wiley and Sons Limited, Chichester. Reproduced with permission.*

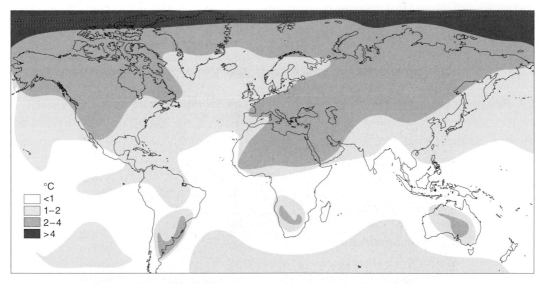

Figure 12.7 *General circulation model scenario of global warming for doubled CO_2 concentrations compared with pre-industrial levels*
Source: *after Beniston (2000), reprinted by permission of Hodder Arnold.*

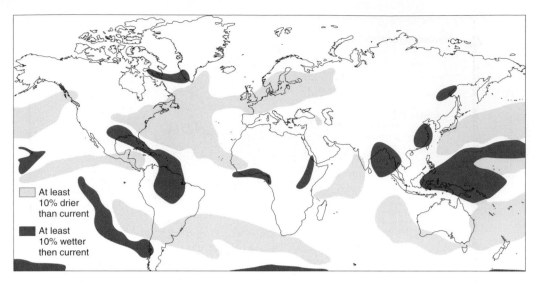

Figure 12.8 *General circulation model scenario of precipitation change for doubled CO_2 concentrations compared with pre-industrial levels*
Source: *after Beniston (2000), reprinted by permission of Hodder Arnold.*

Hence, anticipated climate change will vary between and within mountain ranges. The geomorphic response, which is also temporally lagged with respect to a climatic perturbation, becomes doubly difficult to predict (cf. Slaymaker, 1990).

Land-use changes in mountain regions today are more evident than any climatic changes (Slaymaker and McPherson, 1977; Slaymaker, 2001) and Deforestation and biomass burning (in central America, Papua New Guinea and the East African Highlands) and soil degradation (in the Andes, Himalayas, central America, China and the USA) are a result of intensifying mountain agriculture, where indigenous populations are fighting for survival (Beniston, 1994; see also Owens and Slaymaker, this volume). Tourism, hydropower development and mining also contribute locally and cumulatively to global environmental change (Figure 12.9).

6 Anthropogenic processes in mountains and the problem of complexity

6.1 Revisiting landscape resistance, vulnerability and community resilience

When anthropogenic processes are added to the biogeochemical processes, the assessment of vulnerability becomes more complex than that discussed in Section 5.1 (Liverman, 1994). Brunsden's (1993) framework relies on 'system state resistance' to take care of the human impact. We will come back to this but for the moment it is helpful to look to the concept of an adaptive cycle (Holling, 1986) which is viewed by many as the most successful way of looking at complexity and which tries to give equal prominence to ecological and socio-economic factors.

6.2 The concept of an adaptive cycle

Gunderson and Holling (2002) propose the term 'panarchy' for a structure in which systems of Nature and People as well as combined People–Nature systems and socio-ecological systems are interlinked in never-ending adaptive cycles of growth, accumulation, restructuring and renewal. These transformational cycles occur in nested sets at scales ranging from a leaf to the biosphere over periods

ENVIRONMENT

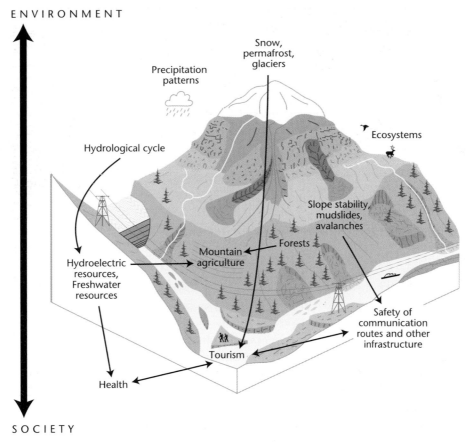

Figure 12.9 *Links between global environmental change and socio-economic activity in mountains*
Source: *modified from Beniston (2000).*

from days to geological epochs, and from the scale of a family to a socio-political region over periods from years to centuries. The adaptive cycle has three properties: wealth (potential for change), sensitivity (controllability) and adaptive capacity (resilience) (Holling, 2001). If we can understand these cycles and their associated scales we should be able to identify points at which a system is vulnerable and points at which a system is able to absorb positive change. Although this discussion has no special relevance to mountains, the strong reflexivity between ecological and socio-economic systems in mountains makes it important to include this framework for consideration.

6.3 Constraint and vulnerability matrices

In a less holistic framework than that above we can examine community resilience to disturbance or global environmental change in two stages: (a) both socio-economic and ecological constraints offer resistance to change – they are essentially a wealth parameter; and (b) both socio-economic and ecological vulnerability to change are a combination of sensitivity and resilience – this parameter indicates how well a community can absorb change. Mountain regions outside the Tropics have severe biogeochemical constraints for development by comparison with lowland regions. Simply put, they have poor market access and low potential agricultural productivity, because of their poor biogeochemical endowment. They also tend to have high vulnerability (Khor, 1997) in the face of global change. They are at the mercy of globalization processes over which they have no control and they experience global warming or desertification with limited adaptive capacity. We can apply this framework to many extra-tropical mountain regions such as Tibet, the Caucasus region and the Pyrenees. Within the Tropics, where mountain environments often have the highest agricultural productivity and good *local* market access, development potential is more variable regionally within mountain zones. For example, in the East African Highlands of Burundi, Ethiopia, Kenya, Rwanda, northern Tanzania and Uganda, a mountain region of 90 million people, there is a wide variation in constraints and vulnerability to change. Parts of Ethiopia close to Addis Ababa, the central and western Kenyan Highlands, the Lake Victoria crescent and much of the Highlands of Uganda are the most favoured areas (Pender and Hazell, 2000). They have fewer constraints and less vulnerability to change than the rest of the region. By contrast, those areas with low agricultural potential, poor access to local markets and high vulnerability to change would probably benefit from investment in human capital to permit migration.

6.4 *Environmental security as a component of human security*

Human security is a political concept that traditionally refers to threats to human lives and livelihoods in time of conflict or war (Department of Foreign Affairs and International Trade, 1999). Since the end of the Cold War there has been pressure from academics and others to expand this definition to encompass a broader range of threats to human well-being, including that of the environment and global environmental change (Lonergan, 1999). Case studies suggest that introducing environmental security into the debate on human security can provide a new imperative for environmental policy. In the Coast Temperate Forest Biome of British Columbia, the case of Clayoquot Sound represented a paradigm shift from environmental policy to the economic and human security portfolios (Clayoquot Sound Scientific Panel, 1995). Policy reform thereby became a higher priority (Kasperson and Kasperson, 2001). The case of world food security can also be seen to be closely related to questions of global environmental change and, more specifically, global climate change (Downing, 1995).

6.5 *Varieties of land use in mountain regions*

Increasing pressure on the land, the increasing desire for better living conditions, higher standards of living and the search for land to inhabit have resulted in human interventions that have caused land to degrade. The following types of causative factors are implicated: (a) deforestation or natural vegetation removal (Slaymaker, 2000; Figure 12.10; Figure 1.7 of Owens and Slaymaker, this volume); (b) overgrazing; (c) agricultural activities; (d) overexploitation of vegetation for domestic use; and (e) bio-industrial and industrial activity (including mining) (Oldeman *et al.*, 1991). To this list we could add urbanization and infrastructure development. Vitousek *et al.* (1997) made the case that human alteration of the Earth is substantial and growing. Between one-third and one-half of the land surface has been transformed by human action. More than half of all accessible freshwater is put to use by humanity and about one-third of the bird species on Earth have been driven to extinction. In light of the magnitude of these human impacts, Lubchenco (1998) has directed attention to the intimate connections between ecological systems and human health, the economy, social justice and national security. Our planet has become a human-dominated planet and there is urgency in unravelling these complex interconnections at cumulative as well as systemic scale (Roots, 2000). We do not know what is the relative contribution of cumulative compared with systemic processes in global environmental change. The expertise of the mountain geomorphologist is geared towards cumulative processes. The transformation of the land, the exploitation of freshwater and the accelerated nitrogen fixation by human action are all examples of important themes in geomorphology: themes of relief development, variable runoff sources, and biogeochemical and rock cycles.

Figure 12.10 *Clearcut logging in British Columbia, Canada*
Photo: P.N. Owens.

7 Monitoring and institutional mechanisms needed in order to prepare for future global environmental change in mountains

7.1 What to monitor

In mountain regions, geomorphic, cryospheric, climatic, hydrologic, ecological and socio-economic conditions change rapidly over relatively short vertical and horizontal distances (see Figure 1.1 of Owens and Slaymaker, this volume). It would be important to stratify any monitoring programme between (a) mountain environments not directly affected by human activity and (b) anthropogenically influenced mountain environments. Under (a) biodiversity tends to be high, and characteristic sequences of geo-ecosystems are found along mountain slopes. The boundaries between these systems (e.g., ecotones, snowlines and glacier boundaries) are the first to be affected by environmental changes and should be used as indicators of change (Troll, 1973). Vanishing glaciers, degrading permafrost, changing magnitude and frequency of geomorphic events, water quantity and quality changes, and adjustments in the position of treeline, snowline and glaciation limit are all responses to climate change. Unfortunately, critical information on many of these parameters is lacking. A number of scientific initiatives, of which the Global Change and Mountain Regions Research Initiative is one (Becker and Bugmann, 2001), have been developed. In order to understand and predict the effects of recent and future global environmental changes it will be necessary to: (a) assemble information on contemporary processes and their response to change and to set up comprehensive monitoring programmes; (b) assemble information on responses to past environmental changes over periods of time long enough to record land-use/cover and climatic changes; and (c) develop coupled models sufficient to cope with multiple, complex components and which can use the information provided by (a) and (b) above (Becker and Bugmann, 2001). Contemporary monitoring programmes need to be conducted along altitudinal and latitudinal gradients. The former are needed to assemble data on geomorphic responses in different altitudinal zones, while the latter are needed to provide information on geomorphic responses in different mountain types (e.g., arctic, temperate, arid, tropical). In addition, geomorphic responses to changes in land cover, use and management are likely to be different from those resulting from climate change. There will therefore be a need for nested studies at scales ranging from plot scale through to hillslope and catchment scales. To the authors' knowledge, the first appropriately scaled approach to the question of variations in sediment yield as an indicator of environmental change was carried out in Canada (Church et al., 1999). This latter study demonstrates that specific sediment yield is scale dependent (nothing new in this) and that the precise scale dependence differs from region to region. In five out of seven Canadian regions, specific sediment yield increases with area; in one region it decreases with basin area and one region shows no significant scale dependence. The practical implication of these findings is that comparisons of the effects of change (whether climate or land use induced) on specific sediment yield can only be made between basins of similar size.

As described in Section 5.5, many components of cryospheric systems are particularly sensitive to changing environmental conditions, especially ambient temperature. Thus many studies have documented changes in the extent of glaciers, changes in the temperature of mountain permafrost, changing thermokarst development, and changing slope stability leading to intensification of mountain natural hazards (e.g., Harris et al., 2001; see also Hewitt, this volume).

Table 12.2 lists some potential hazards associated with the degradation of mountain permafrost. The European Union Permafrost and Climate in Europe (PACE) Project is a good example of a long-term measurement, monitoring and modelling programme aimed at investigating the effect of changes

Table 12.2 – Matrix of potential hazards from mountain permafrost degradation

Slope class (degrees)	Bedrock			Sediment	
	Noncompetent lithologies (shales, soft mudstones, etc.)	Competent well-jointed lithologies	Competent massive lithologies	Fine-grained (silts, clays, some tills)	Coarse-grained (screes, gravels, sands)
>75	Rock fall	Rock fall	Occasional rock fall		
30–74	Debris flows and landslides (including deep-seated failures)	Rockslides, debris flows		Debris flows	Debris flows
15–29	Landslides, thaw subsidence	Rockslides		Landslides/mudflow	Accelerated permafrost creep (rock glaciers)
<15	Thaw settlement			Thaw subsidence, solifluction, mudslides on steeper slopes	Accelerated permafrost creep (rock glaciers)
0	Thaw settlement			Thaw settlement	

Source: from Harris, C., Haeberli, W., Muhll, D.V. and King, L., Permafrost monitoring in the high mountains of Europe: the PACE Project in its global context. Permafrost and Periglacial Processes, 12, 3–11, 2001. © John Wiley and Sons Limited. Reproduced with permission.

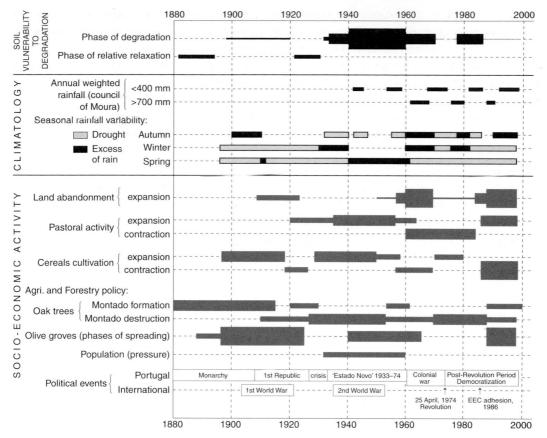

Figure 12.11 *Socio-economic and climatic context of soil degradation in Portugal*
Source: *modified from Brum Ferreira (2000).*

in climate on mountain permafrost degradation and associated geotechnical hazards (cf. Table 12.2). The programme consists of key permafrost monitoring sites in contrasting mountain environments in Europe running along a transect from Spain to Svalbard (Harris *et al.*, 2001).

In order to understand the response of geomorphic processes to environmental changes there is also a need to assemble proxy records of past changes. Such records are required for a number of reasons: (a) placing the contemporary measurements within a historical or longer temporal context so as to check on trends; (b) establishing environmental change–response relations and associated recovery times; and (c) providing the necessary long-term empirical information for models. Figure 12.6 illustrates results based on deep boreholes in permafrost on Svalbard as part of the PACE project. Where anthropogenic processes are significant, the question of what to monitor becomes immediately more complex (Strömquist *et al.*, 1998). One of the more interesting studies known to the authors is recorded by Brum Ferreira from the Iberian Mountains (Brum Ferreira, 2000). She attempts to plot the socio-economic and climatic context of the Inner Alentejo region, Portugal, from 1880 to 2000. She shows the necessity to incorporate political developments, population pressure, agricultural and forestry policy, climatic variables and estimates of soil vulnerability in order to explain the patterns of observed land degradation in this mountainous environment (Figure 12.11).

This example underlines the need for monitoring standard indicators that are also integrative and holistic. There is a burgeoning literature on the use of human development indicators in mountain regions (Price, 1999) and an equally lively debate on the need for parallel indicators of ecological and social conditions (Goudie, 1994; Prescott-Allen, 2001). As always, the debate centres on the validity of indicators deriving from general theories in natural science *vis-à-vis* indicators that more faithfully reflect individual decision-making. Given that indicators at the individual, or even at the individual household, scale do not exist for whole mountain regions, the search for sensitive and sensible surrogate variables is ongoing (Kreutzmann, 2001).

7.2 Mechanisms and institutions for overseeing global environmental change in mountain environments

In the later part of the twentieth century, increasing global attention was directed towards mountain regions. Chapter 13 of Agenda 21 (Table 12.3) was a landmark event in conjunction with the 1992 UN Conference on Environment and Development (Stone, 1992). In spite of this, there has been a

Table 12.3 – Summary research recommendations from Agenda 21, Chapter 13

Two research areas A and B

A. Generating and strengthening knowledge about the ecology and sustainable development of mountain ecosystems.

   a) Survey of soils, forest, water use, crop, plant and animal resources of mountain ecosystems.

   b) Maintain and generate database and information systems to facilitate the integrated management and environmental assessment of mountain ecosystems.

   c) Improve and build the existing land/water ecological knowledge base regarding technologies and agricultural and conservation practices in the mountain regions of the world, with the participation of local communities.

   d) Create and strengthen the communications network and information clearing-house for existing organizations concerned with mountain issues.

   e) Improve coordination of regional efforts to protect fragile mountain ecosystems through the consideration of appropriate mechanisms, including regional legal and other instruments.

   f) Generate information to establish databases and information systems to facilitate an evaluation of environmental risks and natural disasters in mountain ecosystems.

B. Promoting integrated watershed development and alternative livelihood opportunities.

   a) Develop appropriate land-use planning and management for both arable and nonarable land in mountain-fed watershed areas to prevent soil erosion, increase biomass production and maintain the ecological balance.

   b) Promote income-generating activities, such as sustainable tourism, fisheries and environmentally sound mining, and improve infrastructure and social services, in particular to protect livelihoods of local communities and indigenous people.

   c) Develop technical and institutional arrangements for affected countries to mitigate the effects of natural disasters through hazard-prevention measures, risk zoning, early-warning systems, evacuation plans and emergency supplies.

Source: from United Nations (1992).

concurrent loss of control by many mountain populations of their own mountain lands and, in a sense, a loss of control over their own futures. Mountain ecosystems and biodiversity have deteriorated and resources on which mountain people depend have been reduced. The situation is strikingly different in mountain regions of the North and of the South. In the North, mountain environments are being changed by decreased human demand and natural forest regeneration and active afforestation pro-grammes. Decision-making with respect to European mountains is likely to be increasingly made in Brussels, the seat of decision-making in the European Union.

In the South, particularly in the Tropics, mountain regions are often far more attractive for the primary sector than neighbouring regions. Rapidly growing populations and intensifying mountain agri-cultural systems are leading to overexploitation of land and soil resources. Land degradation is often the result (Stone, 1992; Messerli and Ives, 1997). A number of international agreements exist which go some way towards providing a legal and economic framework to tackle major environmental issues. An adequate set of policy measures, which must be developed in cooperation with the local indigenous people, and a reasonable balance between economic interests and environmental sustain-ability, would help to reduce the vulnerability of mountains and mountain people to global environ-mental change (Price, 1999).

7.3 The mutual vulnerability of the South and North and of mountains and lowlands

The mutual vulnerability of people from the South and North across the so-called North–South divide (Brandt Commission, 1980) seems to be as difficult to appreciate as the mutual vulnerability of ecosystems and human society or the critical interdependence of mountains and lowlands (often expressed as the highland–lowland interaction problem). Mountain environments in the South and North incorporate all of the above vulnerabilities and mutual dependences. They are, therefore, a microcosm of this global problem so brilliantly described by Head (1991). The extent to which humanity internalizes the universality of the global community, both biogeochemical and socio-economic, and the extent to which the South–North divide can be bridged, may determine the long-term future of humanity. This book, which emphasizes the biogeochemical realities of mountain geomorphology, will have served its purpose if it contributes to our understanding of mountain geo-morphology. More important, we submit, is an appreciation of the ways in which global environmen-tal change in mountain regions forces investment of effort into unravelling the linkages between society and the mountain environment. The future of our mountain world may depend on a deeper understanding of reflexivity and mutual vulnerability.

8 Conclusion

In terms of global environmental change, the historical geomorphology of mountains provides a long-term perspective on the significance of anthropogenic processes in the mountains, as well as providing a check against the enthusiastic claims of researchers of contemporary processes that they can extrap-olate their findings into the remote past and into the immediate future. The functional geomorphology of mountains provides understanding of the ways in which contemporary mountain systems function and as such establish their constraints and vulnerabilities in the face of global environmental change. The applied geomorphology of mountains raises questions of sustainable development, of risk (and hazard) assessment and acceptance, and the institutional and ethical obligations of mountain manage-ment (Slaymaker, 1999). We are convinced that, important as the study of climate change certainly is, the way in which we internalize our collective dependence on one another and on the land which we are disrupting in the legitimate cause of human development is the most fundamental challenge of the twenty-first century. Mountain geomorphology has much to say about our chances of success.

Acknowledgement

We would like to thank Eric Leinberger for producing the diagrams.

References

Ager, D.G., 1981. *The nature of the stratigraphic record.* New York: Halsted Press.

Ballantyne, C.K., 2002. Paraglacial geomorphology. *Quaternary Science Reviews,* **21**: 1935–2017.

Barrow, C.J., 1995. *Developing the environment: problems and management.* London: Longman.

Barsch, D., 1996. *Rockglaciers: indicators of the present and former geoecology of high mountain environments.* Heidelberg: Springer-Verlag.

Becker, A. and Bugmann, H., 2001. *Global change and mountain regions.* International Geosphere–Biosphere Programme (IGBP) Report 49, Stockholm.

Beniston, M. (ed.), 1994. *Mountain environments in changing climates.* London: Routledge.

Beniston, M., 2000. *Environmental change in mountains and uplands.* London: Arnold.

Benn, D.I. and Evans, D.J.A., 1998. *Glaciers and glaciation.* London: Arnold.

Bowler, J.M., Hope, G.S., Jennings, J.N., Singh, G. and Walter, D., 1976. Late Quaternary climates of Australia and New Zealand. *Quaternary Research,* **6**: 359–99.

Bradley, R.S. and Jones, P.D., 1992. *Climate since 1500 AD.* London: Routledge.

Brandt Commission, 1980. *North–South: a programme for survival.* London: Pan Books.

Bray, J.R., 1977. Pleistocene volcanism and glacial initiation. *Science,* **197**: 251–54.

Brum Ferreira, D., 2000. Environmental impact of land use change in the inner Alentejo (Portugal) in the twentieth century. In Slaymaker, O. (ed.), *Geomorphology, human activity and global environmental change.* Chichester: Wiley, 249–68.

Brunsden, D., 1980. Applicable models of landform evolution. *Zeitschrift für Geomorphologie, Supplementband,* **36**: 16–26.

Brunsden, D., 1993. The persistence of landforms. *Zeitschrift für Geomorphologie, Supplementband,* **93**: 13–28.

Brunsden, D. and Thornes, J.B., 1979. Landscape sensitivity and change. *Transactions of the Institute of British Geographers,* **4**: 463–84.

Chang, J.C. and Slaymaker, O., 2002. Frequency and spatial distribution of landslides in a mountainous drainage basin: Western Foothills, Taiwan. *Catena,* **46**: 285–307.

Church, M. and Ryder, J.M., 1972. Paraglacial sedimentation: a consideration of fluvial processes conditioned by glaciation. *Geological Society of America Bulletin,* **83**: 3059–72.

Church, M. and Slaymaker, O., 1989. Disequilibrium of Holocene sediment yield in glaciated British Columbia. *Nature,* **337**: 452–54.

Church, M., Ham, D., Hassan, M. and Slaymaker, O., 1999. Fluvial clastic sediment yield in Canada: scaled analysis. *Canadian Journal of Earth Sciences,* **36**: 1267–80.

Clague, J.J., 1986. The Quaternary stratigraphic record in British Columbia. *Canadian Journal of Earth Sciences,* **23**: 885–94.

Clayoquot Sound Scientific Panel, 1995. *Sustainable ecosystem management in Clayoquot Sound.* Reports 1–5. Victoria: Cortex Consultants.

CLIMAP Project Members, 1976. The surface of the Ice Age Earth. *Science,* **191**: 1131–37.

CLIMAP Project Members, 1981. *Seasonal reconstruction of the Earth's surface at the Last Glacial Maximum.* Geological Society of America, Map Chart Series, MC-36.

COHMAP Members, 1988. Climatic changes of the last 18,000 years: observations and model simulations. *Science*, **241**: 1043–52.

Deming, D., 1995. Climatic warming in North America: analysis of borehole temperatures. *Science*, **268**: 1576–77.

Department of Foreign Affairs and International Trade, 1999. *Human security: safety for people in a changing world*. Ottawa: Government of Canada.

Desloges, J.R. and Gilbert, R., 1998. Sedimentation in Chilko Lake: a record of the geomorphic environment of the eastern Coast Mountains of British Columbia. *Geomorphology*, **25**: 75–91.

Dickinson, W.R., 1995. The times are always changing: the Holocene saga. *Geological Society of America Bulletin*, **107**: 1–7.

Downing, T. (ed.), 1995. *Climate change and world food security*. Heidelberg: Springer-Verlag.

Fitzharris, B.B., 1996. The cryosphere: changes and their impacts. In Watson, R.T., Zinyowera, M.C., Moss, R.H. and Dokken, D.J. (eds), *Climate change 1995: impacts, adaptation and mitigation*. Cambridge: Cambridge University Press, 243–65.

Ford, D.C., 1987. Effects of glaciation and permafrost upon the development of karst in Canada. *Earth Surface Processes and Landforms*, **12**: 507–21.

French, H.M., 1996. *The periglacial environment*. Harlow: Addison Wesley Longman.

Gage, M., 1970. The tempo of geomorphic change. *Journal of Geology*, **78**: 619–25.

Goudie, A.S., 1994. The nature of physical geography: a view from the drylands. *Geography*, **79**: 194–209.

Grove, J.M., 1988. *The Little Ice Age*. London: Routledge.

Gunderson, L. and Holling, C.S. (eds), 2002. *Panarchy: understanding transformations in human and natural systems*. Washington DC: Highland Press.

Haeberli, W. and Beniston, M., 1998. Climate change and its impact on glaciers and permafrost in the Alps. *Ambio*, **27**: 258–65.

Hallet, B., Hunter, L.E. and Bogen, J., 1996. Rates of erosion and sediment evacuation by glaciers: a review of field data and their implications. *Global and Planetary Change*, **12**: 213–35.

Harris, C., Haeberli, W., Muhll, D.V. and King, I., 2001. Permafrost monitoring in the high mountains of Europe: the PACE Project in its global context. *Permafrost and Periglacial Processes*, **12**: 3–11.

Head, I., 1991. *On a hinge of history: the mutual vulnerability of south and north*. Toronto: University of Toronto Press.

Holling, C.S., 1986. The resilience of terrestrial ecosystems: local surprise and global change. In Clark, W.C. and Munn, R.E. (eds), *Sustainable development of the biosphere*. Cambridge: Cambridge University Press, 292–320.

Holling, C.S., 2001. Understanding the complexity of economic, ecological and social systems. *Ecosystems*, **4**: 390–405.

International Geosphere–Biosphere Programme, 1990. *Past global changes project in the IGBP: a study of global change. The initial core project*. Stockholm: IGBP.

Isaksen, K., Holmlund, P., Sollid, J.L. and Harris, C., 2001. Three deep alpine permafrost boreholes in Svalbard and Scandinavia. *Permafrost and Periglacial Processes*, **12**: 13–25.

Kasperson, J.X. and Kasperson, R.E. (eds), 2001. *Global environmental risk*. Tokyo: United Nations University Press.

Khor, M., 1997. *Is globalization undermining the prospects for sustainable development?* 5th Annual Hopper Lecture. Guelph, Ontario: University of Guelph.

Kreutzmann, H., 2001. Development indicators for mountain regions. *Mountain Research and Development*, **21**: 132–39.

Lamb, H.H., 1995. *Climate history and the modern world*. London: Routledge.

Land Use and Cover Change Scientific Steering Committee, 1999. *Implementation plan for land use and cover change (LUCC)*. IGBP Report 48. Stockholm: IGBP.

Leonard, E.M., 1997. The relation between glacial activity and sediment production: evidence from a 4450 year varve record of Neoglacial sedimentation in Hector Lake, Alberta. *Journal of Paleolimnology*, **17**: 319–30.

Liverman, D.M., 1994. Vulnerability to global environmental change. In Cutter, S. (ed.), *Environmental risks and hazards*. Englewood Cliffs NJ: Prentice-Hall, 326–42.

Lonergan, S.C. (ed.), 1999. *Environmental change, adaptation and security*. Dordrecht: Kluwer Academic Press.

Lubchenco, J., 1998. Entering the century of the environment: a new social contract for science. *Science*, **279**: 491–97.

Luckman, B.H., 1990. Mountain areas and global change: a view from the Canadian Rockies. *Mountain Research and Development*, **10**: 171–82.

Luckman, B.H., 1996. Dendrochronology and global change. In Dean, J.S., Meko, D.M. and Swetman, D.W. (eds), *Tree rings, environment and humanity. Proceedings of a 1994 conference held in Tucson*. Tucson AZ: Radiocarbon, Department of Geosciences, University of Arizona, 3–24.

Luckman, B. and Kavanagh, T., 2000. Impact of climate fluctuations on mountain environments in the Canadian Rockies. *Ambio*, **29**: 371–80.

MacAyeal, D.R., 1993. Binge-purge oscillations of the Laurentide Ice Sheet as a cause of the North Atlantic's Heinrich events. *Palaeoceanography*, **8**: 775–84.

Marshall, S.J. and Clarke, G.C., 1999. Modelling North American freshwater runoff through the last glacial cycle. *Quaternary Research*, **52**: 300–15.

Messerli, B. and Ives, J.D. (eds), 1997. *Mountains of the world: a global priority*. London: Parthenon.

Milankovitch, M., 1920. *Theorie mathematique des phenomenes thermiques produits par la radiation solaire*. Paris: Gauthier-Villars.

Molnar, P. and England, P., 1990. Late Cenozoic uplift of mountain ranges and global climate change: chicken or egg? *Nature*, **346**: 29–34.

Oldeman, L.R., Hakkeling, R.T.A. and Sombroek, W.G., 1991. *World map of the status of human-induced soil degradation: an explanatory note*. International Soil Reference and Information Centre. Nairobi: United Nations Environment Programme.

Oldfield, F., 1977. Lakes and their drainage basins as units of sediment-based ecological study. *Progress in Physical Geography*, **1**: 460–504.

Ollier, C.D. and Pain, C.F., 2000. *The origin of mountains*. London: Routledge.

Östrem, G., 1966. The height of the glaciation limit in southern BC and Alberta. *Geografiska Annaler*, **48A**: 126–38.

Östrem, G., 1972. Height of the glaciation level in northern BC and southeastern Alaska. *Geografiska Annaler*, **54A**: 76–84.

Owens, P.N. and Slaymaker, O., 1994. Post-glacial temporal variability of sediment accumulation in a small alpine lake. In Olive, L.J., Loughran, R.J. and Kesby, J.A. (eds), *Variability in stream erosion and sediment transport*, IAHS Publication 224. Wallingford: IAHS Press, 187–95.

Pender, J. and Hazell, P., 2000. *Promoting sustainable development in less-favoured areas*. Washington, DC: International Food Policy Research Institute.

Petit, J.R., Mounier, L., Jouzel, J., Korotkevitch, Y.S., Kotlyakov, V.I. and Lorius, C., 1990. Paleoclimatological and chronological implications of the Vostok core dust record. *Nature*, **343**: 56–58.

Porter, S.C., 1979. Hawaiian glacial ages. *Quaternary Research*, **12**: 161–87.

Prescott-Allen, R., 2001. *The well-being of nations: a country by country index of quality of life and the environment.* Ottawa and Victoria: IDRC/Island Press.

Price, M.F., 1999. *Chapter 13 in action, 1992–1997.* Rome: Food and Agriculture Organization of the United Nations.

Rajaee, F., 2000. *Globalization on trial: the human condition and the information civilization.* Ottawa: International Development Research Centre.

Rind, D., 1988. Dependence of warm and cold climate depiction on climate model resolution. *Journal of Climate*, **1**: 965–97.

Roots, E.F. (ed.), 2000. *Landscape changes at Canada's biosphere reserves.* Toronto: Environment Canada.

Ruddiman, W.F., 1984. The Last Interglacial Ocean. CLIMAP Project Members. *Quaternary Research*, **21**: 123–224.

Ruddiman, W.F. and Raymo, M.E., 1988. Northern hemisphere climatic regimes during the past 3 Ma: possible tectonic connections. *Philosophical Transactions of the Royal Society of London*, **B318**: 411–30.

Ryder, J.M., 1971. The stratigraphy and morphology of paraglacial alluvial fans in south-central British Columbia. *Canadian Journal of Earth Sciences*, **8**: 279–98.

Schumm, S.A., 1973. Geomorphic thresholds and complex response of drainage systems. In Morisawa, M. (ed.), *Fluvial geomorphology.* Publications in Geomorphology 3, Binghamton, NY: SUNY Press, 299–310.

Schumm, S.A. and Lichty, R.W., 1965. Time, space and causality in geomorphology. *American Journal of Science*, **263**: 110–19.

Slaymaker, O., 1990. Climate change and erosion processes in mountain regions of western Canada. *Mountain Research and Development*, **10**: 183–95.

Slaymaker, O. (ed.), 1991. Focus: global change. *The Canadian Geographer*, **35**: 83–92.

Slaymaker, O., 1996. Geomorphology and global sustainability. *Zeitschrift für Geomorphologie, Supplementband*, **104**: 1–11.

Slaymaker, O., 1999. Natural hazards in BC: an interdisciplinary and interinstitutional challenge. *International Journal of Earth Sciences*, **88**: 317–24.

Slaymaker, O., 2000. Assessment of the geomorphic impacts of forestry in British Columbia. *Ambio*, **29**: 381–87.

Slaymaker, O., 2001. Why so much concern about climate change and so little attention to land use change? *The Canadian Geographer*, **45**: 71–78.

Slaymaker, O. and French, H.M., 1993. Cold environments and global change. In French, H.M. and Slaymaker, O. (eds), *Canada's cold environments.* Montreal: McGill-Queens University Press, 313–34.

Slaymaker, O. and McPherson, H.J., 1977. An overview of geomorphic processes in the Canadian Cordillera. *Zeitschrift für Geomorphologie*, **21**: 169–86.

Slaymaker, O. and Spencer, T., 1998. *Physical geography and global environmental change.* Harlow: Addison Wesley Longman.

Smith, G.I. and Street-Perrott, F.A., 1983. Pluvial lakes of the western United States. In Porter, S.C. (ed.), *Late Quaternary environments of the United States.* Harlow: Longman, 190–212.

Souch, C.J. and Slaymaker, O., 1986. Temporal variability of sediment yield using accumulations in small ponds. *Physical Geography*, **7**: 140–53.

Starkel, L.A., 1987. Long-term and short-term rhythmicity in terrestrial landforms and deposits. In Rampiro, M.R., Sanders, J.E., Newman, W.S. and Konigsson, L.K. (eds), *Climate: history, periodicity and predictability*. New York: Van Nostrand Reinhold, 323–32.

Stone, P. (ed.), 1992. *The state of the world's mountains: a global report*. London: Zed Books.

Strömquist, L., Yanda, P., Msemwa, P., Lindberg, C. and Simonson-Forsberg, L., 1998. Utilizing landscape information to analyze and predict environmental change: the extended baseline perspective. *Ambio*, **28**: 436–43.

Summerfield, M.A., 1991. *Global geomorphology*. Harlow: Longman Scientific and Technical.

Summerfield, M.A. (ed.), 2000. *Geomorphology and global tectonics*. Chichester: Wiley.

Syvitski, J.P.M., Burrell, D.C. and Skei, J.M., 1987. *Fiord processes and products*. New York: Springer-Verlag.

Troll, C., 1973. High mountain belts between the polar caps and the equator: their definition and lower limit. *Arctic and Alpine Research*, **5**: 19–27.

Trottier, D., 2001. *Water management: an uphill battle in the Andes*. Science from the Developing World Reports. 30 April, 2001. Ottawa: International Development Research Centre.

Turner, B.L., Kasperson, R.E., Meyer, W.B., Dow, K., Golding, D., Kasperson, J.X., Mitchell, R.C. and Ratick, S.J., 1990. Two types of global environmental change: definitional and spatial scale issues in their human dimensions. *Global Environmental Change: Human and Policy Dimensions*, **1**: 14–22.

United Nations, 1992. *Earth Summit: Agenda 21*. Rio de Janeiro: UNCED.

Vitousek, P.M., Mooney, H.A., Lubchenco, J. and Melillo, J.M., 1997. Human domination of Earth's ecosystems. *Science*, **277**: 494–99.

Westman, W.E., 1978. *Ecology, impact assessment and environmental planning*. New York: Wiley.

Index